4th International Conference on Agricultural and Biological Sciences (ABS 2018)

AA006104

IOP Conference Series: Earth and Environmental Science
Volume 185

Hangzhou, China
26 - 29 June 2018

Editor:

Mohammad H. Golabi

ISBN: 978-1-5108-7024-6
ISSN: 1755-1307

Printed from e-media with permission by:

Curran Associates, Inc.
57 Morehouse Lane
Red Hook, NY 12571

Some format issues inherent in the e-media version may also appear in this print version.

Printed by Curran Associates, Inc. (2018)

For permission requests, please contact the Institute of Physics
at the address below.

Institute of Physics
Dirac House, Temple Back
Bristol BS1 6BE UK

Phone: 44 1 17 929 7481
Fax: 44 1 17 920 0979

techtracking@iop.org

Additional copies of this publication are available from:

Curran Associates, Inc.
57 Morehouse Lane
Red Hook, NY 12571 USA
Phone: 845-758-0400
Fax: 845-758-2633
Email: curran@proceedings.com
Web: www.proceedings.com

Table of contents

Volume 185

4th International Conference on Agricultural and Biological Sciences

26–29 June 2018, Hangzhou, China

Accepted papers received: 7 August 2018
Published online: 30 August 2018

Preface

Papers

IOP Conf. Series: Earth and Environmental Science **185** (2018) 011001 doi:10.1088/1755-1315/185/1/011001

Preface

Thanks to the support and contributions from participants of more than 30 countries, *the 4th International Conference on Agricultural and Biological Sciences (ABS 2018)* was held successfully from June 26th to June 29th in Hangzhou, Zhejiang Province, China.

The International Conference on Agricultural and Biological Sciences (ABS) is convened annually to gather related experts to exchange the newest ideas and experiences in Agricultural and Biological Sciences. The previous three ABS Conferences brought together a forum of renowned personalities from the fields of agriculture and biology from the leading research institutes to most of the universities making it a successful international gathering. Based on ABS2017, the main topics of ABS2018 will still be five parts with more specified sessions: Rice and Crop Productions and Physiology Studies; Plant Sciences including Metabolites, Stress Response, Systematics, Physiology and Application, Environmental Sciences and Toxicants, Diseases and Environmental Biotechnology etc.

More than 300 papers were submitted to ABS2018, and 71 submitted manuscripts met the scope of IOP Conference Series: Earth and Environmental Science (EES). After pre-review on the originality and language, the peer review process was arranged by the editorial committee. The editorial board was led by the Guest Editor Prof. Mohammad Golabi and reviewers like Dr. Wahyu Nurbandi, Dr. Ajin R S, Dr. Erryana Martati etc and the 40 best manuscripts were selected after the pre-review and peer review, and those manuscripts provide some insights in the field of soil, environmental and biological sciences.

On behalf of the conference organizers, we would like to take this opportunity to express our sincere thanks to guest editor and all the reviewers for their tremendous efforts and dedication to the conference, to all the authors for their trust and contributions to the conference, as well as all the colleagues from IOP Publishing for their support and their endless efforts towards the publication of the proceedings for the Conference. We believe that with their earnest support and contributions, future ABS Conferences would scale new heights.

Editor

Prof. Mohammad H. Golabi

University of Guam,

Guam-USA

Below are some interesting pictures from ABS2018:

IOP Conf. Series: Earth and Environmental Science **185** (2018) 011001 doi:10.1088/1755-1315/185/1/011001

Peer review statement

All papers published in this volume of *IOP Conference Series: Earth and Environmental Science* have been peer reviewed through processes administered by the proceedings Editors. Reviews were conducted by expert referees to the professional and scientific standards expected of a proceedings journal published by IOP Publishing.

IOP Conf. Series: Earth and Environmental Science **185** (2018) 012001 doi:10.1088/1755-1315/185/1/012001

TNF-family member Receptor Activator of NF-κB (RANK) and RANK-Ligand (RANKL) in bone remodelling

W J Quax[1], A H G van Assen and Y Z Wang

Department of Chemical and Pharmaceutical Biology, University of Groningen, Antonius Deusinglaan 1, 9713 AV, Groningen, The Netherlands

E-mail: w.j.quax@rug.nl

Abstract. Receptor Activator of NF-κB (RANK) and RANK-Ligand (RANKL) are members of the Tumour Necrosis Factor (TNF)-superfamily involved in bone homeostasis. In a tightly controlled interplay of this receptor and ligand, bone is continuously being remodelled. Several pathological conditions resulting from a misbalance between bone resorption and bone formation have been documented. Most frequently resorption gets the overhand resulting in a lower Bone Mineral Density (BMD), increased fracture risk and reduced mobility of patients. RANKL is expressed on bone producing osteoblasts whereas RANK is expressed on pre-osteoclasts, which can develop in bone resorbing osteoclasts. The binding of RANKL to RANK functions as a trigger for the formation of these bone resorbing osteoclasts. With the development of a monoclonal antibody directed against RANKL a new therapeutic strategy to interfere with bone remodelling has become available some years ago. In this manuscript we discuss the prospective of interfering with the RANK/RANKL pathway as a therapeutic target for bone diseases. We discuss the role of the soluble receptor osteoprotegerin (OPG) as a therapeutic in bone diseases. Then we focus on the possibility to develop antagonistic and agonistic variants of RANKL based on computational protein design and we discuss the development of antagonistic RANKL variants by changing the stoichiometry of the RANKL molecule.

1. Introduction

The Tumor Necrosis Factor (TNF) - superfamily is a group of cytokines consisting of ligands and receptors whose interactions regulate cell differentiation and controlled cell death (apoptosis) [1]. Thanks to their essential role in cell survival, many TNF-superfamily members are being investigated as potential therapeutic targets for diseases as diverse as cancer [2] and rheumatoid arthritis [3]. TNF-superfamily members Receptor Activator of NF-κB (RANK), RANK Ligand (RANKL) and osteoprotegerin (OPG) have a role in proliferation of bone cells as they control the balance between bone producing osteoblasts and bone resorbing osteoclasts, thereby forming the human skeleton [4].

From its appearance the skeleton seems to be a static structure, but bone tissue is continuously being remodelled. The complete human skeleton is renewed in approximately 10 years [5]. Remodelling is necessary to adapt to changes due to mechanical loads [6]. These lead to micro damage, which needs to be repaired to maintain the strength of the skeleton [7]. Differences in inter individual pressure on the skeleton lead to differences in bone remodelling. Individuals who exercise more and have more physical activity have a lower risk of fractures. Age related thinning of the bone is associated with an increase in fragility of bone, which may result in fractures upon falling. By aging of our population bone diseases become more frequent [8].

Bone diseases can be classified as syndromes with a lowered bone mass, with an increased bone mass and as bone tumours. In osteoporosis, there is an imbalance between osteoblastic and osteoclastic activity in favor of osteoclastic activity, leading to bone fragility and an increased risk of fractures [9]. Paget's disease is the most common bone disease in which there is an increased bone mass. It is characterized by deformation of the bone. Usually there are one or more focal areas of disorganized bone remodelling. Following osteoporosis, it is the most frequent bone disease [10]. Multiple myeloma is the most common primary malignancy of bone with marrow infiltration of the skeleton, which produces widespread osteolytic bone damage. In multiple myeloma, the balance between osteoblastic and osteoclastic activity is disturbed in favor of osteoclastic activity [11].

Recently, the first biological -denosumab, an antibody directed against RANKL- was introduced and accepted by legal authorities for treating osteoporosis. The monoclonal antibody (mAB) interferes with the RANK/RANKL signalling and thereby inhibits not only osteoclast activity [12] but also bone tumours [13]. Recent discoveries such as the structure of RANKL bound to RANK and OPG have given further insight into the molecular mechanisms of RANK/RANKL and OPG in bone remodelling, elucidating key residues interaction of the ligand and the receptors [14-17].

This review will focus on strategies to combat bone diseases making use of the RANK/RANKL pathway. We will discuss the process of RANK/RANKL controlled bone remodelling in the next paragraph and will thereafter discuss mechanisms of action of (new) therapeutic options to treat bone diseases interfering with this pathway.

2. RANK/RANKL/OPG controlled bone remodelling

Osteoblasts originate from pluripotent mesenchymal cells. Osteoblast cells are responsible for the formation, deposition and mineralization of bone tissue [18]. Bone is an important source for calcium and phosphate and as such important for mineral homeostasis [19]. Parathyroid Hormone (PTH) is a key regulator of calcium metabolism. When blood calcium levels are reduced, PTH is produced and secreted thereby increasing osteoclast mediated bone resorption which increases calcium levels [20].

RANK and its corresponding ligand RANKL play a role in bone homeostasis by tightly controlling the balance between bone resorbing osteoclasts and bone producing osteoblasts. Upon activation by PTH, RANKL is presented on the surface of osteoblasts [5]. RANK is presented on the surface of pre-osteoclasts upon binding of a paracrine factor, Macrophage Colony Stimulating Factor (M-CSF) [21]. Osteoblasts secrete M-CSF, which in turn binds to its corresponding receptor M-CSFR on pre-osteoclasts activating the RANK/RANKL pathway [22].

Interaction between RANKL and RANK leads to fusion of the osteoclast precursors resulting in multinucleated mature osteoclasts having the capability to resorb bone. Upon RANKL binding to RANK, TNF receptor associated factor-1 (TRAF1), -2, -3, -5, and -6 are recruited to the intracellular domain of RANK [23]. TRAF6 seems to be the most important of these as recruitment leads to the activation of nuclear factor κβ (NF-κB) [24]. NF-κB in turn regulates transcriptional activity of osteoclast-specific proteins [25]. Osteoprotegerin (OPG) acts as a decoy receptor for RANKL thus preventing interaction between RANK and RANKL resulting in less fusion of the OPCs [26]. The RANK-RANKL pathway is important in bone development in both physiological and pathological situations. Therefore, this signal pathway can be considered as a promising target for bone-related diseases [27].

3. Interfering with the RANK/RANKL pathway in combatting bone diseases

3.1. Antagonizing RANKL

The natural antagonist for suppressing RANKL titers is OPG. This soluble receptor of the TNF family is present at high concentrations in bone and it plays a natural role in protecting bone mass, hence, its name osteoprotegerin. OPG and RANK bind to slightly different regions of RANKL. RANK distributes its contacts evenly around the DE-loop and A'A''-loop of RANKL. In contrast the contacts made by OPG are limited to the E-strand and DE-loop region of RANKL, almost completely avoiding the A'A''-loop region [14]. OPG exerts its function as decoy receptor by blocking residues essential for RANKL to bind RANK [28]. Initial trials to use OPG as a therapeutic have failed due to the poor pharmacokinetic and pharmacodynamic properties and due to interference with the heparin binding domain in mice [29]. As a result, numerous fusion proteins containing a variety of residues of human OPG fused to human Fc fragments were tested and some were found to be 200 times more active than full-length OPG in vivo. A recombinant Fc-OPG, AMGN-0007, was in phase I clinical study for osteolytic bone metastasis and was discontinued due to the potential safety risk of immune response [30]. The human RANKL binding domain of RANK fused Fc fragment (RANK-Fc) was also found to be more effective than native OPG. However, in primates, following repeated dosing of human RANK-Fc, inactivating autoantibody titers against RANK were detected and the development was discontinued. Subsequently, monoclonal antibodies were selected and one monoclonal antibody obtained from immunization of the IgG2 XenoMouse emerged as a potent inhibitor of RANKL induced osteoclast formation. The resulting monoclonal antibody (an IgG2κ antibody) was originally designated AMG 162 and is now known as denosumab. Alternatively, peptide mimics of OPG (OP3-4 peptide) [31], RANK [32] and the tumour necrosis factor (TNF) receptor (WP9QY peptide) [33] were also developed. Although they showed inhibitory activity against the RANKL-induced osteoclastogenesis, further research of them is needed before these can be applied for therapeutic treatment.

The use of OPG in bone malignancies is limited as OPG also functions as a decoy receptor for Tumour Necrosis Factor Related Apoptosis Inducing Ligand (TRAIL) as well, preventing the apoptosis of tumour cells[34]. Therefore, structure-based OPG mutants have been designed that still do antagonize RANKL inhibiting osteoclastogenesis, but which do not bind TRAIL and therefore do not interfere with tumour apoptosis [34]. TRAIL cancer therapy has been shown to be effective. Recombinant human TRAIL variants with a decreased affinity for OPG have been investigated and this research was done to overcome TRAIL resistance when applied in the bone environment [35].

3.2. Antagonizing RANK

One other approach to target bone diseases is to target and antagonize RANK directly. The advantage of targeting the receptor over targeting the ligand RANKL can be found in the direct inactivation of the receptor. Although denosumab titrates away RANKL, it does not block the receptor. Furthermore the administration of antibodies such as denosumab carries the risk of immune reactions [36]. Compared with antibody, administration of structure-based protein mutants are most likely to trigger the auto immune system with less impact, as the mutations in the proteins involve less residues. Therefore, the design of antagonistic proteins based on the RANKL seems to be an attractive approach.

The feasibility to design variants of TNF-ligands with altered binding properties was first shown for TRAIL[37]. In many cancer cells, apoptosis seems to occur primarily via only one of the two TRAIL death receptors and the introduction of receptor selectivity is a way to create more potent TRAIL agonists. With computational protein design it is possible to predict protein segments and amino acids important for binding, affinity and strength of the ligand-receptor complexes [38]. All residues in the receptor binding interface can be mutated into any possible other amino acid. A set of predicted energetic values for the wild type TRAIL - receptor complex formation was compared with the obtained energetic values after mutagenesis. The change in binding energy at every residue can be used as a prediction for improved or decreased binding. Afterwards predicted variants were expressed, purified and screened for receptor selectivity by surface plasmon resonance (SPR). TRAIL variants selective for a specific TRAIL death receptor could be obtained this way [37, 39-40].

Due to the high homology among TNF ligands and their cognate receptors, an alternative strategy to predict variants with agonistic or antagonistic properties is to compare different TNF-superfamily members; interaction domains and interacting residues do show similarity among TNF ligand-receptor complexes. One of these interaction loops is the so called DE-loop. There is a conserved tyrosine in the DE-loop of all TNF-ligands except CD40L and RANKL. Shibata and others found that mutations in this hydrophobic residue of the DE-loop in TNF-α yielded TNF-receptor 1 selective antagonist [41]. The 3D structure of the complex between RANK and RANKL suggests that hydrophilic interactions are predominant among all the surface interactions between the DE-loop of RANKL and RANK. However, RANKL mutation at the equivalent residue 248 from isoleucine to aspartate resulted in an eightfold lower activity [16]. Then, the effect of further mutations at position I248 in the DE-loop of murine RANKL on the interaction of RANKL with RANK was investigated. RANKL I248Y and I248Kwere found to compete with wild-type RANKL binding to RANK while reducing wild-type RANKL-induced osteoclastogenesis [42]. These mutations suggest that the DE-loop of RANKL is important for osteoclastogenesis and two newly formed mutants, RANKL I248Y and I248K, can offer a starting point for novel therapeutic proteins in new osteoporosis treatments (Figure 1)

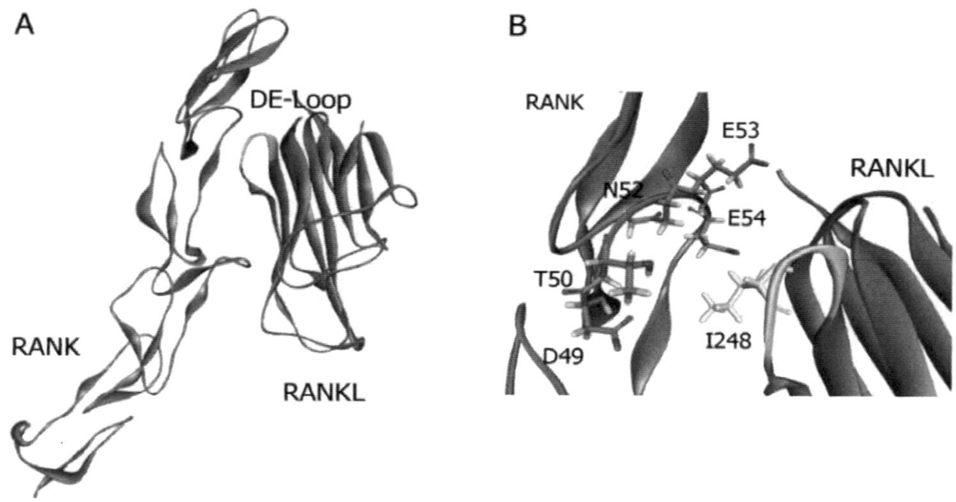

Figure 1. Structural overview of the area surrounding the DE loop of (A) Refined structure of RANKL (orange) -RANK (green) complex. The DE loop of RANKL is shown in yellow. The figure was derived from the known structure of RANK: RANKL complex (PDB ID: 4GIQ). (B) Close-up view of residues near I248 in the DE-loop of RANKL Changes in the stoichiometry of the receptor.

3.3. Changes in the stoichiometry of the receptor

An important characteristic of the TNF superfamily is the multimeric organization of receptors and ligands. For RANK and RANKL a trimeric RANK receptor binds to a trimeric RANKL ligand; this is needed for the biological effect [43]. RANK trimerizes through hydrophobic interactions at the core of the trimer, leading to assembly of a homotrimer [16]. However, TNF ligands can also bind to their corresponding monomeric and dimeric receptor subunits. It was found that there is a high affinity for the first binding to a monomeric subunit of the receptor. Using Surface Plasmon Resonance a high affinity binding between RANKL and the first monomeric subunit of RANK was demonstrated. A dissociation constant (K_D) of 0.67 nM with k_a=4.9 x 10^5 $M^{-1}s^{-1}$ and k_d=3.3 x 10^{-4} $M^{-1}s^{-1}$ could be calculated [44].

The development of RANKL variants occupying only one monomer of the RANK receptor obstructing the binding to three monomeric RANK receptors will have an antagonistic effect on osteoclast formation [45]. By creating RANKL variant that only has a single intact receptor binding interface, one RANK receptor might still be bound but additional receptor molecules can no longer be recruited to form the biologically active trimeric receptor complex. This leaves less receptor molecules available to trimerize. Thus, these variants do not only block RANK-RANKL interaction by competition with trimeric RANKL but also leave less RANK receptor molecules available on the cell surface.

Interestingly, monoclonal antibodies, such as denosumab, bind to trimeric TNF family members mostly in a 3:2 complex, meaning 3 denosumab antibody molecules bind two RANKL molecules (trimeric), resulting in complicated kinetics [46]. Compared with the antibodies, structure-based variants binding in a 1:1 complex might exhibit simpler kinetics resulting in a reduction of the risk of side effects.

4. Conclusion

The TNF-superfamily members RANKL and RANK are important for bone remodelling. Their roles in osteoclast derived bone resorption have been the subject of study for a long time. With the development of a monoclonal antibody derived against RANKL, a new therapeutic approach to combat bone diseases became available. In the review we presented three ways to develop new possible therapeutics to interfere with RANK/RANKL signalling. First, we focused on the possibility to use OPG, the natural decoy receptor for RANKL. Then we discussed the development of RANKL variants with antagonistic properties, based on computational design using conserved domains in the TNF-superfamily. Finally, we discussed antagonistic molecules interfering with the trimerization of RANK, a conformation essential for signal transduction. Interfering with the RANK/RANKL pathway is a promising strategy to develop therapeutics to combat bone diseases.

References

[1] Aggarwal B B, 2003, Signalling pathways of the TNF superfamily: A double-edged sword, *Nature Reviews Immunology*, **3**: 745–56

[2] Azijli K, Weyhenmeyer B, Peters G J, De Jong S and Kruyt F A E, 2013, Non-canonical kinase signaling by the death ligand TRAIL in cancer cells: Discord in the death receptor family, *Cell Death Differ.*, **20**: 858–868

[3] Vinay D S and Kwon B S, 2012, Targeting TNF superfamily members for therapeutic intervention in rheumatoid arthritis, *Cytokine*, **57**: 305–312

[4] Yasuda H, 2013, RANKL, a necessary chance for clinical application to osteoporosis and cancer-related bone diseases, *World J. Orthop.*, **4**: 207

[5] Manolagas S C, 2000, Birth and death of bone cells: basic regulatory mechanisms and implications for the pathogenesis and treatment of osteoporosis., *Endocr. Rev.*, **21**:115–37

[6] Cameron J A, Milner D J, Lee J S, Cheng J, Fang N X and Jasiuk I M, 2013, Employing the biology of successful fracture repair to heal critical size bone defects, *Curr. Top. Microbiol. Immunol.*, **367**:113–32

[7] Eriksen E F, 2010, Cellular mechanisms of bone remodeling, *Rev. Endocr. Metab. Disord.*, **11**: 219–27

[8] Brent Richards J, Zheng H F and Spector T D, 2012, Genetics of osteoporosis from genome-wide association studies: Advances and challenges, *Nat. Rev. Genet.*, **13**:576–88

[9] Davidson M R, 2003, Pharmacotherapeutics for osteoporosis prevention and treatment, *J. Midwifery Womens. Health*, **48**: 39–52

[10] Ralston S H, 2013, Paget's Disease of Bone, *N. Engl. J. Med.*, **368**: 644–650

[11] Tosi P, 2013, Diagnosis and treatment of bone disease in multiple myeloma: spotlight on spinal

involvement, *Scientifica (Cairo).*, **2013**: 1–12

[12] Brown J P, Reid I R, Wagman R B, Kendler D, Miller P D, Jensen J E B, Bolognese M A, Daizadeh N, Valter I, Zerbini C A F and Dempster D W, 2014, Effects of Up to 5 years of denosumab treatment on bone histology and histomorphometry: The FREEDOM study extension, *J. Bone Miner. Res.*, **29**: 2051–6

[13] Shapiro C L, Moriarty J P, Dusetzina S, Himelstein A L, Foster J C, Grubbs S S, Novotny P J , and Borah B J, 2018, Cost-effectiveness analysis of monthly zoledronic acid , zoledronic acid every 3 months, and monthly denosumab in women with breast cancer and skeletal metastases : CALGB 70604 (Alliance), *J. Clin. Oncol.*, **35:** 3949–3956

[14] Nelson C A, Warren J T, Wang M W, Teitelbaum S L, Fremont D H and Engage R, 2012, RANKL employs distinct binding modes to enage RANK and the osteoprotegerin decoy receptor, *Structure*, **20:** 1971–82.

[15] Liu C, Walter T S, Huang P, Zhang S, Zhu X, Wu Y, Wedderburn L R, Tang P, Owens R J, Stuart D I, Ren J and Gao B, 2010, Structural and functional insights of RANKL-RANK interaction and signaling, *J. Immunol.*, **184:** 6910–9.

[16] Lam J, Nelson C A, Ross F P, Teitelbaum S L and Fremont D H, 2001, Crystal structure of the TRANCE/RANKL cytokine reveals determinants of receptor-ligand specifity, *J. Clin. Invest.*, **108:** 971–979.

[17] Luan X, Lu Q, Jiang Y, Zhang S, Wang Q, Yuan H, Zhao W, Wang J and Wang X, 2016, Crystal structure of human RANKL complexed with its decoy receptor opsteoprotegerin, *J. Immunol.*, **189:** 245–52.

[18] Phan T C, Xu J and Zheng M H, 2004, Interaction between osteoblast and osteoclast: impact in bone disease., *Histol Histophathol*, **19:** 1325–1344.

[19] Bergwitz C and Jüppner H, 2010, Regulation of phosphate homeostasis by PTH, Vitamin D, and FGF23, *Annu. Rev. Med.*, **61:** 91–104.

[20] Gaffney-Stomberg E, Lutz L J, Rood J C, Cable S J, Pasiakos S M, Young A J and McClung J P, 2014, Calcium and vitamin D supplementation maintains parathyroid hormone and improves bone density during initial military training: a randomized, double-blind, placebo controlled trial, *Bone*, 46–56.

[21] Lagasse E and Weissman I L, 1997, Enforced expression of Bcl-2 in monocytes rescues macrophages and partially reverses osteopetrosis in op/op mice., *Cell*, **89**: 1021–1031.

[22] Boyce B F, Rosenberg E, de Papp A E, and Duong L T, 2012, The osteoclast, bone remodelling and treatment of metabolic bone disease, *Europ. J. Clin. Investig.*, **42:** 1332–41.

[23] Galibert L, Tometsko M E, Anderson D M, Cosman D and Dougall W C, 1998, The involvement of multiple tumor necrosis factor receptor (TNFR)- associated factors in the signaling mechanisms of receptor activator of NF-kappaB, a member of the TNFR superfamily, *J Biol Chem*, **273:** 34120–7.

[24] Darnay B G, Besse A, Poblenz A T, 2007, Lamothe B and Jacoby J J, TRAFs in RANK signaling, *Adv. Experi. Medici. Biol.*, **597:** 152–159.

[25] Honma M, Ikebuchi Y, Kariya Y and Suzuki H, 2014, Regulatory mechanisms of RANKL presentation to osteoclast precursors, *Curr. Osteoporos. Rep.*, **12:** 115–20.

[26] Simonet W S, Lacay D L, Dunstan C R, Kelley M, and Chang M S, 1997, Lüthy, R.,. Osteoprotegerin: a novel secreted protein involved in the regulation of bone density., *Cell*, **89:** 09–319.

[27] Tanaka S, Nakamura K, Takahasi N and Suda T, 2005, Role of RANKL in physiological and pathological bone resorption and therapeutics targeting the RANKL-RANK signaling system., *Immunol. Rev.*, **208:** 30–49.

[28] Luan X, Lu Q, Jiang Y, Zhang S, Wang Q, Yuan H, Zhao W, Wang J and Wang X, 2012, Crystal structure of human RANKL complexed with its decoy receptor osteoprotegerin, *J. Immunol.*, **189:** 245–52.

[29] Bekker P J, Holloway D, Nakanishi A, Arrighi M, Leese P T and Dunstan C R, 2001, The effect

of a single dose of osteoprotegerin in postmenopausal women., *J. Bone Miner. Res.*, **16:** 348–60.

[30] Body J J, Greipp P, Coleman R E, Facon T, Geurs F, Fermand J P, Harousseau J L, Lipton A, Mariette X, Williams C D, Nakanishi A, Holloway D, Martin S W, Dunstan C R and Bekker P J, 2003, A phase I study of AMGN-0007, a recombinant osteoprotegerin construct, in patients with multiple myeloma or breast carcinoma related bone metastases., *Cancer*, **97:** 887–92.

[31] Heath D J, Vanderkerken K, Cheng X, Gallagher O, Prideaux M, Murali R and Croucher P I, 2007, An osteoprotegerin-like peptidomimetic inhibits osteoclastic bone resorption and osteolytic bone disease in myeloma, *Cancer Res.*, **67:** 202–8.

[32] Ta H M, Nguyen G T T, Jin H M, Choi J, Park H, Kim N, Hwang H Y and Kim K K, 2010, Structure-based development of a receptor activator of nuclear factor-B ligand (RANKL) inhibitor peptide and molecular basis for osteopetrosis, *Proc. Natl. Acad. Sci.*, **107:** 20281–6.

[33] Aoki K, Saito H, Itzstein C, Ishiguro M, Shibata T, Blanque R, Mian A H, Takahashi M, Suzuki Y, Yoshimatsu M, Yamaguchi A, Deprez P, Mollat P, Murali R, Ohya K, Horne W C and Baron R, 2006, A TNF receptor loop peptide mimic blocks RANK ligand-induced signaling, bone resorption, and bone loss, *J. Clin. Invest.*, **116:** 1525–34.

[34] Higgs J T, Jarboe J S, Lee J H, Chanda D, Lee C M, Deivanayagam C, and Ponnazhagan S, 2015, Variants of osteoprotegerin lacking TRAIL binding for therapeutic bone remodeling in osteolytic malignancies, *Mol. Cancer Res.*, **33:** 395–401.

[35] Bosman M C J, Reis C R, Schuringa J J, Vellenga E and Quax W J, 2014, Decreased affinity of recombinant human tumor necrosis factor-related apoptosis-inducing ligand (rhTRAIL) D269H/E195R to osteoprotegerin (OPG) overcomes trail resistance mediated by the bone microenvironment, *J. Biol. Chem.*, **289:** 1071–8.

[36] Hansel T T, Kropshofer H, Singer T, Mitchell J A and George A J T, 2010, The safety and side effects of monoclonal antibodies, *Nat. Rev. Drug Discov.*, **9:** 325–38.

[37] Van der Sloot A M, Tur V, Szegezdi E, Mullally M M, Cool R H, Samali A, Serrano L and Quax W J, 2006, Designed tumor necrosis factor-related apoptosis-inducing ligand variants initiating apoptosis exclusively via the DR5 receptor., *Proc. Natl. Acad. Sci. USA*, **103:** 8634-9.

[38] Van der Sloot A M and Quax W J, 2011, Computational design of TNF ligand-based protein therapeutics, **691:** 521–34.

[39] Reis C R, Van Der Sloot A M, Szegezdi E, Natoni A, Tur V, Cool R H, Samali A, Serrano L and Quax W J, 2009, Enhancement of antitumor properties of rhTRAIL by affinity increase toward its death receptors, *Biochemistry*, **48**.

[40] Reis C R, Van Der Sloot A M, Natoni A, Szegezdi E, Setroikromo R, Meijer M, Sjollema K, Stricher F, Cool R H, Samali A, Serrano L and Quax W J, 2010, Rapid and efficient cancer cell killing mediated by high-affinity death receptor homotrimerizing TRAIL variants, *Cell Death Dis.*, **1**.

[41] Shibata H, Yoshioka Y, Ohkawa A, Minowa K, Mukai Y, Abe Y, Taniai M, Nomura T, Kayamuro H, Nabeshi H, et al., 2008, Creation and X-ray structure analysis of the tumor necrosis factor receptor-1-selective mutant of a tumor necrosis factor-alpha antagonist, *J. Biol. Chem.*, **283:** 998–1007.

[42] Wang Y, van Assen A H G, Reis C R, Setroikromo R, van Merkerk R, Boersma Y L, Cool R H, and Quax W J, 2017, Novel RANKL DE-loop mutants antagonize RANK-mediated osteoclastogenesis, *FEBS J.*, **284:** 2501–12.

[43] Douni E, Rinotas V, Makrinou E, Zwerina J, Penninger J M, Eliopoulos E, Schett G and Kollias G, 2012, A RANKL G278R mutation causing osteopetrosis identifies a functional amino acid essential for trimer assembly in RANKL and TNF, *Hum. Mol. Genet.*, **21:** 784–798.

[44] Reis C R, van Assen A H G, Quax W J and Cool R H, 2011, Unraveling the binding mechanism of trivalent tumor necrosis factor ligands and their receptors, *Mol. Cell. Proteomics*, **10:**

IOP Conf. Series: Earth and Environmental Science **185** (2018) 012001 doi:10.1088/1755-1315/185/1/012001

M110.002808.

[45] Warren J T, Nelson C A, Decker C E, Zou W, Fremont D H, and Teitelbaum S L, 2014, Manipulation of receptor oligomerization as a strategy to inhibit signaling by TNF superfamily members, *Sci. Signal.*, **7:** ra80.

[46] Arthur K K, Gabrielson J P, Hawkins N, Anafi D, Wypych J, Nagi A, Sullivan J K, and Bondarenko P V, 2012, In vitro stoichiometry of complexes between the soluble RANK ligand and the monoclonal antibody denosumab, *Biochemistry*, **51:** 795–806.

Stability and reusability of cyclodextrin glucanotransferase immobilized on hollow fiber membrane

S Suhaimi[1], R Che Man[1,3], N Jamil[1], Z I Mohd Arshad[1], S Md Shaarani[1], S K Abdul Mudalip[1], S Z Sulaiman[1] and A N M Ramli[2]

[1] Faculty of Chemical and Natural Resources Engineering, Universiti Malaysia Pahang, Lebuhraya Tun Razak, 26300, Kuantan, Pahang, Malaysia

[2] Faculty of Industrial Sciences and Technology, Universiti Malaysia Pahang, Lebuhraya Tun Razak, 26300, Kuantan, Pahang, Malaysia

E-mail: rohaida@ump.edu.my

Abstract. Cyclodextrin glucanotransferase (CGTase) is a starch degrading enzyme to produce cyclodextrin (CD), which can encapsulate compounds to remove flavors and aroma. However, the use of the enzyme is often limited due to the instability of the enzyme resulting in low production of CD. Immobilization technique has greatly improved the characteristics of the enzyme during the reaction. In this study, CGTase was immobilized on polyvinylidene difluoride (PVDF) hollow fiber membrane by adsorption method. The stability and reusability of the immobilized CGTase were studied and compared with free form of the CGTase. Thermal stability of the immobilized CGTase able to retain 50% of the initial residual activity at temperature up to 70°C. While there was an improvement in pH stability with a wider pH range from pH 5 to pH 8. The reusability of the immobilized CGTase was able to retain up to 40% of the initial CD production after repeatedly used for 10 cycles. Hence, the immobilization of CGTase shows a good physical and chemical resistance, which may be advantageous to be applied in industry.

1. Introduction

Cyclodextrin glucanotransferase is a type of enzyme that belongs to amylase family [1]. The CGTase has the capability to catalyze glycosidic bonds and due to the amylolytic in nature, the CGTase is able to convert starch into cyclodextrin (CD) [2-3]. The CD can be divided into three types: α, β and γ-CD that contain six, seven and eight glucose residues, respectively. The CD is a cyclic oligosaccharide, which consists of hydrophilic outer surface and hydrophobic inner surface. With this properties, the CD can form an inclusion complex with guest molecules to minimize the side effect of drug formulation and to stabilize volatile substances [4]. Apart from food and pharmaceutical industry, the CD also has been widely used in cosmetic, chemical and agricultural industry [5-7]. Due to this, the demand of the CD has increased about 20 to 30% per annum [8].

In industrial application, the CD is generally produced in batch system with free CGTase used directly in the reaction mixture. This often leads to problem such as low stability of the enzyme during the enzymatic reaction [9]. Besides, long time process and high cost of separation between the enzyme and product are the main drawback in the production of CD [10]. Hence, the enzyme immobilization technique is introduced as an alternative to improve the enzymatic properties. This strategy can enhance the enzyme stability and allows the enzyme to be easily recovered and repeatedly used which makes the process more economically feasible. The enzyme has been immobilized by using various technique such as adsorption [11], entrapment [12], covalent binding [13] and cross-linking [14]. Among the immobilization technique proposed, adsorption is widely used due to the useful physical interactions created between the enzyme and support that does not require chemical alteration. Other than that, this technique is relatively simple and inexpensive [15-16].

Different support offers different physical and chemical properties that can affect the enzyme immobilization and its catalytic properties. Thus, the characteristic of the support is important in determining the performance of the enzyme immobilization. For example, immobilized enzyme on nanoporous materials showed remarkably high enzyme loading and enzyme recovery [17]. However, the nonporous materials usually limit the substrate diffusion due to the entrapment of the enzyme in the porous support [18]. Other than that, the used of carrageenan as a support in enzyme immobilization was proven to be ineffective due to weak mechanical stability resulting in enzyme leakage even though the support is inexpensive [19].

In the present study, the polyvinylidene difluoride (PVDF) membrane was employed for the CGTase immobilization. The hollow fiber membrane has a high surface to volume ratio, high mechanical strength and high operational durability [20]. Besides, the hollow fiber membrane shows a great chemical resistance, lack of toxicity and also biodegradable [21]. Most studies for the immobilization of CGTase thus far have focused on the attachment to commercial resin [22]; entrapment on the sodium alginate beads and agarose [22-23] or covalent binding on magnetic carrier and Eupergit C [24-25]. To this date, the immobilized CGTase on hollow fiber membrane by adsorption has not been studied. Therefore, the stability and reusability of the immobilized CGTase to enhance the production of CD was examined. The results presented here suggested that the immobilization method is a promising approach for CD production with high CGTase stability.

2. Materials and method

2.1. Materials

Polyninylidene difluoride (PVDF) hollow fiber membrane was synthesized in Faculty of Chemical and Natural Resources Engineering laboratory, Universiti Malaysia Pahang. Commercial CGTase from *Bacillus licheniformis* (Toruzyme 3.0) was purchased from Novozymes A/S (Bagsvaerd, Denmark). Standard α-cyclodextrin (98%) was purchased from Next Gene Scientific Sdn Bhd (Selangor, Malaysia). Glycine and soluble potato starch were purchased from Chemolab Supplies Sdn Bhd (Selangor, Malaysia). Methyl orange, hydrochloric acid (37%) and acetonitrile HPLC grade were purchased from Merck Sdn Bhd (Selangor, Malaysia).

2.2. Immobilization of CGTase

PVDF membrane was cut into 3 cm and transferred into solution containing 1 mL of enzyme and 10 mL of 0.05 M of sodium phosphate buffer (pH6). The sample was incubated at 25°C for 24 h with agitation of 100 rpm. Then, the membrane was collected and washed thoroughly with sodium phosphate buffer to eliminate the non-immobilized enzyme.

2.3. Thermal stability of the immobilized and free CGTase

The thermal stability of the immobilized CGTase was determined at various temperatures of 30, 40, 50, 60, 70 and 80°C. The immobilized CGTase was transferred into two different test tube containing 0.05M sodium phosphate buffer (pH 6). The test tubes containing immobilized CGTase was immersed in water bath for 30 min. Then, CGTase assay was conducted to determine the residual activity of the enzyme.

Thermal stability for free enzyme was compared with the immobilized enzyme. About 0.3 mL of free enzyme undergoes similar condition applied in the immobilized enzyme.

2.4. pH stability of the immobilized and free CGTase

The pH stability of the immobilized and free CGTase was investigated at pH 4, 5, 6, 7, 8 and 9. The buffer solutions used were citrate phosphate (pH 4 and pH 5), sodium phosphate (pH 6 and pH 7) and glycine-NaOH (pH 8 and pH 9). The test tube containing immobilized CGTase was immersed in water bath (40°C) for 30 min. Then, CGTase assay was conducted to determine the residual activity of the enzyme.

pH stability for free enzyme was compared with the immobilized enzyme. About 0.3 mL of the free enzyme undergoes the similar condition applied in the immobilized enzyme.

2.5. *Reusability of the immobilized CGTase*

The reusability of the immobilized CGTase was evaluated using different cycles. The immobilized enzyme was incubated in 20 mL of 3% w/v of soluble potato starch with 0.05M sodium phosphate buffer (pH 6). The reaction was performed at temperature of 40°C with 150 rpm of agitation rate for 1h. After 1 h, the immobilized enzyme was separated from the reaction mixture and rinsed with 0.05M sodium phosphate buffer to eliminate any unwanted substrate or product remained on the hollow fiber membrane. Then, the immobilized CGTase was re-suspended again in a newly prepared reaction mixture for the next new cycle. The amount of CD produced were determined after every cycle. The percentage of reduction was determined by equation (1).

$$\text{Retained activity } (\%) = \frac{C_o}{C_i} \times 100\% \quad (1)$$

whereby C_o is the concentration of CD after each cycle and C_i is the highest concentration of CD. All experiments were conducted in triplicate.

2.6. *Analytical analysis*

2.6.1. CGTase assay. A 0.1mL of enzyme was added into 0.9mL of 3% (w/v) soluble potato starch in 0.05M of sodium phosphate buffer solution (pH 6). The reaction mixture was set at 40°C for 10 min. The reaction was terminated immediately by addition of 1mL of HCl (1.0M) and 0.1mL of methyl orange (0.01M). The mixture was then transferred into water bath at 16°C for 20 min. The absorbance was determined using UV vis spectroscopy at 505nm. Single unit of enzyme was described as the amount enzyme that produced 1 micromol of α-CD per min under standard conditions [26].

2.6.2. High Performance Liquid Chromatography (HPLC). High Performance Liquid Chromatography was used to determine the production of CD as described by Sakinah *et al.* [27] with some modification. The column used in this study was Agilent Eclipse Plus C18 and a mixture of acetonitrile: water (60:40) was used as the mobile phase with flowrate of 1 mL/min. Reflective index detector (RID) was used to detect the CD. Samples were centrifuged for 10 min at 5000 rpm. The sample was then filtered using Whatman® nylon membrane.

3. Results and discussion

3.1. *Thermal stability of immobilized and free CGTase*

High thermal stability is one of the most important criteria for the immobilized enzyme. In the present study, the thermal stability of free and immobilized CGTase were conducted by varying the incubation temperature ranging from 40°C to 90°C. Figure 1 reveals that both free and immobilized CGTase were positively stable at temperature up to 60°C. Nevertheless, the immobilized CGTase was significantly more resistant to heat treatment at higher temperature (70°C to 90°C), in comparison with the free CGTase. A significant different of heat resistance by the immobilized CGTase was clearly showed at temperature 70°C. The immobilized CGTase was able to remain active with 50% of the residual activity at 70°C while the free CGTase was only about 30%. Meanwhile, at temperature 80°C and 90°C the immobilized CGTase managed to retain 15% and 8% of activity, respectively.

The result in this study indicated that the thermal stability of the CGTase was enhanced by the immobilization method. This could be due to the rigidity of the enzyme structure and preservation of the active site from denaturation upon immobilization. Thus, the binding of the CGTase on the membrane would limit the movement of the enzyme molecules at various temperatures. However, for the free CGTase system, the enzyme molecules would gain more kinetic energy resulting in rigorous movement and eventually leads to enzyme inactivation [28]. A research performed by Chiou *et al.* [29] found that lipase from *Candida rugose* immobilized on chitosan was thermally stable up to 40°C, while the free lipase was stable at 30°C.

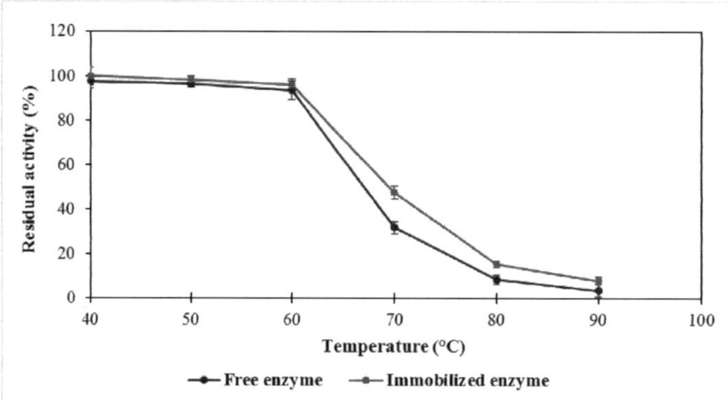

Figure 1. Comparison of thermal stability of free and immobilized CGTase. The residual activity was measured after incubated at different temperature for 30 min.

A similar finding was reported previously in the immobilization of CGTase on magnetic carrier whereby, there was a significant loss recorded by the free CGTase (95%) compared to the immobilized CGTase (50%) at 80°C [30]. Therefore, the improvement of the thermal stability of the immobilized CGTase might be advantageous in the industrial application, which required high reaction temperature [26].

3.2. *pH stability of immobilized and free CGTase*

pH stability also should be taken into consideration in order to determine the enzymatic properties of the immobilized CGTase. The pH stability of the enzyme in this study was investigated by varying the pH ranging from pH 4 to 10. As shown in Figure 2, the maximum stability achieved by the free CGTase was from pH 6 to pH 8. Interestingly, the immobilized CGTase showed a wider pH stability which was from pH 5 to pH 8. The different of the enzyme activity between the free and immobilized CGTase was significant especially at pH 5. The free enzyme has lost about 60% of the initial activity while the immobilized enzyme has loss only 20% of the initial activity. The present study showed that the immobilized CGTase has a higher stability in wider pH range which able to produce higher amount of CD in comparison with the free CGTase.

The immobilized enzyme has a higher stability in different pH value was due to the ability to exist in stable conformation and preservation of the enzyme structure from inactivation. According to Manas *et al.* [31], acidic reaction condition was found to be favourable for the enzymatic reaction. Thus, the ability of the immobilized CGTase to remain active in acidic condition would be beneficial in order to improve the production of CD.

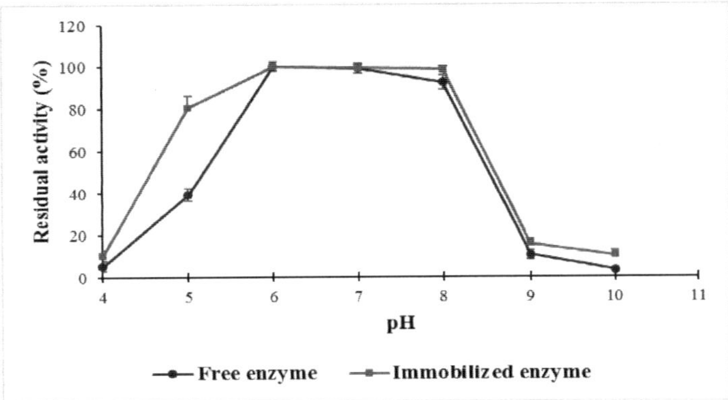

Figure 2. Comparison of pH stability for free and immobilized CGTase. The residual activity for was measured after incubated at different pH value for 30 min.

The finding in the present study was consistent with a research done by Abdel-Naby *et al.* [32] whereby the immobilized CGTase on Duolite XAD 761 resin managed to retain high activity at wider acidic range (pH 4 to 6.8) compared to the free enzyme (pH 5.5 to 6.5). This was due to the cationic characteristic of the resin as a support which resulted in the stabilization of the immobilized enzyme in acidic medium. However, the result showed in the present study was in contrast with a study conducted by Kim *et al.* [22], whereby there was no significant changes in pH stability (pH 4 to pH 8) between the free and immobilized CGTase on Amberlite IRA-900. The insignificant changes in pH stability after the immobilization process was probably due to the support which unsuccessfully change the microenvironment of the enzyme, subsequently resulted in enzyme denaturation [33].

3.3. Reusability of the immobilized CGTase

For industrial application, the cost of enzyme and downstream process can be reduced greatly if the enzyme can be reused for several times. To ensure a cost-effective production, the reusability of immobilized CGTase was studied by determining the amount of CD produced. The amount of CD in the first batch was taken as a reference (100 percentage of reduction). Based on Figure 3, the immobilized CGTase in the present study was capable to retain up to 50% and 40% of the initial CD production when repeatedly used for fifth and tenth successive batch reactions, respectively. A total of 26 mg/mL of CD was successfully produced by the immobilized CGTase in the 10 cycles.

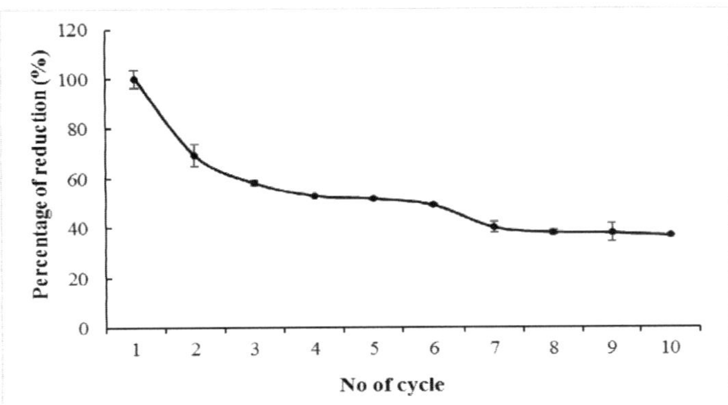

Figure 3. Reusability of immobilized CGTase on hollow fiber membrane in repeated batch reaction. CD production was measured for every cycle (1 h reaction time) until 10 cycles.

In addition, as the number of cycle increases, the reduction percentage of the immobilized CGTase also decreases. This phenomenon was due to the desorption of the immobilized enzyme upon washing in each cycle [34]. The finding in the present study was consistent with the findings by Ivanova [31] who investigated the reusability of immobilized CGTase on magnetic carrier treated with polyethyleneimine (PEI). In the study, the immobilized CGTase undergoes repeated batch reaction for 7 cycles. Based on the result, the production of CD decreased as the number of cycle increased. Similar finding was also portrayed by different enzyme such as a study conducted by Hung *et al.* [35] who studied on the reusability of immobilized lipase on chitosan for the production of p-nitrophenol. The results clearly showed decrement of p-nitrophenol in the 10 cycles which concluded that the immobilized enzyme often contributed to the decrease in enzymatic activity after repeatedly used for severeal times.

The ability of the immobilized CGTase to retain up to 40% of the reduction suggested that the CGTase was stable and remains active throughout the 10 cycles. Interestingly, the present study recorded relatively higher stability compared to the CGTase immobilized on nanofiber activated with dodecanediamine derivative, where 40% of initial activity was maintained after only 7 cycles [36]. Therefore, the reusability of the immobilized CGTase presented in the present study might applicable for industrial scale since it can be reused with high enzyme stability.

4. Conclusion

The immobilization of CGTase on hollow fiber membrane has been successfully conducted via adsorption method without any chemical alteration. The results showed an improvement in resistance against thermal and pH denaturation compared to the free CGTase. Besides, the immobilized CGTase can be used repeatedly with high enzyme stability for the production of CD. This indicated that the enzyme immobilization using PVDF hollow fiber membrane as a support is a promising strategy and is capable to be applied in scale-up process.

Acknowledgment

This research was funded by Universiti Malaysia Pahang (Grant No: PGRS 170396). The authors also would like to acknowledge Faculty of Chemical and Natural Resources Engineering for the facilities.

References

[1] Bernardi N Z, Blanco K C, Monti R and Contiero J 2015 *J. Food Ind. Microbiol.* **1** 1–5
[2] Qi Q and Zimmermann W 2005 *Appl. Microbiol. Biotechnol.* **66** 475–485
[3] Coelho S L A, Magalhães V C, Marbach P A S and Cazetta M L 2016 *Brazilian J. Microbiol.* **47** 120–128
[4] Uyar T, Havelund R, Hacaloglu J, Zhou X, Besenbacher F and Kingshott P 2009 *Nanotechnology* **20**
[5] Suhaimi S, Man R C, Jamil N, Arshad Z I M, Shaarani S, Sulaiman S Z, Mudalip S K A, Ramli A N M 2018 *J. Teknol.* **3** 165–171
[6] Andrea R, Gastón J, Szerman N, Costa H, Krymkiewicz N and Ferrarotti S A 2009 *Enzyme Microb. Technol.* **45** 36–41
[7] Jamil N, Man R C, Shaarani S M, Sulaiman S Z, Mudalip S K A and Arshad Z I M 2017 *Indian J. Sci. Technol.* **10** 1–5
[8] Saallah S, Naim M N, Lenggoro I W, Mokhtar M N, Abu Bakar N F and Gen M 2016 *Biotechnol. Reports.* **10** 44–48
[9] Matte C R, Nunes M R, Benvenutti E V, Schoffer J N, Ayub M A Z and Hertz P F 2012 *Journal Mol. Catal. B, Enzym.* **78** 51–56
[10] Ravinder K, Prabhakar T and Prasanthkumar K 2012 *Adv. Appl. Sci. Res.* **3** 2288–98
[11] Abdulla R, Sanny S A and Derman E 2017 *IOP Conf. Ser. Mater. Sci. Eng.* **206** 012032
[12] Yildiz H B, Kiralp S, Toppare L, and Yağci Y 2005 *J. Appl. Polym. Sci.* **96** 502–507
[13] Atacan K, Çakıroğlu B and Özacar M 2017 *Int. J. Biol. Macromol.* **97** 148–155

[14] Sojitra U V, Nadar S S and Rathod V K 2017 *Carbohydr. Polym.* **157** 677–685
[15] Salleh S, Serri N A, Hena S and Tarajudin. H A 2016 *IOP Conf. Ser. Earth Environ. Sci.* **36** 012026
[16] Zucca P and Sanjust E 2014 *Molecules* **19** 14139–94
[17] Lei C, Soares T a, Shin Y, Liu J and Ackerman E J 2008 *Nanotechnology* **19** 125102
[18] Kim J, Grate J W and Wang P 2006 *Chem. Eng. Sci.* **61** 1017–26
[19] Tan I S and Lee K T 2014 *Bioresour. Technol.* **184** 386–394
[20] Che Man R, Ismail A F, Fuzi S F Z M, Ghazali N, and Illias R M 2016 *Process Biochem.* **51** 474–483
[21] Algieri C, Donato L and Giorno L *Biotechnol. Appl. Biochem.* 1–8
[22] Kim P, Shin H, Park J and Lee Y 2000 *Biotechnol. Bioprocess Eng.* **5** 174–180
[23] Arya S K and Srivastava S K 2006 *Enzyme Microb. Technol.* **39** no 507–510
[24] Vieira M F, Vieira A M S, Tardioli P W, Zanin G M and Guisán J M 2009 *N. Biotechnol.* **25S** S169
[25] Blanco K C, Santos F J d, Bernardi N S, Júnior M J, Monti R and Contiero J 2013 *Enzym. Eng.* **02** 111
[26] Ding R, Li Z, Chen S, Wu D, Wu J and Chen J 2010 *Process Biochem.* **45** 880–886
[27] Sakinah A M M, Ismail A F, Illias R M, Zularisam A W, Hassan O and Matsuura T 2014 *Sep. Purif. Technol.* **124** 61–67
[28] Silva A M d, Tavares A P M, Rocha C M R, Cristóvão R O, Teixeira J a and MacEdo E a 2012 *Process Biochem.* **47** 1095–1101
[29] Chiou S H and Wu W T 2004 *Biomaterials* **25** 197–204
[30] Ivanova V 2010 *Biotechnol. Biotechnol. Equip.* 516–528
[31] Manas N H, Illias R M and Mahadi N M 2017 *Crit. Rev. Biotechnol.* 1–22
[32] Abdel-Naby M A, Fouad A and Reyad R M 2015 **116** *J. Mol. Catal. B Enzym.* 140–147
[33] Lee J M, Xu G R, Kim B K, Choi H N and Lee W Y 2011 *Electroanalysis*. **23** 962–970
[34] Silva A L P, Nascimento R G, Arakaki L N H, Arakaki T, Espínola J G P and Fonseca M G 2013 *J. Non. Cryst. Solids*. **376** 139–144
[35] Hung T C, Giridhar R, Chiou S H and Wu W T 2003 *J. Mol. Catal. B Enzym.* **26** 1–2 69–78, 2003.
[36] Sulaiman S, Mokhtar M N, Naim M N, Baharuddin A S, Salleh M A M and Sulaiman A *Sains Malaysiana*. **44** 1541–50

IOP Conf. Series: Earth and Environmental Science **185** (2018) 012003 doi:10.1088/1755-1315/185/1/012003

Production of bioactive trastuzumab and chimeric anti-VEGF antibody in the cytoplasm of *Escherichia coli*

P Charoenpun[1], J Leelawattanachai[2], A Meechai[1,3] and D Waraho-Zhmayev[1,4]

[1] Biological Engineering Program, Faculty of Engineering, King Mongkut's University of Technology Thonburi, Bangkok 10140, Thailand

[2] Nano-Molecular Target Discovery Laboratory, National Nanotechnology Center, National Science and Technology Development Agency, Pathum Thani 12120, Thailand

[3] Department of Chemical Engineering, Faculty of Engineering, King Mongkut's University of Technology Thonburi, Bangkok 10140, Thailand

E-mail : dujduan.war@mail.kmutt.ac.th

Abstract. Monoclonal antibody (mAb) is a very useful tool in not only medical field but also in research field. However, the current cost of its production is very high as most mAbs are produced in mammalian cell system. *Escherichia coli* is a good alternative but has many bottlenecks. Recently, it was demonstrated that biological active full-length Immunoglobulin G (IgG) could be produced in the reductive cytoplasm of an engineered *E. coli* strain called SHuffle. In this study, we investigated the effect of temperature and induction level on the production of full-length humanized anti-Her2 and chimeric anti-VEGF IgGs expressed as cyclonal. The solubility as well as the assembly of the IgG molecules was examined by western blot analysis. It was found that expression at 30 °C with 0.1 mM IPTG induction was the most suitable for trastuzumab. In contrast, the full-length expression of chimeric anti-VEGF required induction with 1 mM IPTG at 30 °C., however, 1 mM IPTG induction resulted in more full-length IgG. In conclusion, the present study indicated that E coli could be successfully used for the full-length expression of IgG by suitably modifying the expression conditions.

1. Introduction

Currently, there are more than 300 biopharmaceutical products in the market with combined market share of more than USD 100 billion [1]. Therapeutic monoclonal antibody (mAb) represents a large class of biopharmaceutical products. They are used for many indications majority of them are used in various types of cancer and autoimmune disease. As new products are being developed and many are already in clinical trial, unsurprisingly, their revenue is continuing to grow and the combined global market is projected to be nearly USD 125 billion by 2020 [2]. In fact, the use of mAbs is not limited to therapeutic purposes but also as diagnostic and research tools due to their ability to recognize and bind specifically with respective antigens.

Currently, the most popular method for production of mAbs involves hybridoma cell line that expresses mAb in a bioreactor. It allows asparagine-linked (N-linked) glycosylation, which is important for antibody function, high level mAb production and high stability [3]. However, the major drawbacks of mammalian expression system for recombinant protein production include low productivity, low growth rate, potential human pathogenic viral contamination and the use of serum in the cell culture [4]. Moreover, in terms of bioprocessing and scaling up, long production cycle due to slow growth rate, high capital investment as well as expensive medium and serum consequently result in much higher production cost of mAbs [5-6]. Therefore, microbial system such as *E. coli* is a good alternative. Previous attempts were based on expression of mAbs in the naturally oxidative periplasmic compartment of *E. coli*, where it has similar conditions to the endoplasmic reticulum of mammalian cells, allowing for correct disulfide bond formation which is crucial to antibody's function. Nonetheless, periplasmic mAb production has limitations such as small periplasmic volume when compared to cytoplasmic volume leading to lower protein yield, lack of adenosine triphosphate (ATP)-dependent molecular chaperones that aid in antibody folding and efficiency of IgG heavy and light chain secretion through the tightly sealed cytoplasmic membrane [7].

Recently, Successful expression of bioactive full-length IgG was recently reported by using an engineered oxidative cytoplasm of an E coli strain called SHuffle T7 express [8]. They proved that the yield of the cytoplasmic IgG, termed 'cyclonal' was significantly more than the yield of typical IgG produced in the periplasm of *E. coli.* Furthermore, they also showed that the produced cyclonal could be redirected to newer antigens by swapping variable regions of the scaffold cyclonal with the new variable heavy chain (VH) and variable light chain (VL) domains with specificity for certain antigens [7]. These findings showed potentials in creating on-demand IgGs because of the ease with which new genes can be designed, cloned and expressed, in *E. coli.* This technique is a powerful tool for designing and producing humanized antibodies to reduce innate human immunogenic response against the conventional to animal-derived antibodies.

The high cost of pharmaceuticals, has become an important issue for ever-increasing healthcare costs. Here, we demonstrated the effect of temperature and inducer concentration on the production of trastuzumab and chimeric anti-vascular endothelial growth factor (VEGF) in the cytoplasm of *E.coli* Shuffle T7 express strain. Trastuzumab is a humanized anti-human epidermal growth factor receptor 2 (HER2) mAb used to treat breast cancer. It is sold under the trade name Herceptin by Roche and was the world's eighth-largest-selling pharmaceutical product in 2015 (USD 6,603 million) [9]. While another product from Roche, humanized anti-VEGF or bevacizumab, sold under the trade name Avastin, was the world's seventh-largest-selling pharmaceutical product in 2015 (USD 6,751 million) [9]. Both of these mAbs are currently produced in mammalian cell system. In view of above, the current study might be helpful to understand factors that affect cytoplasmic IgG production in *E. coli*, which can be optimized further for large scale production. In the future, *E. coli* may replace the current mammalian cell system and can reduce high cost of mAbs such that therapeutic antibodies can be more affordable for patients around the world.

2. Materials and methods

2.1. Bacterial strains and growth conditions

E. coli SHuffle T7 express strain (NEB) was used for all experiments expect for cloning, where DH5α strain was used instead. Typically, SHuffle T7 Express cells transformed with one of the pMAZ360–cIgG expression vectors were grown in Luria–Bertani (LB) medium, supplemented with 100 μg/ml ampicillin and 25 μg/ml spectinomycin, overnight at 37 °C. The next day, cells were subcultured into fresh LB supplemented with antibiotics and grown at 23, 30 or 37 °C until the absorbance at 600 nm reached 0.7 then protein expression was induced by addition of 0.01, 0.1 or 1.0 mM isopropyl b-D-thiogalactopyranoside (IPTG), after which cells were incubated an additional 16 h at indicated temperatures.

2.2. Construction of recombinant plasmid

Chimeric anti-VEGF was created by replacing Fv of anti-HER2 (trastuzumab) with Fv of anti-VEGF antibody. To generate pMAZ360-anti-VEGF (mouse/human chimeric Ab), *E. coli* codon optimized V_H and V_L domains of anti-VEGF murine mAb (muMAb A.4.6.1) [10] were chemically synthesized by GenScript (USA) and cloned into pMAZ360-Herceptin. Briefly, murine anti-VEGF V_L domain was PCR amplified and religated into *Nde*I-*Not*I digested pMAZ360-Herceptin, thus replacing the original anti-HER2 V_L domain creating pMAZ360-anti-VEGF V_L/Herceptin plasmid. To replace anti-HER2 V_H domain, murine anti-VEGF V_H domain was PCR amplified using primers that appended a 5' RBS sequence with *Hind* III cut site flanking on both ends and religated into *Hind*III treated pMAZ360-anti-VEGF V_L/Herceptin plasmid to obtain pMAZ360 anti-VEGF (mouse/human chimeric Ab). Sequencing analysis confirmed correct sequences and directions of both V_H and V_L.

2.3. Protein analysis

SHuffle cells expressing cyclonals were harvested 16 hours after induction by centrifugation at 5,000 rpm, at 4 °C for 10 min. Cell pellets were resuspended in lysis buffer (1 mg/mL lysozyme, 50 mM Tris-HCl, 1 mM EDTA and 25 mM NaCl) and sonicated for 10 s 3 times at 40% amplitude on a Branson 150 sonifier (Branson Ultrasonics Co., Danbury, CT). The supernatants containing soluble cytoplasmic proteins were recovered as cell lysates containing soluble proteins. SDS-PAGE analysis was performed under non-reducing conditions. Samples were diluted 1:1 in 2 x Laemmli sample buffer without addition of 2-mercaptoethanol and heated at 100 °C for 10 min. All samples were normalized by total protein, as determined by the Bradford assay and loaded on 10% Tris-HCl gels. Western blotting was performed as previously described [7]. Herceptin or anti-VEGF (mouse/human chimeric Ab) were probed with 1:15,000-diluted anti-human IgG (H+L)-HRP conjugate (Abcam). The result was obtained using a Bio-Rad ChemiDoc™ MP Imaging System (USA).

2.4. Relative quantification of IgG fragments

The program on Bio-Rad ChemiDoc™ MP Imaging System (USA) was used for densitometric analysis of the protein bands. Solubility of each sample was calculated as total IgG of each sample by addition of all the bands for each sample. On the other hand, full-length IgG yield was determined by considering only the band representing constructs with 2 heavy and 2 light chains. Relative solubility and relative full-length IgG yield were calculated by comparing data of each sample with the data of sample incubated at 30 °C with 1 mM IPTG induction. This condition was used as the reference because it was the condition reported previously [7]. % assembly efficiency was the ratio of full-length IgG to total IgG.

3. Results and discussion

3.1. Humanization of anti-VEGF using variable-domain grafting technique

For clinical use, animal-derived antibodies can cause immunogenicity in patients. With advances in recombinant DNA technology, therapeutic antibodies are now undergone 'humanization', which involves combining the antigen biding portions of a mouse antibody with the constant regions of a human antibody. In the current study, we tried to create a chimeric antibody by grafting murine anti-VEGF variable fragment (Fv) [11] on to the humanized anti-HER2 antibody or trastuzumab scaffold [7], The result showed that we can create on-demand antibody that can be produced in *E. coli*, thus can reduce the production cost greatly compared to the mammalian cell systems used currently. It is noteworthy that this trastuzumab version contains E382V and M428I mutations, which allow aglycosylated IgG expressed from *E. coli* to bind effector FcγRI thus potentiating the killing of Her2 oeverexpressing tumor cells with dendritic cells as effecotrs [12]. Figure 1 shows the comparison between the amino acid sequences of Fv of anti-HER2 (trastuzumab) and the Fv of murine anti-VEGF named A.4.6.1, which was used to replace the Fv of trastuzumab and expressed as a mouse/human chimeric antibody.

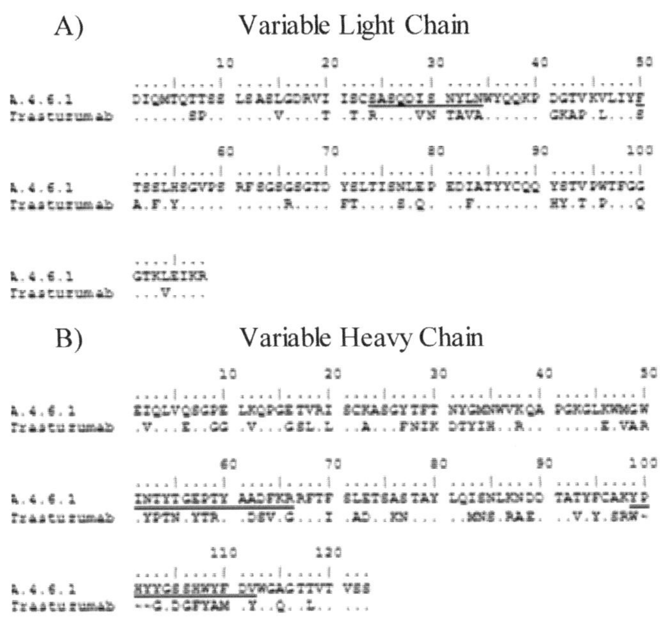

Figure 1. Amino acid sequences of variable light and heavy domains of murine anti-VEGF monoclonal antibody (A.4.6.1) and anti-HER2 scaffold. CDRs are underlined.

3.2. Comparison of cytoplasmic IgG expressions at various temperatures

We first investigated the effect of temperature on the cytoplasmic IgG expression of trastuzumab and the chimeric anti-VEGF using the same condition previously reported for trastuzumab which included induction with 1 mM IPTG for 16 h of protein expression [7]. In addition, the expression was carried out at 23, 30 and 37 °C. In both cases, it was found that the higher the temperature the less amount of soluble IgG fragments were observed (Figure 2A). It is clear that chimeric anti-VEGF can be expressed but, unfortunately, the amount of the soluble anti-VEGF is less than the trastuzumab used as scaffold for chimeric antibody grafting. Moreover, full-length IgG assembly was observed in the case of trastuzumab expressed at all temperatures investigated.

Only small amount of full-length anti-VEGF was observed when expressed at 30 °C (Figure 2A). This is not surprising because it has been reported that the yield of anti-PA-63, an antibody against anthrax protective antigen, when expressed in the cytoplasm of *E. coli* as heterotetrameric IgG in mFab/hFc format, where the whole Fab is replaced, is about 15 fold higher than of the chimeric format, where only Fv region was replaced (17.0 mg/L vs. 1.1 mg/L) [7]. In order to obtain the highest yield of full-length IgG, it is crucial to balance the expression conditions for high soluble protein expression and high assembly efficiency.

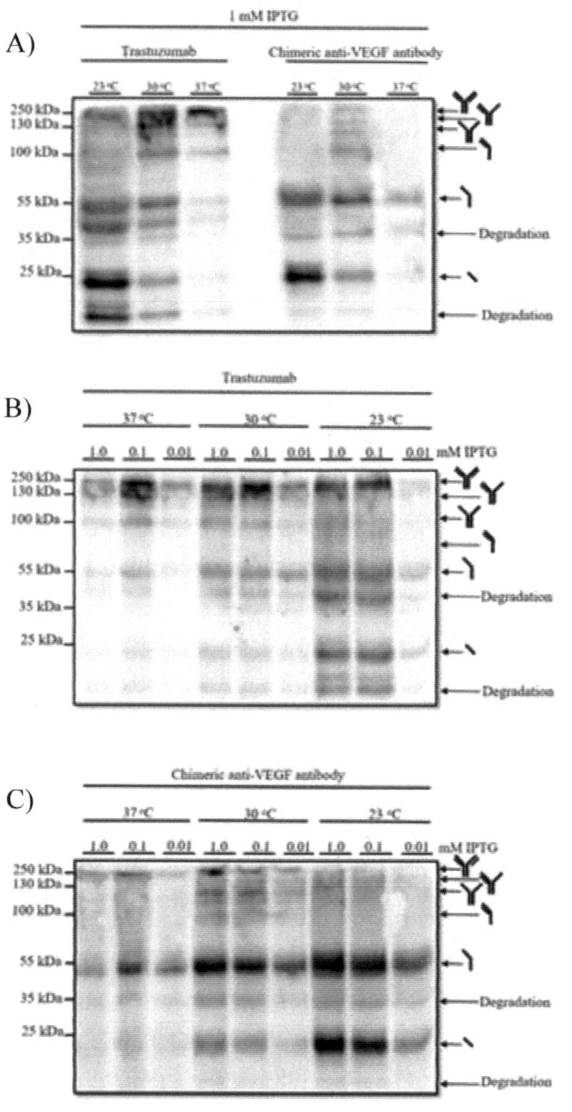

Figure 2. Expressions of trastuzumab and chimeric anti-VEGF antibodies in SHuffle T7 cells. Representative non-reducing western blot of A) both IgGs expressed at 23 °C, 30 °C, and 37 °C with 1 mM IPTG after 16 h of induction. B) Trastuzumab and C) chimeric anti-VEGF antibody with varying amount of inducers in SHufffle cells are shown here.

3.3. Investigation of the effect of temperature and induction level

Temperature and induction level are crucial parameters that can greatly effect a solubility of recombinant proteins expressing in *E. coli* system as they can affect translation rate. Typically, proteins of mammalian origin are expressed as soluble form when the translation rate is low, which could be easily achieved by growing the cells at lower temperature or inducing with lower IPTG levels [13]. Therefore, a full factorial experiment was performed in order to study the interaction effect of both factors. Three temperatures, i.e. 23 , 30 and 37 °C, and three induction levels, i.e. 0.01, 0.1 and 1 mM IPTG, were investigated. Figure 2B and 2C show the expression of trastuzumab and chimeric anti-VEGF analyzed by immunoblotting. IgGs were detected using polyclonal anti-human antibodies that detect both heavy and light chains such that various types of assembly can be detected. Bio-Rad ChemiDoc™ MP Imaging

System was used to quantify the results. It was confirmed that low temperature was better forexpression of the IgGs since, the highest yield of soluble IgG was obtain when cells were grown at 23 °C. as quantified by band intensity (Figure 3A). Among the various IPTG levels tested, induction with 0.01 mM IPTG induction was too low for all temperatures while 0.1 and 1 mM IPTG seemed to have similar yield of IgG fragments at 23 °C and 30 °C while at 37 °C induction with 1 mM IPTG seemed to be too high resulting in less soluble IgG. The possible explanation might be that at high induction level, the rate of translation is too high resulting in partially folded protein having hydrophobic regions that can interact and form large aggregates. Eventhough the amount of full-length trastuzumab was similar when expressed at 23 °C and 30 °C, the assembly efficiency was better at 30 °C, which would ease purification process of full-length antibody in industrial scale. Currently, the majority of mAb recovery process employs Protein A chromatography following harvest [14]. Protein A binds strongly to Fc region of the heavy chain. If the assembly efficiency is low, it can result in high amount of unwanted product variant species. Moreover, as induction with 0.1 or 1 mM IPTG gave rise to similar amount of soluble IgG, it would be more economical to use lower IPTG concentration in industrial scale. Similar to trastuzumab, low temperature was better for expression of the chimeric anti-VEGF as more amount of soluble IgG was observed (Figure 3D). However, as % assembly of chimeric anti-VEGF at 23 °C was too low, 30 °C is the most suitable expression temperature for this IgG. Chimeric anti-VEGF showed highest relative full-lengh IgG yield following induction with 1 mM IPTG (Figure 3F) but the amount is still quite low ; therefore, further optimization maybe required.

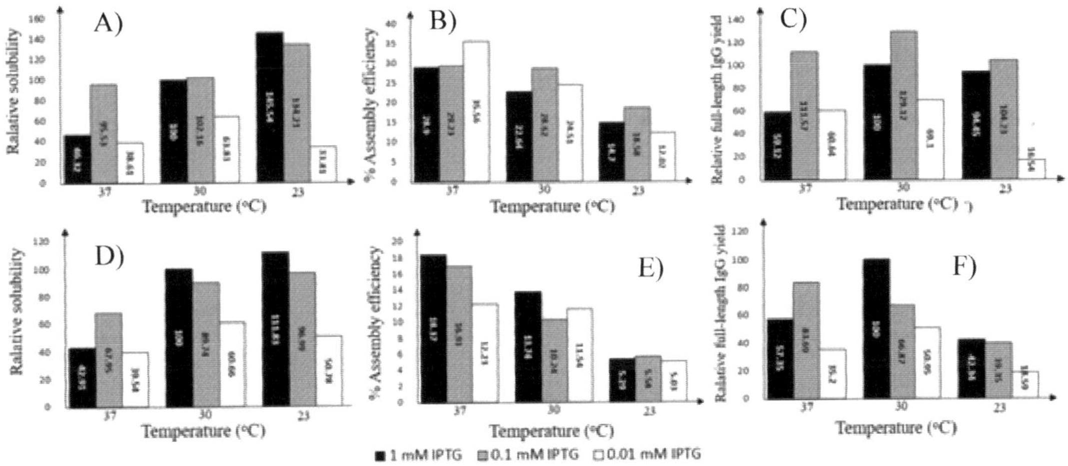

Figure 3. Relative quantification of IgG expressions. Band intensities were quantified using Bio-Rad ChemiDoc™ MP Imaging System to calculate A) and D) relative solubility, B) and E) % assembly efficiency, and C) and F) relative full-length IgG yield for trastuzumab and chimeric anti-VEGF, respectively.

4. Conclusions

A method to produce mAb in the cytoplasm of *E. coli* has emerged in recent years which could revolutionize the industrial production of mAb as it will not only reduce the cost of production but also allow for mAb engineering as genetic manipulation can be performed easily using *E. coli* as a host. In this study, we investigated the effect of temperature and induction level which affect the translation rate of the IgG. In conclusion, full-factorial experiment showed that expression at 30 °C with 0.1 mM IPTG induction was the most suitable for trastuzumab. Expression of chimeric anti-VEGF was also most suitable at 30 °C, however, higher induction level resulted in more full-length IgG. Since high assembly efficiency of chimeric anti-VEGF was still not achieved, other factors or different grafting methods could be investigated in the future.

Acknowledgment
This work was financially supported by KMUTT Research Fund M2009-014207-078172).

References

[1] Nielsen J 2013 Production of biopharmaceutical proteins by yeast: advances through metabolic engineering *Bioengineered* **4** 207-11
[2] Ecker D M, Jones S D and Levine H L 2015 The therapeutic monoclonal antibody market *mAbs* **7** 9-14
[3] Li F, Vijayasankaran N, Shen A Y, Kiss R and Amanullah A 2010 Cell culture processes for monoclonal antibody production *mAbs* **2**, 466-79
[4] Spadiut O, Capone S, Krainer F, Glieder A and Herwig C 2014 Microbials for the production of monoclonal antibodies and antibody fragments *Trends Biotechnol* **32**, 54-60
[5] Chartrain M and Chu L 2008 Development and production of commercial therapeutic monoclonal antibodies in Mammalian cell expression systems: an overview of the current upstream technologies *Curr Pharm Biotechnol* **9**, 447-67
[6] Farid S S 2007 Process economics of industrial monoclonal antibody manufacture *J Chromatogr B Analyt Technol Biomed Life Sci* **848**, 8-18
[7] Robinson M P, Ke N, Lobstein J, Peterson C, Szkodny A, Mansell T J, Tuckey C, Riggs P D, Colussi P A, Noren C J, Taron C H, DeLisa M P and Berkmen M 2015 Efficient expression of full-length antibodies in the cytoplasm of engineered bacteria *Nat Commun* **6**, 8072
[8] Lobstein J, Emrich C A, Jeans C, Faulkner M, Riggs P and Berkmen M 2012 SHuffle, a novel Escherichia coli protein expression strain capable of correctly folding disulfide bonded proteins in its cytoplasm *Microb Cell Fact* **11** 56
[9] Top drugs by sales revenue in 2015: Who sold the biggest blockbuster drugs? March 10, 2016 Retrieved April 18, 2018
[10] Kim K J, Li B, Houck K, Winer J and Ferrara N 1992 The vascular endothelial growth factor proteins: identification of biologically relevant regions by neutralizing monoclonal antibodies *Growth Factors* **7** 53-64
[11] Presta L G, Chen H, O'Connor S J, Chisholm V, Meng Y G, Krummen L, Winkler M and Ferrara N 1997 Humanization of an anti-vascular endothelial growth factor monoclonal antibody for the therapy of solid tumors and other disorders *Cancer Res.* **57** 4593-9
[12] Jung S T, Reddy S T, Kang T H, Borrok M J, Sandlie I, Tucker P W and Georgiou G 2010 Aglycosylated IgG variants expressed in bacteria that selectively bind FcγRI potentiate tumor cell killing by monocyte-dendritic cells *PNAS* **107** 604-9

[13] Dyson M R, Shadbolt S P, Vincent K J, Perera R and McCafferty J 2004 Production of soluble mammalian proteins in Escherichia coli: identification of protein features that correlate with successful expression *BMC Biotechnol* **4**:32

[14] Liu H F, Ma J, Winter C and Bayer R 2010 Recovery and purification process development for monoclonal antibody production *mAbs* **2**:5 480-499

IOP Conf. Series: Earth and Environmental Science **185** (2018) 012004 doi:10.1088/1755-1315/185/1/012004

Application of ligninolytic enzyme of *Lentinus polychrous* on synthetic dye decolorization

J Isanapong[1] and S Mataraj

Department of Agro-Industrial, Food, and Environmental Technology, Faculty of Applied Science, King Mongkut's University of Technology North Bangkok, Bangkok, Thailand

E-mail: Jantiya.i@sci.kmutnb.ac.th

Abstract. Contamination of water resources by synthetic dyes causes many environmental and health problems. Fungal ligninolytic enzymes have been applied for dye decolorization due to its ability in degrading a wide variety of recalcitrant substances. Extracellular ligninolytic enzyme production by white rot fungus, *Lentinus polychrous*, grown in glucose containing medium supplemented with rice straw powder and soybean pomace was monitored. Optimum condition for remazol brilliant blue R (RBBR) decolorization was studied by varying initial RBBR concentrations, pH values, and initial crude enzyme concentrations. The result revealed that the ligninolytic enzyme dominantly produced was laccase with an activity of 0.095 U/ml on day 15, and a small amount of manganese peroxidase was also detected. The crude laccase produced from *L. polychrous*, effectively decolorized the dye within a short period of time, that was approximately 50% of 20 mg/l RBBR was decolorized within the first 2 hours. The optimal pH for RBBR decolorization was 3.0 which had an efficiency of 87% within 6 hours after incubation. Moreover, the best redox mediator of laccase was $CuSO_4$, whose decolorization efficiency was over twice than that of samples without this mediator. The decolorization intensified with increase of $CuSO_4$ concentration but higher concentrations of chromium tended to suppress decolorization.

1. Introduction

Synthetic dyes have been widely used in textile, paper, cosmetic, food, and pharmaceutical industries. Textile industry consumes large amounts of water supply and produces tremendous amounts of highly colored effluent containing up to 30% of unfixed dyes and other organic compounds, that is released during the dyeing process [1]. Based on different structures in chromophores of the synthetic dyes, they are classified into several groups, like azo, anthraquinone, indigo, triphenylmethyl and phthalocyanine dyes, containing complex aromatic and heterocyclic structures. These complex structures make the dyes toxic, impermeable to light, and barely biodegradable when discharged into the environment [2]. Furthermore, the dyes hold carcinogenic and mutagenic properties which multiple adverse effects on the environment and health. Dye-containing effluent commonly has high biological oxygen demand (BOD), chemical oxygen demand (COD), and high salt concentrations [1]. The dye-containing effluent accumulates in the water body obstructs sunlight penetrating through the water, thereby reducing photosynthesis of aquatic plants and resulting in lower oxygen concentration in the water. Moreover, as microbes also require dissolved oxygen to biologically degrade excess amounts of contaminated dyes, these dyes in the water, eventually cause oxygen depletion, with negative environmental consequences.

Biological decolorization is a promising approach due to its advantages like being an environmentally friendly process, lower chemical and energy consumptions, and reduced cost in breakdown of organic compounds [3]. Several studies have demonstrated that fungi are able to degrade complex organic substances using extracellular ligninolytic enzymes like laccase, manganese peroxidase, and lignin peroxidase [4]. Ability of the fungi to produce each ligninolytic enzyme is different among fungal species due to differences in genetic materials and culture conditions. Some fungi can produce laccase and manganese peroxidase, while others can produce only laccase [5, 6]. In general, application of fungi or their extracellular enzymes are more favorable than those of other organisms since fungi can decolorize several types of synthetic dyes having different chemical structures [6], and ligninolytic enzymes originating from different fungal species can breakdown the same dye at different levels [7]. Therefore, investigation of organisms useful in degrading several complex and recalcitrant pollutants is an important approach in bioremediation.

Laccase is a copper-containing oxidoreductase enzyme recognized for its broad specificity in catalyzing a wide range of substrates like lignin, phenolic and non-phenolic molecules, synthetic dyes and aromatic amines coupled with reduction of oxygen [8]. It is evident that laccase production can be enhanced by addition of aromatic inducers like benzyl alcohol, guaiacol, veratryl alcohol, and some heavy metals [3, 9]. Laccase from several fungal species has been studied for its ability in decolorization. Most of them are *Trametes* sp., *Phanerochaete* sp., *Funalia* sp., *Pleurotus* sp. [10] such as *Trametes versicolor*, *Trametes trogii*, *Phanerochaete chrysosporium*, *Aspergillus* sp., *Pleurotus ostreatus*, *Coriolus versicolor* and *Funalia trogii* [5, 8, 11-13]. In this study, a member of the relatively under-investigated *Lentinus* sp., which is *Lentinus polychrous*, was studied for its decolorizing performances using extracellular ligninolytic enzymes. *L. polychrous* is a famous edible wood-degrading white rot fungus commonly found in northern and northeastern regions of Thailand. This research aimed to monitor ligninolytic enzyme activity produced from *L. polychrous* and apply the crude enzyme for decolorization of remazol brilliant blue R (RBBR). RBBR is a sulfonated anthraquinone dye used as a starting substance for polymeric dye in textile industry [14] and considered as a recalcitrant organopollutant containing complex aromatic structure. In this study, several parameters like pH values, initial dye concentrations and initial enzyme concentrations, were varied to determine the optimal condition for RBBR decolorization, as well as effects of mediators and heavy metal ions on decolorization.

2. Materials and methods

2.1. Materials
RBBR (reactive blue 19), 2,2-azino-bis (3-ethylbenzothiazoline-6-sulfonic acid) (ABTS), and hydroxybenzotriazole hydrate (HBT) were purchased from Sigma-Aldrich (St. Louis, MO, USA). Bromophenol blue was obtained from LabChem. Methyl orange, chromium (III) chloride, and copper (II) sulphate were obtained from M&B, QReC, and Univar, respectively.

2.2. Culture condition for ligninolytic enzyme production
The fungus, *L. polychrous*, was grown on freshly prepared potato dextrose agar (PDA) for one week to obtain the active fungal mycelium. Then, three mycelial bits (7 mm each) were transferred to 250 ml Erlenmeyer flasks containing 100 ml potato dextrose broth (PDB) supplemented with 0.1 g of rice straw powder and soybean pomace. The cultures were incubated at 27°C, 110 rpm for ligninolytic enzyme production and enzyme activity in the supernatant was monitored every three days until day 15. Thereafter, the culture was centrifuged at 12,000 rpm for 10 min to obtain the mycelial free crude enzyme, which would subsequently be tested for dye decolorization.

2.3. Ligninolytic enzyme activity determination
Enzyme activity of the three main ligninolytic enzymes, which are laccase, manganese peroxidase, and lignin peroxidase were monitored every three days. Laccase activity was determined using 2,2'-azino-

bis (3-ethylbenzothiazoline-6-sulphonic acid, ABTS) in acetate buffer, pH 4.5. Oxidation of ABTS was spectrophotometrically determined by an increase in absorbance at 420 nm ($\varepsilon_{420} = 0.036\ \mu M^{-1}cm^{-1}$) [4]. Measurement of manganese peroxidase and lignin peroxidase was modified from Bholay *et al.*, 2012. Manganese peroxidase was monitored using 0.1 mg/ml phenol red in the presence of 0.1 mM manganese sulfate, 1 mg/ml of bovine serum albumin, 0.1 mM hydrogen peroxide, 25 mM sodium lactate, and 0.5 ml culture filtrate [15]. Oxidation of phenol red was monitored at 610 nm ($\varepsilon_{610} = 0.0446\ \mu M^{-1}cm^{-1}$) [16]. Lignin peroxidase was determined by monitoring the demethylation of methylene blue to azure C [15]. The reaction contained 32 μM methylene blue in 50 mM sodium tartarate buffer (pH 4.0), 10 μl of culture filtrate and 0.1 ml of 0.1 mM H_2O_2. The reaction was incubated at room temperature for 1 hour and lignin peroxidase activity was measured at A_{650nm}. Percentage of methylene blue decolorization was calculated as (A_{650} of control – A_{650} of test/A_{650} of control) x 100. All enzyme activity reactions were stopped by adding 100 μl of H_2SO_4. One unit of enzyme activity (U) was defined as the amount of enzyme required for oxidation of 1 μmol substrate per minute.

2.4. *Effects of different dye concentrations, pH values, and enzyme concentrations on decolorization*

Different RBBR concentrations of 20, 40, 60, 80 and 100 mg/l were prepared using citrate phosphate buffer (pH 7.0). Each reaction contained 25 ml of RBBR and 2 ml of crude enzyme. To determine the optimum pH, decolorization of RBBR was performed within the pH ranges of 3.0 – 7.0 and the desired pH of the dye solution was prepared in citrate phosphate buffer. Each reaction contained 25 ml of 20 mg/l dye solution and 2 ml of crude enzyme. To determine the effect of enzyme concentrations on decolorization, crude enzyme amounts of 0.5, 1, 1.5, 2 and 2.5 ml, having laccase activity of 0.095 U/ml, were tested for ability in decolorizing 20 mg/l RBBR. Each reaction contained 25 ml of dye solution and different amounts of enzymes. All reactions were incubated at room temperature without shaking, and decrease in absorbance was spectrophotometrically monitored at 0, 1, 2, 4, and 6 hours at the absorbance of 592 nm. The reactions were tested in triplicate, and the experimental control was the sample without the crude enzyme, performed under the same conditions.

2.5. *Effect of mediators and heavy metal on dye decolorization*

$CuSO_4$ (10, 20, 30 mg/l) and HBT (1, 5, 10 mM) were used as mediators on RBBR decolorization. $CrCl_4$ (10, 20, 30 mg/l) was used to study effects of heavy metals on RBBR decolorization at the optimum pH. Each mediator and heavy metal was added to 100 mg/l RBBR in citrate phosphate buffer (pH 3.0). Each reaction contained 25 ml of dye solution and 2.5 ml of the crude enzyme. Monitoring of dye decolorization was performed as described above. Absorbance spectra of the untreated and treated samples were scanned using UV-VIS 1800 spectrophotometer (Shimadzu).

2.6. *Percentage of dye decolorization*

Percentage of dye decolorization was calculated using the following formula.

$$\text{Decolorization efficiency } (\%) = [(A_I - A_T)/ A_I] \times 100$$

where, A_I is initial absorbance of the heat inactivated crude enzyme, A_T is specific time absorbance measured after incubation with the enzyme. The values shown were mean of triplicate with standard deviation.

3. Results and discussion

3.1. *Ligninolytic enzyme production*

Ligninolytic enzyme production was studied by culturing the active mycelium of *L. polychrous* in PDB medium supplemented with rice straw and soybean pomace. The results revealed that extracellular ligninolytic enzymes secreted by *L. polychrous* were predominantly laccase, along with low levels of manganese peroxidase. Figure 1 shows that the maximum laccase activity of 0.095 U/ml was obtained on day 15, and the activity was around six folds higher than that of manganese peroxidase. It is possible that laccase activity might increase if longer incubation time is provided, and addition of laccase inducers would stimulate fungal growth and enzyme production. Lignin peroxidase was undetectable under the culture condition, which could be because the fungus lacks genes responsible for lignin peroxidase production or the gene expression is suppressed [17]. For these reasons, laccase was the major enzyme that played a vital role in decolorizing RBBR in this study.

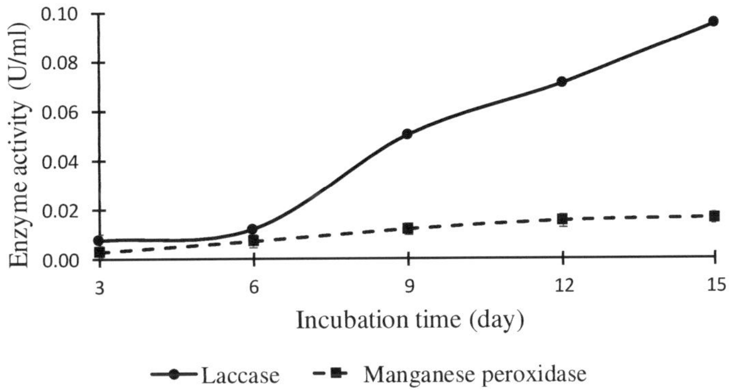

Figure 1. Ligninolytic enzyme production by *L. polychrous*.

Ligninolytic enzyme production shown in the present study was in agreement with Wangpradit *et al.*, 2014 who reported that ligninolytic enzyme produced by *L. polychrous* was only laccase and manganese peroxidase [6]. Another research showed that laccase production of *L. polychrous* on solid substrates, corn husk, rice bran and rice husk was 0.145 U/ml on day 12, which was 1.5 times higher than our study [18]. The different types and compositions of the culture medium as well as culture conditions contribute to different amounts of enzyme production. It has been reported that ligninolytic enzyme production by white-rot fungi substantially depends on medium compositions such as carbon and nitrogen contents [9, 19]. When applying different types of carbon sources, *L. crinitus* grown in 10 g/l glucose decolorized the highest levels of RB220 dye up to 85%, while the culture with glycerol and starch yielded low destaining activity of only 5-10% [19]. This indicated that different types of carbon sources resulted in variable amounts of ligninolytic enzyme production, thus influencing decolorization performance. Apart from *Lentinus* sp., other white rot fungi, *P. ostreatus* strain 32 and *P. ostreatus* HAUCC 162 produced only laccase, while lignin peroxidase and manganese peroxidase were undetectable [8, 20]. They revealed that laccase activity substantially increased with organic nitrogen sources like peptone and yeast extract [8], and several metal ions, especially Cu^{2+}, could promote laccase production [20]. Ligninolytic enzyme of *Polyporus* sp. S133 was mostly laccase, and production of lignin peroxidase and manganese peroxidase was relatively low [21].

3.2. Effect of RBBR concentrations on decolorization

Effect of RBBR concentrations on decolorization efficiency was performed at 20 - 100 mg/l dye concentrations at pH 7.0. After 2 hours, 20 mg/l RBBR was decolorized up to 50%, and more than 40% decolorization was obtained with 40 and 60 mg/l RBBR as shown in Figure 2. It was observed that destaining activity barely increased after 2 hours of incubation, possibly because the enzyme becomes saturated with the substrate or enzyme denaturation occurs during a long incubation period. Decolorization of 100 mg/l dye concentration was only 23% after 6 hours of incubation. The results demonstrated that decolorization efficiency decreased with increasing dye concentrations. Our results are similar to those of Mechichi *et al.*, 2006 and Hadibarata *et al.*, 2011, who revealed that dye decolorization decreased with increasing concentrations of RBBR, suggesting that the rate of reaction increased with the substrate concentrations until saturation [12, 21]. Also, adding high concentrations of dye as a substrate directly to the culture could be toxic to fungal growth, and the growth of *P. ostreatus* was inhibited when initial RBBR concentration increased to 100 mg/l demonstrating the inhibitory effects of the dye on fungal growth [5].

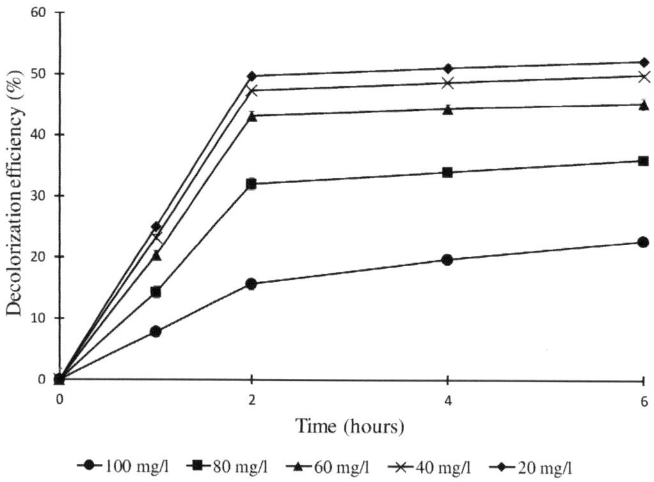

Figure 2. Decolorization of different RBBR concentrations.

3.3. *Effect of pH values on decolorization*

RBBR decolorization was performed at different pH values from 3.0 to 7.0 in citrate phosphate buffer containing 20 mg/l RBBR and 2 ml of crude enzyme. Among five tested pH values, the maximum decolorization was obtained at pH 3.0 where majority of RBBR decolorization occurred at 78% within 2 hours of incubation and the maximum decolorization of 86% was obtained within 4 hours as shown in Figure 3. It was obvious that the enzyme could effectively decolorize RBBR at low pH ranges from 3.0 to 5.0, and the dye was distinctly degraded within the first 2 hours of incubation, after that the percentage of decolorization barely increased. The decolorization efficiency was greatly decreased at pH 6.0 and 7.0 with a removal efficiency of only 59% and 47% at 6 hours, respectively. It was evident that dye decolorization largely depended on pH levels and the decolorization was more favorable in acidic environment.

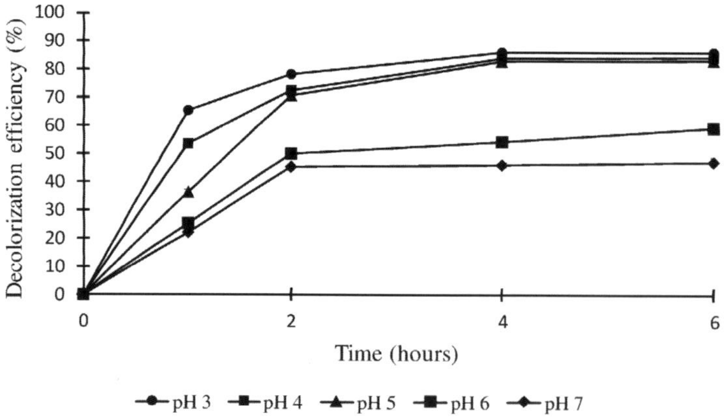

Figure 3. Optimum pH for decolorization of RBBR.

The maximum decolorization efficiency at pH 3.0 was in accordance with Sarnthima *et al.*, 2009 who reported that decolorization of 20 mg/l RBBR at pH 3.0 was 58% within 1 hour [22]. They also reported using 0.2 U/ml of crude laccase which was almost twice than that of our study. Another observation was that *L. polychrous* decolorized 66% of RBBR at pH 4.0 within 3.5 hours, whereas the optimum pH on ABTS substrate was 3.0, and relative activity continuously decreased with increasing pH values from 3.0 to 7.0, and the remaining activity of 16% at pH 7.0 [23]. The higher catalytic activity at low pH ranges might contribute to higher performance of RBBR decolorization in acidic environment and small decolorization at neutral pH. It was reported that most laccase produced from fungi have the optimum pH in acidic ranges [14]. Laccase from *Lentinus* sp. showed the optimum pH at 2.5, 3.5, and 5.0 on the substrates ABTS, 2,6-DMP, and guaiacol respectively, indicating that the optimum pH mainly depended on the substrates [14]. It is noticeable that the optimum pH values were distinct in the acidic ranges, where laccase had a higher catalytic performance. Hadibarata *et al.*, 2011 found that the optimal RBBR decolorization rate of laccase purified from *Polyporus* sp. S133 was at pH 5.0 and that the crude laccase was unable to decolorize RBBR at pH 7.0 [21].

According to dye's chemical structure and enzyme conformation, electrical charge of the dye and enzyme plays an important role in decolorization. RBBR contains sulfonic ($–SO^{-}_{3}–$) functional group which is negatively charged, while electrical charge at the active site of the enzyme is largely dependent on the pH of the solution. At lower pH values with high concentration of H^{+}, the active site has a strong positively charged facilitating binding with anionic sulfonic group of the RBBR dye; thus, high removal efficiency is achieved at low pH values. On the other hand, an active site at higher pH values contains less positively charged H^{+}, therefore ionic attraction between the dye and enzyme is less, and a reduction in decolorization efficiency is thus observed [24].

To determine the optimum pH of other synthetic dyes, decolorization of bromophenol blue and methyl orange was studied in pH ranges from 3.0 to 7.0 (data not shown). The results showed that both dyes had the optimum pH at 4.0 with maximum decolorization efficiency of 73% and 71% for bromophenol blue and methyl orange, respectively (Table 1). Similar observation was found by Rakrudee 2009 showing that the optimum pH for bromophenol blue decolorization was at pH 4.0 with 37% decolorization efficiency [22]. The dye decolorization in this study indicated that the optimal pH of laccase was substrate dependent and that the enzyme was more active in the acidic range. It was also noteworthy that bromophenol blue and methyl orange decolorization was less than 5% at pH 7.0. However, crude laccase produced from *L. crinitus* was active in a wide range of pH values from acidic to basic conditions and was able to decolorize more than 70% of reactive blue 220 from pH 3.5 to 8.0. [19].

Table 1. Decolorization of other synthetic dyes by *L. polychrous*.

Dyes	Types	Optimum pH	Time (h)	Decolorization (%)
Bromophenol blue	Triphenylmethane	4.0	2 h	70
			12 h	73
Methyl orange	Azo	4.0	2 h	37
			12 h	71

In general, different dye structures affect decolorization capacity and laccase decolorizes anthraquinone dye more effective than other dye groups [18], and the most resistant dyes are triphenylmethane and azo dyes [25]. Our result found that the most degradation-resistant group was azo (methyl orange), followed by triphenylmethane (bromophenol blue), while anthraquinone (RBBR) was the most degradable dye. However, the decolorization efficiency of bromophenol blue was almost as high as that of RBBR, as crude laccase also decolorized bromophenol blue as effectively. Similarly, crude enzyme from *Paraconiothyrium variabile* also had a good capacity in decolorizing the triphenylmethane dye [7]. This could be because different isoforms of crude laccase may be effective in decolorizing different types of dyes and some isoforms are capable of decolorizing triphenylmethane or azo dyes [20]. Unlike other dyes which were decolorized to reach maximum efficiency within 2 to 4 hours, degradation of methyl orange took 12 hours to reach maximum degradation of 72%, while decolorization at 2 hours was only 32%. This indicated that methyl orange was more resistant than other groups and longer degradation time was required to reach maximum decolorization. It was noteworthy that three types of dyes tested in the present study had optimal pH in acidic ranges, which might contribute to high catalytic performance of the crude laccase in low pH ranges.

3.4. Effect of enzyme concentrations on decolorization

Effect of enzyme concentrations on decolorization varied from 0.5 to 2.5 ml (with laccase activity of 0.095 U/ml). The results revealed that decolorization efficiency increased with increasing enzyme concentrations. The difference in decolorization was obvious within the first two hours after incubation where higher enzyme concentration resulted in higher decolorization rate. The maximum decolorization of 89% was obtained when using 2.5 ml of crude enzyme, and the lowest decolorization of 68% was obtained when using 0.5 ml of crude enzyme as shown in Figure 4. It was noticeable that more than 80% of the dye was decolorized within the first 2 hours when using more than 2 ml of crude enzyme.

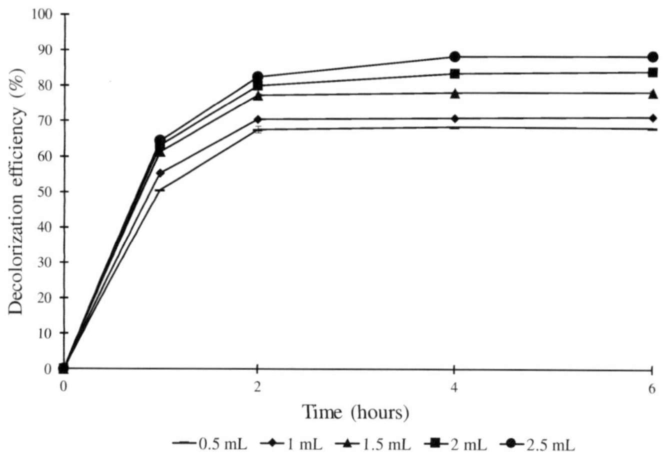

Figure 4. Effect of different enzyme concentrations on decolorization.

Enzyme activity was an important factor in decolorization. The decolorization efficiency mainly increased with enzyme concentrations [18]. Hsu *et al.*, 2012 found that laccase from *Lentinus* sp. decolorized RBBR only 29% with 1 U/ml; however, by increasing laccase activity to 20 U/ml, the decolorization efficiency increased to 88% [14]. Moreover, laccase from *Lentinus* sp. showed better decolorization efficiency than that of a commercial *T. versicolor*. It was observed that high decolorization efficiency could be achieved if higher amount of laccase was used [22]; nevertheless, adding excess amount of crude enzyme for effluent decolorization in practical application might pose another environmental problem. This is because high protein contents or organic matters in the crude enzyme might require further wastewater treatment. Therefore, an appropriate amount of enzyme concentration should be applied.

3.5. Effect of mediators and metal ions on decolorization

As decolorization efficiency of 100 mg/l RBBR was relatively low, this initial concentration was chosen to test the effect of the mediators $CuSO_4$ and HBT, on improving decolorization. In addition, chromium, which has usually been found in wastewater of textile industry, was selected to test the effect of metal ion on decolorization [1]. The result showed that addition of HBT (1, 5, 10 mM), $CuSO_4$ and $CrCl_4$ (10, 20, 30 mg/l) could increase decolorization efficiency, and $CuSO_4$ was better than HBT and $CrCl_4$ in enhancing decolorization, which could be because laccase is a copper-containing oxidase enzyme requiring Cu^{2+} to perform its active function. The maximum decolorization efficiency of the sample with $CuSO_4$ was 74% when using 30 mg/l $CuSO_4$, which was more than twice than that of the sample without mediator (Figure 5). Addition of HBT increased decolorization and 1 mM HBT was the best concentration to enhance decolorization up to 50%. However, decolorization efficiency did not correspond with increasing concentrations of HBT to 5 and 10 mM, which might be attributed to the inhibitory effect of excessive concentrations of mediators. Also, chromium could enhance decolorization up to 65% using 10 mg/l $CrCl_4$ and using higher concentrations of Cr^{3+} tended to decrease decolorization efficiency (Figure 6). Decolorization of the samples with 20 and 30 mg/l Cr^{3+} was 57% and 54%, respectively, at 6 hours of incubation. This indicated that decolorization efficiency was potentially affected by high concentration of Cr^{3+}, this could be because of the interaction of the high metal ion concentration interfered with the electron transport system of laccase contributing to its instability [26, 27]. Absorbance spectra of the samples with and without the mediators showed the peak at maximum wavelength around 592 nm (Figure 7). Degradation of the sample with $CuSO_4$ was the most effective in decolorizing RBBR; showing the lowest peak compared to the absorbance spectrum of the sample before decolorization.

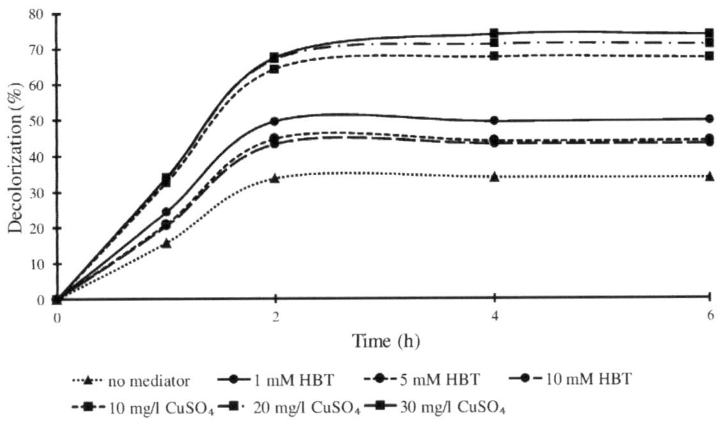

Figure 5. Effect of mediators, $CuSO_4$ and HBT on decolorization.

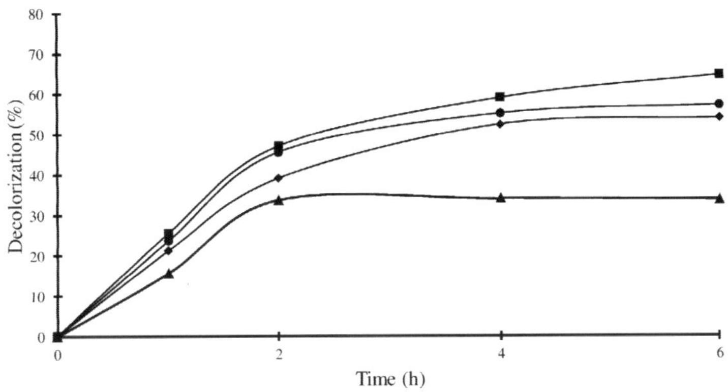

Figure 6. Effect of heavy metal, Cr^{3+} on decolorization.

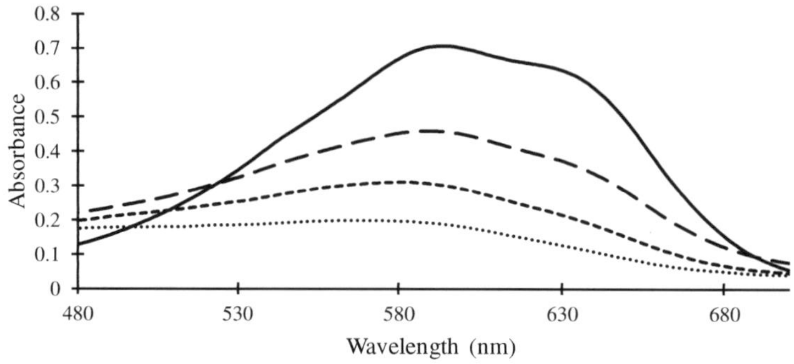

Figure 7. Absorbance spectra of the native RBBR and the decolorized RBBR in the presence and absence of mediators at 24 hours of incubation.

In the presence of mediators, laccase enzyme oxidizes low molecular weight mediators to stable radicals or redox mediators, then the radicals oxidize non-laccase substrates. Therefore, the mediator that has high affinity to the substrate, will enhance a laccase-mediator system increasing decolorization efficiency [26]. It has been reported that using redox mediators to enhance RBBR decolorization by adding Cu^{2+} to the culture medium, a maximum RBBR decolorization of 79% was obtained, compared to other metal ion mediators [26]. In addition, crude laccase (30 U/ml) produced from *P. ostreatus* strain 32 was able to decolorize 70% of remazol brilliant blue SN4R and the decolorization efficiency increased up to 90% in the presence of 0.16% ABTS mediator. However, inhibitory effect in decolorization appeared when an excessive amount of ABTS was applied [8]. Apart from using fungi for decolorization, cyanobacteria were also able to perform dye decolorization. Afreen *et al.*, 2017 found that laccase from cyanobacteria, *Spirulina platensis* CFTRI, could decolorize 47% of 100 mg/l RBBR within 48 hours and decolorization efficiency increased almost twice in the presence of redox mediator syringaldehyde [3]. On the other hand, laccase produced from *L. polychrous* in the present study rapidly decolorized the dye up to 67% in the presence of $CuSO_4$ in a much shorter time which was within 2 hours, indicating high performance of the enzyme produced from the fungi. Among ligninolytic enzymes produced from the fungus, *Funalia trogii*, only laccase was responsible for RBBR decolorization, and decolorization was achieved without addition of hydrogen peroxide or redox mediator [5]. Similarly, RBBR decolorization in this study occurs without addition of H_2O_2 or $MnSO_4$, the main cofactor and substrate for lignin peroxidase and manganese peroxidase, indicating that laccase is the enzyme that plays a vital role in RBBR decolorization.

4. Conclusion

Ligninolytic enzyme produced by *L. polychrous* is mainly laccase, and it is responsible for RBBR decolorization. The crude laccase is effective in decolorization of RBBR at pH 3.0, while small decolorization is obtained at pH 6.0 and 7.0. At this optimum pH, around 78% of 20 mg/l RBBR is decolorized within 2 hours, which is relatively fast compared to some of the other fungal species or cyanobacteria, and decolorization decreases with increasing dye concentrations from 20 to 100 mg/l. By increasing concentrations of crude enzymes, the decolorization efficiency increases. In addition, decolorization efficiency increases more than twice in the presence of the mediator $CuSO_4$. However, increasing concentration of chromium tends to inhibit dye decolorization, thus contamination of chromium in the effluent of textile industry might decrease decolorization efficiency of the crude laccase. Using crude extract produced by *L. polychrous* to decolorize synthetic dyes is a simple, environmentally friendly and ecofriendly method which has high potential in the degradation of several synthetic dyes and could also be applied for remediation of other aromatic and xenobiotic pollutants.

Acknowledgments
The authors gratefully acknowledge the Faculty of Applied Science, King Mongkut's University of Technology North Bangkok, Thailand for the financial support of this research project under Grant No. 5944109.

References

[1] Babu B R, Parande A K, Raghu S and Kumar T P 2007 *J Cotton Sci* **11** 141-53
[2] Afreen S, Anwer R, Singh R K and Fatma T 2016 (in press) *Saudi J Biol Sci* (https://doi.org/10.1016/j.sjbs.2016.01.015) 1-8
[3] Afreen S, Bano F, Ahmad N and Fatma T 2017 *Biocatal Agric Biotechnol* **10** 403-10
[4] Sing N N, Husaini A, Zulkharnain A and Roslan H A 2017 *Biomed Res Int* **2017** 1-8
[5] Erkurt E A, Ünyayar A and Kumbur H 2007 *Process Biochem* **42** 1429-35
[6] Wangpradit R and Chitprasert P 2014 *Int Biodeterior Biodegradation* **93** 168-76
[7] Forootanfar H, Moezzi A, Aghaie-Khozani M, Mahmoudjanlou Y, Ameri A, Niknejad F and Faramarzi M A 2012 *Iranian J Environ Health Sci Eng* **9**:27 1-10
[8] Hou H, Zhou J, Wang J, Du C and Yan B 2004 *Process Biochem* **39** 1415-9
[9] Isanapong J, Krailoekpaiboon T, Noiniyom W and Panchal S 2017 *KMUTNB: IJAST* **10** 239-44
[10] Sen S K, Raut S, Bandyopadhyay P and Raut S 2016 *Fungal Biol Rev* **30** 112-33
[11] Sasmaz S, Gedikli S, Aytar P, Gungormedi G, Cabuk A, Hur E, Unal A and Kolankaya N 2011 *Appl Biochem Biotechnol* **163** 346-61
[12]]Mechichi T, Mhiri N and Sayadi S 2006 *Chemosphere* **64** 998-1005
[13] Soares G M, Costa-Ferreira M and Pessoa de Amorim M T 2001 *Bioresour Technol* **79** 171-7
[14] Hsu C A, Wen T N, Su Y C, Jiang Z B, Chen C W and Shyur L F 2012 *Environ Sci Technol* **46** 5109-17
[15] Bholay A D, Borkhataria Bhavna V, Jadhav Priyanka U, Palekar Kaveri S, Dhalkari Mayuri V and Nalawade P M 2012 *Univers J Environ Res Technol* **2** 58-64
[16] Silva M L C, Souza V B d, Santos V d S, Kamida H M, Vasconcellos-Neto J R T d, Góes-Neto A and Koblitz M G B 2014 *Adv Biosci Biotechnol* **05** 1067-77
[17] Budda W, Sarnthima R, Khammuang S, Milintawisamai N and Naknil S 2012 *J Biol Sci* **12** 25-33
[18] Ratanapongleka K and Phetsom J 2014 *Int J Chem Eng Appl* **5** 26-30
[19] Niebisch C H, Malinowski A K, Schadeck R, Mitchell D A, Kava-Cordeiro V and Paba J 2010 *J Hazard Mater* **180** 316-22
[20] Zhuo R, Yuan P, Yang Y, Zhang S, Ma F and Zhang X 2017 *Biochem Eng J* **117** 62-72
[21] Hadibarata T, Mohd Yusoff A R and Ayu R 2011 *Water Air Soil Pollut* **223** 933-41
[22] Sarnthima R, Khammuang S and Svasti J 2009 *Biotechnol Bioprocess Eng* **14** 513-22
[23] Khammuang S and Sarnthima R 2007 *Biotechnology* **6** 408-13
[24] Kumar Sen S, Raut S, Bandyopadhyay P and Raut S 2016 *Fungal Biol Rev* **30** 112-33
[25] Eichlerová I, Homolka L and Nerud F 2006 *Bioresour Technol* **97** 2153-9
[26] Hadibarata T, Yusoff A R M and Kristanti R A 2012 *Water Air Soil Pollut* **223** 4669-77
[27] Rodríguez Couto S, Sanromán M and Gübitz G M 2005 *Chemosphere* **58** 417-22

Comparison of antioxidant activity of ginseng root extracts obtained by pulsed electric field and hydrolytic enzyme processing

C W Lu[1] and Y G Yin[2, 3]

[1] College of Life Science, Changchun Normal University, Changchun 130032, China

[2] College of Biological and Agricultural Engineering, Jilin University, Changchun, 130032, China

E-mail: biofoodyyg@hotmail.com

Abstract. The antioxidant activity and biological compounds of ginseng root extracts were compared during pulsed electric field (PEF) and hydrolytic enzyme (HE) processing. The total ginsenoside content of ginseng root extracts by PEF (29.51±1.67 mg/g) was higher than cellulase enzyme (HEC 23.74±1.85 mg/g) and *β*-glucosidase enzyme (HEG 25.31±2.34 mg/g) treatment. The ginseng root extracts obtained by PEF contained more total polyphenols (15.62±1.07 mg/g) and flavonoids (16.58±1.5 mg/g) than HEC (polyphenols 10.76±1.22 mg/g; flavonoids 13.2±1.63 mg/g) and HEG (polyphenols 14.09±0.39 mg/g; flavonoids 15.37±1.05 mg/g) treatment. Moreover, PEF processing showed more powerful scavenging activities against DPPH, ABTS radicals and FRAP activity of ginseng root extracts. Scanning electron microscopic examination of the ginseng root tissue with PEF processing provided direct evidence for the disruption of structure integrity. These results indicated that bioactive ingredients extraction with higher antioxidant activity could be obtained via PEF processing.

1. Introduction

As the healthy food and traditional medicine, ginseng was consumed widely in Asian countries. So far, more than 60 kinds of ginsenosides have been identified from ginseng plant [1]. Ginsenosides included 20(*S*)-protopapaxdiol and 20(*S*)-protopapaxtriol categories according to sugar moieties. For example, ginsenoside Rb1, Rc and Rg3 belonged to 20(*S*)-protopapaxdiol; ginsenoside Re, Rh1 and Rg1 belonged to 20(*S*)-protopapaxtriol [2]. The ginsenosides had various effects for human health, including antioxidant, immunomodulatory, neuro-protective, anticancer and antifatigue [3-5].

There are many methods for extracting ginsenoside from ginseng root, such as heat reflux [6], ultrasound extraction [7], microwave extraction [8], supercritical carbon dioxide extraction [9] and ultrahigh pressure extraction [10]. Pulsed electric field (PEF) belonged to a non-thermal technique for food processing [11]. The electrical damage of PEF caused reversible or irreversible pores formation of cell membrane. With the increasing intensity of PEF processing, the reversible pores formation converted into irreversible disruption in membrane. The PEF treatment had applied in the fields of bioactive ingredients extraction [12-14]. The PEF processing can enhance the extraction ration of polysaccharides from corn silk [15]. The maximum polyphenols yield of fresh tea leaves achieved 27% at intensity of electric field 0.9kV/cm [16]. A research reported the different conditions of pulsed electric field on chondroitin sulphate (CS) content in fish bone. These results indicated that the PEF processing enhanced the extraction yield of CS comparison with traditional methods [11]. Recent studies have focused the application of the enzyme hydrolysis on ginsenoside extraction. Lee *et al.* reported that single enzyme of Ultraflo L (UTGL) treatment showed higher levels of ginsenoside content (406.1 g/mg) from ginseng leaf than other enzymes (celluclast 1.5 L, cytolase PCL5 and viscozyme L) [17]. Compared with untreated ginseng leaves, the DPPH and ABTS antioxidant activities were significantly higher than those with UTGL treatment [17].

The ginseng root extract had important potential application on food industry and pharmaceutical industry. In the present study, we compared total ginsenoside content, antioxidant activity and biological compounds under PEF, HEC and HEG extraction.

2. Materials and methods

2.1. Materials

The ginseng roots (four year old) was obtained from Jilin province, dried in vacuum at 60°C and passed through an 80 mesh screen. The cellulase enzyme and β-glucosidase were purchased from Baoman Biology (Shanghai, China). The HPLC solvents were analytical grade and were purchased from J. T. Baker, (PA, USA). All other chemical reagents were analytical grade.

2.2. Instruments and apparatus

The PEF treatment instrument was indicated in Figure 1(a). This apparatus consisted with material chamber, pump, pulse generator and sensing temperature device [12].The following instruments and apparatus: electronic balance (JA10002, Shanghai, China); centrifuge biofuge heraeus (4000-40,000 r/min, Shanghai, China); rotary evaporator (RE-52A, Shanghai, China); the HPLC instrument (Agilent 1100, HP Technologies, USA); spectrophotometer (UV-VIS8500, Beijing, China); steam sterilizer (DSX-280B, Shanghai, China); freeze dryer (ZL-10TD, Shanghai, China); magnetic stirrer (85-2, Jintan, China) were used in the present study.

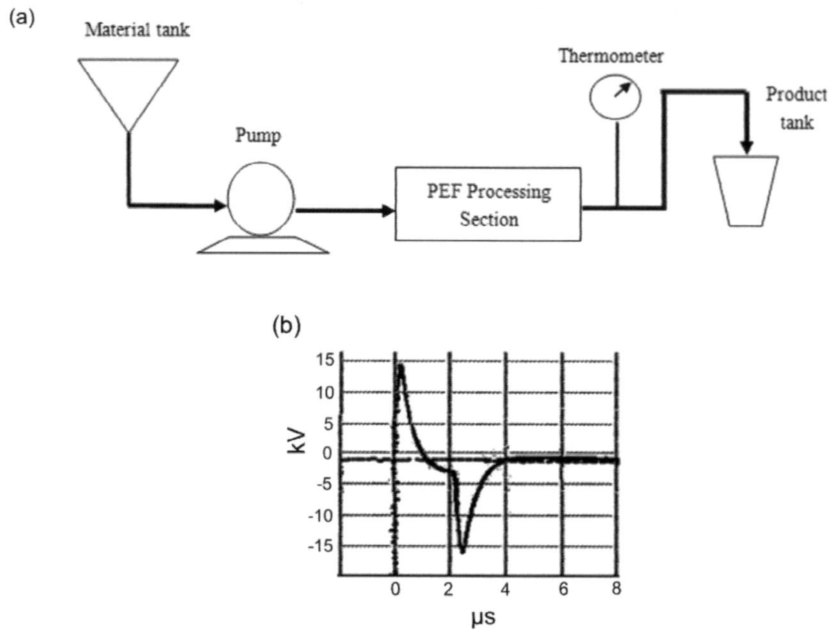

Figure 1. (a) Diagram of high intensity pulsed electric fields processing instrument. (b) The pulsed wave form.

2.3. Methods of extraction of ginsenosides from ginseng root

2.3.1. PEF extraction. The ginseng powder (1g) was treated with 70% ethanol solution (50mL) on the condition of pulsed electric field 10 kV cm^{-1} (PEF1), pulsed electric field 15 kV cm^{-1} (PEF2), pulsed electric field 20 kV cm^{-1} (PEF3) and pulse number 8, respectively [18]. The ginseng root extracts were centrifuged (5000 r/min for 20 min) after treatment. The extracts were freeze-dried and then dissolved in methanol as test solution.

2.3.2. Hydrolytic enzyme extraction. The ginseng powder (1g) was extracted with distilled water (50mL), adding commercial enzyme cellulose and *β*-glucosidase to the solution. The solutions were extracted at 50°C for 4 h (HEC1 and HEG1), 6 h (HEC2 and HEG2) and 8 h (HEC3 and HEG3). To inactivate the enzyme, the treated solutions were boiled for 10 min. Then, ginseng root extracts were centrifuged (5000 r/min for 20min). The freeze-dried extracts were dissolved in methanol as test solution.

2.4. Determination of total ginsenoside content

The HPLC analysis was used Agilent 1100 system (reverse phase C_{18} column) and a UV spectrophotometric detector. The column temperature was maintained at 30℃. The mobile phase included solvent A (water) and solvent B (acetonitrile). The flow rate of solvent was 1 mL/min. The eluate was measured at 203 nm wavelength and 10 μL injection volume.

2.5. *Analysis total polyphenol and flavonoid content*

The modified method from Singleton and Lamuela-Raventos was used to determine total polyphenol content of ginseng root extraction [19]. The colorimetric assay was used to determine total flavonoid content of ginseng root extraction [20-21].

2.6. *Analysis DPPH radical scavenging activity*

The DPPH radical scavenging activity was measured by the method from Yang [22]. The DPPH power (2.5 mg) was dissolved in 10 mL ethanol and mixed completely. DPPH solution in ethanol (2 mL) was added to 2 mL sample solution. The absorbance of the extract was measured at 514 nm at room temperature, using ethanol as the blank.

$$SA(\%) = 1 - \frac{A_I - A_J}{A_0} \times 100$$

A_I indicated absorbance of DPPH and sample; A_J indicated absorbance of sample and ethanol; A_0 indicated absorbance of DPPH and ethanol.

2.7. *Analysis ABTS radical scavenging activity*

The ABTS antioxidant activity was analyzed using the method of Hu and Kitts [23]. The 7 mM ABTS stock solution was mixed with 2.45 mM potassium persulfate as ABTS radical cation in darkness for 14 h. The absorbance of ABTS solution was adjusted to 0.7 with distilled water. The ginsenoside root extract 50 μL was added to 2 mL ABTS radical solution for reacting 6 min. The spectrophotometer was used to determine the ABTS antioxidant activity at 734 nm using trolox as positive control [23].

$$\text{Inhibition ratio } (\%) = (A_0 - A_1)/A_0 \times 100$$

where A_0 is the absorbance of the control, and A_1 is the absorbance of the test sample.

2.8. *FRAP assay*

The FRAP assay was determined following the method from Chen [24]. The ginseng root extracts were mixed with the FRAP solution for 1 h in dark condition. The absorbance was determined at 593 nm against water as a blank. The results were indicated by μmol of Trolox equivalents (TE)/g FW.

2.9. *Analysis structure changes of ginseng root tissue after extraction*

To elucidate the effect of each extraction procedure on structure changes, ginseng root extraction with CPEF processing was analyzed by scanning electron microscopy (SSX-550 SEM). The sample particles were fixed on a specimen holder with aluminum tape and then sputter-coated with gold [10].

2.10. *Statistical analysis*

One way ANOVA and Duncan's multiple range tests were analyzed by SPSS 13.0 software, and statistical significance was showed as $P<0.05$. The experiments were completed in three times and all data were presented as standard deviations (SD).

3. Results and discussion

3.1. Comparison of ginsenoside content between PEF and HE extraction

Total ginsenoside content of ginseng root extraction under PEF and HE treatment was shown in Figure 2. The extraction yield of ginsenoside by PEF2 processing was higher than for HEC2 and HEG2 treatment. Its content was 1.24 times higher than that with HEC2 treatment, and 1.17 times higher than that with HEG2 treatment. These results showed that the PEF processing enhanced the extractability of ginsenoside from ginseng root. Moreover, the extraction time of PEF2 processing was shorter than that with HE treatment. Therefore, the PEF processing was highly efficient to prepare ginsenoside from gingseng root.

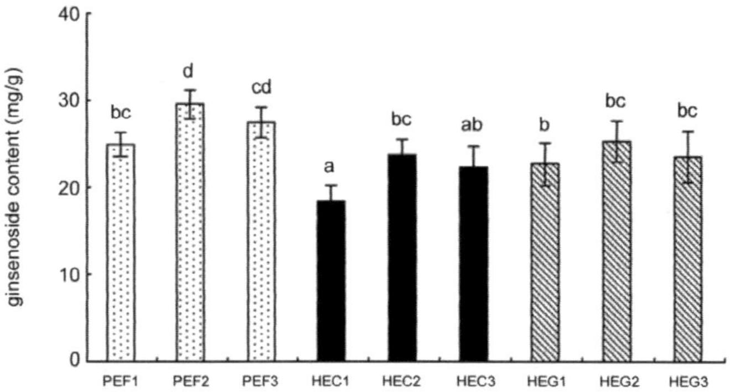

Figure 2. Changes of ginsenoside content in dry mass of ginseng root with different PEF treatment conditions.

3.2. Comparison of total polyphenol content between PEF and HE extraction

As important secondary metabolites in plant, polyphenol compounds had antioxidative effect and preventive effect on cardiovascular diseases [25]. The total phenolic content by PEF2 processing was 15.62 mg/g, which was 45.3 % higher compared to that with HEC2 treatment and 10.86 % higher compared to that with HEG2 treatment (Figure 3). The accumulation of total phenolic compounds may affect the biological activities of ginseng root extraction [26].

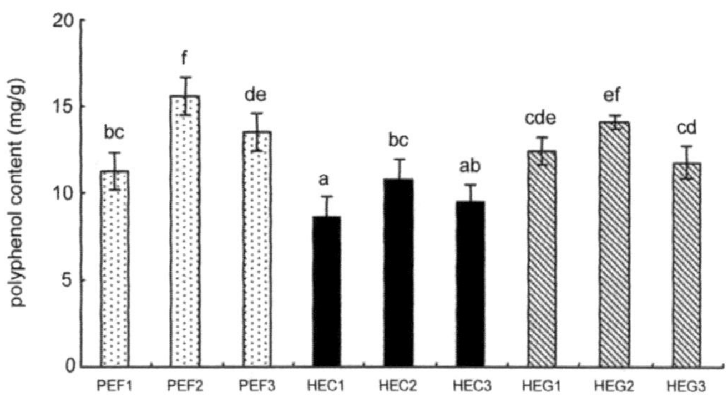

Figure 3. Changes of polyphenol content in dry mass of ginseng root extracts with different PEF treatment conditions.

3.3. Comparison of total flavonoid content between PEF and HE extraction

As complex phenolic molecules, the flavonoid compounds have distinctive antioxidative characteristics in plant [27]. The total flavonoid is important antioxidant ingredients of ginseng root. Total flavonoid contents of ginseng root extract were summarized in Figure 4. The PEF2 treatment contained higher content of total flavonoid (16.58 mg/g) than that with HEC2 (13.19 mg/g) and HEG2 treatment (15.37 mg/g).

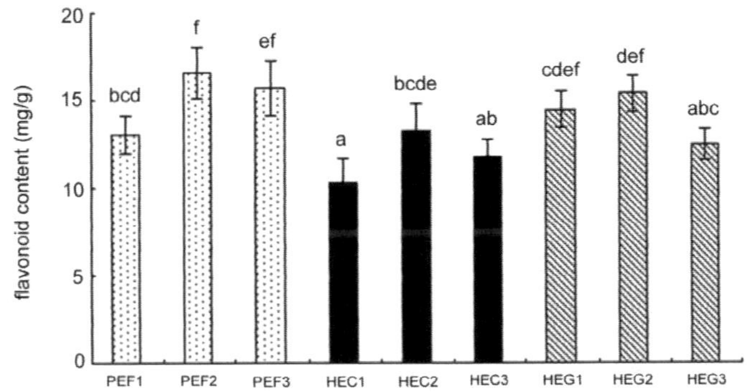

Figure 4. Changes of flavonoid content in dry mass of ginseng root extracts with different PEF treatment conditions.

3.4. Compared with DPPH radical scavenging activity between PEF and HE extraction

There is an interdependent relationship between structure and radical scavenging activity of ginsenoside [28]. We compared the radical scavenging activity of ginseng root extracts obtained with PEF and HE treatment. Figure 5 showed the DPPH radical scavenging activity of ginseng root extracts produced with different methods. Among the different methods, PEF2 method (54.89 %) indicated the maximum efficiency on DPPH scavenging activity, followed by HEG2 (53.59 %) and HEC2 method (46.23 %). Therefore, the ginseng root extracts by PEF processing showed significant capacity of scavenging the free DPPH radicals.

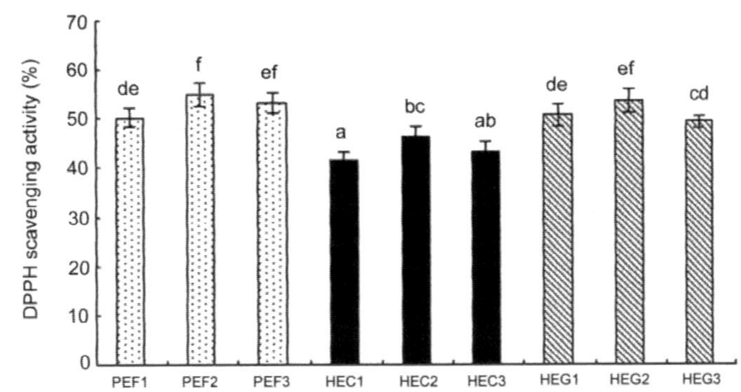

Figure 5. Changes of DPPH scavenging activity of ginseng root extraction with different PEF treatment conditions.

3.5. Comparison of ABTS radical scavenging activity between PEF and HE extraction

The antioxidant capacity of plant can be analyzed by ABTS radical scavenging activity according to the ability of reducing radical cations [29]. The ABTS radical scavenging activity of ginseng root extract was shown in Figure 6. Compared with HEG2 treatment (44.28 %) and HEC2 treatment (40.49 %), the PEF2 processing had significantly higher ABTS radical scavenging activity (45.83 %). From these results, it is confirmed that PEF processing showed higher antioxidant capacity of gradients. These results might be related with the altered position of phenolic group.

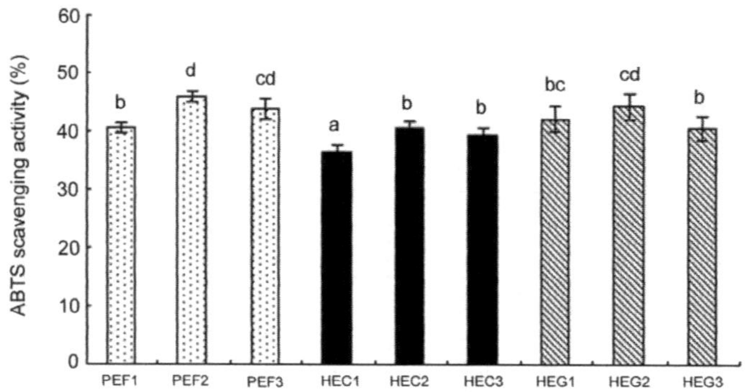

Figure 6. Changes of ABTS scavenging activity of ginseng root extracts with different PEF treatment conditions.

3.6. Comparison of FRAP activity between PEF and HE extraction

Antioxidant capacity of ginseng root extracts as evaluated by FRAP was summarized in Figure 7. The ginseng root extracts via PEF2 processing showed higher FRAP values than that via HEC2 and HEG2 processing. The FRAP values of PEF extraction showed a rapid increase until electric field intensity 15 kV cm^{-1}, then decreased with higher electric field intensity. The maximum FRAP values achieved 7.22 μmol (TE)/g FW at the PEF parameters of electric field intensity 15 kV cm^{-1} and pulse number 8. The analysis clearly demonstrated that PEF processing was capable of improving the antioxidant capacity of ginseng root extracts.

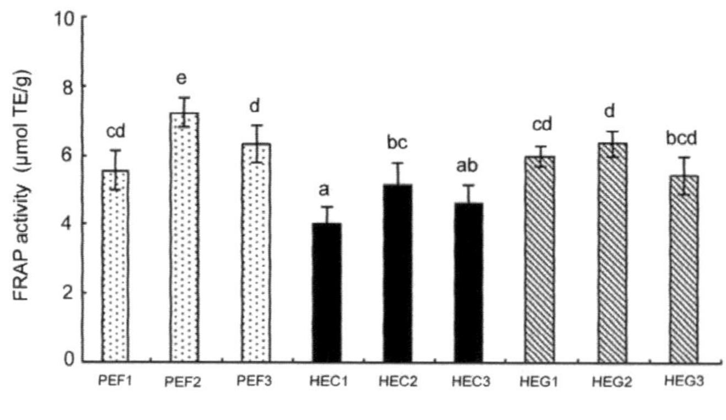

Figure 7. Changes of FRAP activity of ginseng root extracts under different extraction conditions.

3.7. Comparison of structure changes of ginseng particles between PEF and HE extraction

The samples treated with PEF and HE processing were analyzed by scanning electron microscopy. There were significant structural changes of ginseng root powder under PEF and HE treatment. As can be seen from Figure 8a, the ginseng root structures of untreated samples were kept intact, described by SEM micrograph. However, the ginseng root structure of PEF2 processing generated hollow openings and smaller particles (Figure 8b). The samples of HEC2 and HEG2 treatment observed only puny damage of structure (Figure 8c and Figure 8d). These results indicated that the surface tension of ginseng root structure were destroyed by PEF processing and induced many small particles generation.

Figure 8. Electron micrograph of samples. (a) untreated sample; (b) sample treated by PEF (pulsed electric field 10kV/cm, pulse number 8); (c) sample treated by HEC; (d) sample treated by HEG.

4. Conclusion

In this study, total biological compounds and antioxidant activity of ginseng root extractions obtained by PEF, HEC and HEG methods were investigated. Compared with the HEC and HEG methods, the PEF processing had higher ginsenoside extraction efficiency, higher total polyphenol and flavonoid content. The process indicated higher antioxidant activity of ginseng root extractions. The PEF treatment as a non-thermal technology was high yield, high speed and low energy consumption, compared to the traditional methods. Therefore, PEF processing was capable to extract bioactive ingredients from plant materials with high quality.

Acknowledgements

This work was supported by National Natural Science Foundation of China (31601400) and Natural Science Foundation from Jilin Province Department of Science and Technology (20140101040JC).

References

[1] Cheng LQ, Na JR, Bang MH, Kim MK and Yang DC, 2008, Conversion of major ginsenoside Rb1 to 20(S)-ginsenoside Rg3 by Microbacterium sp. GS514. *Phytochemistry*, **69**, 218-24.

[2] Huang D, Li Y, Zhang M, Ruan SL, Zhang HY, Wang YR and Hu P, 2016, Tartaric acid induced conversion of protopanaxadiol to ginsenosides Rg3 and Rg5 and their in situ recoveries by integrated expanded bed adsorption chromatography. *J. Sep. Sci.*, **39**, 2995-3001.

[3] Lee TK, Johnke RM, Allison RR, O'Brien KF and Dobbs LJ, 2005, Radioprotective potential of ginseng. *Mutagenesis*, **20**, 237-43.

[4] Min JK, Kim JH, Cho YL, Maeng YS, Lee SJ and Pyun BJ, 2006, 20(S)-Ginsenoside Rg3 prevents endothelial cell apoptosis via inhibition of a mitochondrial caspase pathway. Biochem. *Bioph. Res. Co.*, **349**, 987-94.

[5] Sun JH, Hu SH and Song XM, 2006, Adjuvant effects of protopanaxadiol and protopanaxatriol saponins from ginseng roots on the immune responses to ovalbumin in mice. *Vaccine*, **25**, 1114-20.

[6] Chen XC, Zhu YG, Zhu LA, Huang C, Chen Y and Chen LM, 2003, Ginsenoside Rg1 attenuates dopamine-induced apoptosis in PC12 cells by suppressing oxidative stress. *Eur. J. Pharmacol.*, **473**, 1-7.

[7] Wu J, Lin L and Chau F, 2001, Ultrasound-assisted extraction of ginseng saponins from ginseng roots and cultured ginseng cells. Ultrason. *Sonochem.*, **8**, 347-52.

[8] Wang YT, You JY, Yu Y, Qu CL, Zhang HR and Ding L, 2008, Analysis of gingsenosides in Panax ginseng in high pressure microwave-assisted extraction. *Food Chem.*, **110**, 161-67.

[9] Wang HC, Chen CR and Chang CJ, 2001, Carbon dioxide extraction of ginseng root hair oil and ginsenosides. *Food Chem.*, **72**, 505-9.

[10] Chen RZ, Meng FL, Zhang SQ and Liu ZQ, 2009, Effects of ultrahigh pressure extraction conditions on yields and antioxidant activity of ginsenoside from ginseng. *Sep. Purif. Technol.*, **66**, 340-6.

[11] He GD, Yin YG, Yan XX and Yu QY, 2014, Optimisation extraction of chondroitin sulfate from fish bone by high intensity pulsed electric fields. *Food Chem.*, **164**, 205-10.

[12] Yin YG, Cui YR and Ding HW, 2008, Optimization of betulin extraction process from Inonotus obliquus with pulsed electric fields. Innov. *Food Sci. Emerg.*, **9**, 306-10.

[13] Sarkis JR, Boussetta N, Tessaro IC, Marczak LDF and Vorobiev E, 2015, Application of pulsed electric fields and high voltage electrical discharges for oil extraction from sesame seeds. *J. Food Eng.*, **153**, 20-7.

[14] Puértolas E, Cregenzán O, Luengo E, Álvarez I and Raso J, 2013, Pulsed electric field assisted extraction of anthocyanins from purple-fleshed potato. *Food Chem.*, **136**, 1330-6.

[15] Zhao WZ, Yu ZP, Liu JB and Yu YD, 2011, Optimized extraction of polysaccharides from corn silk by pulsed electric field and response surface quadratic design. *J. Sci. Food Agric.*, **91**, 2201-9.

[16] Zderic A and Zondervan E, 2016, Polyphenol extraction from fresh tea leaves by pulsed electric field: a study of mechanisms. Chem. *Eng. Res. Des.*, **109**, 586-92.

[17] Lee HJ, Lee HS, Cho HJ, Kim SY and Suh HJ, 2012, Utilization of hydrolytic enzymes for the extraction of ginsenosides from Korean ginseng leaves. *Process Biochem.*, **47**, 538-43.

[18] Lu CW, Yin YG and Yu QY, 2017, Optimized extraction of ginsenosides from ginseng root (Panax ginseng C.A. Meyer) by pulsed electric field combined with commercial enzyme. *J. Food Process. Pres.*, **41**, e12766.

[19] Singleton VL and Lamuela-Raventos RM, 1999, Analysis of total phenols and other oxidation substrates and antioxidants by means of Folin-Ciocalteu reagent. *Method Enzymol.*, **299**, 152-78.

[20] Woisky RG and Salatino A, 1998, Analysis of propolis: Some parameters and procedures for chemical quality control. *J. Agri. Res.*, **37**, 99-105.

[21] Jung CH, Seog HM, Choi IW, Park MW and Cho HY, 2006, Antioxidant properties of various solvent extracts from wild ginseng leaves. *LWT-Food Sci. Technol.*, **39**, 266-74.

[22] Yang B, Wang JS, Zhao MM, Liu Y, Wang W and Jiang YM, 2006, Identification of polysaccharides from pericarp tissues of litchi (Litchi chinensis Sonn.) fruit in relation to their antioxidant activities. *Carbohyd. Res.*, **341**, 634-8.

[23] Hu C and Kitts DD, 2001, Free radical scavenging capacity as related to antioxidant activity and ginsenoside composition of Asian and North American ginseng extracts. *J. Am. Oil Chem. Soc.*, **78**, 249-55.

[24] Chen CO, Ribaya-Mercado JD, Mckay DL, Croom E and Blumberg JB, 2010, Differential antioxidant and quinone reductase inducing activity of American, Asian, and Siberian ginseng. *Food Chem.*, **119**, 445-51.

[25] Yang CS, Landau JM, Huang MT and Newmark HL, 2001, Inhibition of carcinogenesis by dietary polyphenolic compounds. *Annu. Rev. Nutr.*, **21**, 381-406.

[26] Kim BG, Choi SY, Suh HJ and Park HJ, 2011, Bitterness reduction and enzymatic transformation of ginsenosides from Korean red ginseng (Panax ginseng) extract. *J. Food Biochem.*, **35**, 267-82.

[27] Rice-Evans CA, Miller NJ and Pagana G, 1996, Structure- antioxidant activity relationship of flavonoids and phenolic acids. *Free Radical Bio. Med.*, **20**, 933-56.

[28] Liu ZQ, Luo XY, Liu GZ, Chen YP, Wang ZC and Sun YX, 2003, In vitro study of the relationship between the structure of ginsenoside and its antioxidative or prooxidative activity in free radical induced hemolysis of human erythrocytes. *J. Agr. Food Chem.*, **51**, 2555-8.

[29] Kang KS, Yamabe N, Kim HY, Okamoto T, Sei Y and Yokozawa T, 2007, Increase in the free radical scavenging activities of American ginseng by heat processing and its safety evaluation. *J. Ethnopharmacol.*, **113**, 225-32.

IOP Conf. Series: Earth and Environmental Science **185** (2018) 012006 doi:10.1088/1755-1315/185/1/012006

Research on soil quality under traditional drainage and ecological water storage models of saline and alkaline land

J Li[1, 2, 3, 4, 5], D W Hua[1, 2, 3, 4] and Y K Wang[1, 2, 3, 4]

[1] Institute of Land Engineering and Technology, Shaanxi Provincial Land Engineering Construction Group Co., Ltd., 710075, Xi'an, China

[2] Key Laboratory of Degraded and Unused Land Consolidation Engineering, the Ministry of Land and Resources of China, Xi'an 710075, China

[3] Shaanxi Provincial Land Consolidation Engineering Technology Research Center, 710075, Xi'an, Chin

[4] Shaanxi Provincial Land Engineering Construction Group Co., Ltd. 710075, Xi'an, China

Email: lijuan8136@163.com

Abstract. This paper aims to study halide under traditional drainage and ecological water treatment of saline-alkali soil through bulk density, porosity, moisture content, soluble and nutrient characteristics in Shaanxi, to analyze the different governance mode of different soil physical and chemical characteristics to provide scientific basis for saline-alkali land management. Saline-alkali land model test was set in Shaanxi Fuping, respectively set up traditional drainage and ecological water treatment, analysis soil bulk density, porosity, electric conductivity, total salt, organic matter, total N, available K, available P and fractal dimension of 0-30 cm soil layer; Analysis soil moisture content and reservoir capacity of 0-160 cm soil layer, and comprehensive analysis the average fractal dimension of soil and soil nutrient average correlation in 0-30 cm soil depth. Results show that: (1) in 0-30 cm soil depth, two kinds of mode, the soil bulk density and porosity appear consistent, the soil bulk density can be reduced, and ecological water storage mode can effectively increase the soil porousness compare with traditional drainage mode. (2) Under the same amount of water, in the 0-160 cm soil layer, the average soil moisture and average water storage capacity under the ecological water storage treatment were 4.47% and 2.57% higher than the traditional drainage treatment. (3) In the 0-30 cm soil layer, the traditional drainage and ecological water storage treatment significantly reduced the soil pH, conductivity and total salt content before the test, and the ecological water storage treatment decreased significantly; (4) In the 0-30 cm soil layer, the nutrient content of each treatment showed the same trend, which decreased with the increase of soil layer, and the average content of organic matter, total nitrogen, available potassium and available phosphorus under ecological storage treatment was higher than that of traditional drainage treatment. The height was 18.96%, 4.76%, 10.67% and 9.35%, and the difference between treatments was significant ($P<0.05$). (5) In the 0-30 cm soil layer, the fractal dimension of soil aggregates in dry and wet sieves showed opposite trend, and there was a good linear relationship with soil average bulk density, organic matter, total nitrogen, available potassium and available phosphorus (R^2=0.8006~0.9499), the difference was significant ($P<0.05$). In summary, the ecological water storage treatment can effectively improve soil physical and chemical properties, improve soil stability and soil quality, and achieve good salinity treatment.

1. Introduction

There are many Chinese people and few land. The advancement of urbanization has led to a continuous reduction in land area. It is necessary to find effective arable land resources and remediation and utilization as an important way to maintain the dynamic balance of cultivated land [1-2]. Due to China's national conditions and climate distribution characteristics, soil desertification, salinization and uneven distribution of water resources have become important issues in the development of cultivated land and agriculture in China [3-6]. The current research methods for soil desertification are more diversified, such as increasing organic fertilizer, introducing regional pioneer crops, etc. For saline-alkali land, most of them rely on the construction of drainage facilities to reduce the content of salt in the soil, but this technology has great drawbacks. If a large amount of water is to be drained, water resources are wasted, and the migration of saline-alkali water is promoted, so that the pollution of other water bodies is caused, and the soil bulk density is increased, and the soil permeability is reduced, which is not conducive to crop growth [7-10].

Shaanxi lubotan is a heavy saline-alkali land formed by natural and artificial pumping. The soil in the area is mainly composed of moderate saline soil and severely salinized tidal soil, and the soda saline soil is patchy [11].

By analyzing the characteristics of the natural environment in this region, based on the problem of soil salinization, The traditional drainage mode consumes a large amount of manpower, material resources and financial resources, and needs to quote a large amount of fresh water for salt washing, which is easy to cause waste of water resources and soil compaction; the ecological water storage mode can effectively realize the comprehensive utilization of soil and water resources, avoid increasing soil compaction and effectively reducing water waste. On the basis of comprehensive consideration of the traditional drainage mode, the water storage ecological treatment technology is proposed to achieve water retention, water and soil separation, water and land coexistence, and dynamic balance [12-13].

Soil physical and chemical properties are important parameters for measuring soil structure and evaluating soil quality and production performance. Good soil physical and chemical properties are beneficial to crop growth and yield improvement and quality improvement [14-15]. At this stage, the use of water storage mode to control saline-alkali land is mainly for its engineering technology, system and construction regulations, but the improvement of soil physical and chemical characteristics under the water storage mode has rarely been reported. By analyzing the soil bulk density, water, salinity and main nutrient characteristics under the traditional drainage and impounding ecological model, this paper analyzes the differences in soil physical and chemical characteristics under different treatment modes, and provides scientific basis and governance model for the appropriate regional control of saline-alkali land Feasibility and accuracy.

2. Materials and methods

2.1. Site description

The trial was conducted at the Fuping pilot base, located in Ducun Town, Fuping County, Weinan City, Shaanxi Province(109°11'47''E, 34°42'07''N). The area is a warm temperate semi-humid climate zone. The average annual rainfall and evaporation are 472.97 mm and 1213.35 mm respectively. The rainfall is mainly concentrated in July-September, accounting for 49% of the annual rainfall, frost-free period of 225 days, and annual average temperature. At 13.4°C, the total annual light energy radiation is 125.9 kcal/cm^2, which is suitable for the growth of cotton, wheat, corn and other crops.

2.2. Expermental design

The trial was implemented in October 2009. The self-made test device was used to simulate the conditions of the Lubotan land. The length, breadth and height were 23m, 1.5m and 2m, respectively. The bricks and concrete were used for concreting and the water storage and drainage treatment areas were set as Two, a total of four test devices. The test device is composed of a water storage part (drainage part) and a test soil trough. One end is a water storage tank (drainage) and the middle section is an earth tank. In order to prevent soil loss and water mixing, a geotextile is used to separate the two, and the interfaces of the device are treated as waterproof. The test treatment are traditional drainage and ecological storage treatment, that is, water storage treatment is to put quantitative water in the ditch, and use the water-soil dominant role-dispersion to control the movement of soil salinity; drainage treatment The flooding is carried out by flooding with large water, and the water is washed by the convection between the water and the soil, and the amount of water stored in the water storage and drainage is consistent throughout the test period. The salt transport and changes under each treatment are shown in Figure 1.

The test soil samples were selected from the saline-alkaline soil in the Lubotan area, and the unearthed soil in the lubotan area was layered for soiling at intervals of 30 cm. The soil samples were maintained in the original state and backfilled into the self-made test soil tank. According to the test data of the developed area of the test area in the test center of Northwest Agriculture and Forestry University in 2009, the average content of soil organic matter is 0.70%, the average salt content is 4.5g/kg, the average pH value is 9.33, and the average Cl^- content is 0.37%. HCO^{3--} The average content is 0.08%, and the average SO_4^{2-} content is 0.06%.During the experimental period, wheat varieties were tested: Xiaojing 22, 1000-grain weight: 38g, wheat germination rate was 90.1%, wheat sowing rate was 150 kg/hm^{-2}, row spacing was 20cm. Seeding using trench drilling.

The fertilizing amount in the experimental plots was: N = 255 kg/hm^2, P_2O_5 = 180 kg/hm^2, and K_2O = 90 kg/hm^2. Among them, phosphate fertilizer, nitrogen fertilizer and potassium fertilizer were diammonium phosphate, urea and potassium chloride.

In October 2009, when winter wheat was sown, the amount of fertilizing materials was distributed according to the district plan. Through artificial soil preparation and deep testing of the test field, the surface formation and the fertilizer were fully mixed, and the experimental fields were all artificially sowed. Wheat is not fertilized during the growing season and winter wheat is irrigated before delivery.

2.3. Sampling and Measurement

(1) Soil water storage and bulk density measurement [16]. In the spring of 2015 after the corn harvest, using soil drilling, sampling interval is 20 cm, soil moisture content, drying method, and calculate the soil reservoir capacity under various processing; Using ring knife soil method 0 ~ 30 cm soil bulk density, soil sampling interval to 10 cm.

(2) Determination of soil aggregate stability [17]. After year spring corn harvest, select test based on diagonal soil sampling method, set up five sampling points, sampling depth and the sampling interval is 30 cm and 10 cm respectively, the soil sampling undisturbed soil samples used in the determination of soil aggregate related indicators. Soil samples without damage are returned to the laboratory for testing, natural air drying after the stones that exist in the soil sample and cleaned sundries and soil mechanical stability determined by dry sieving method respectively, wet sieve method of determination of soil water stability.

(3) Measurement of soil chemical indicators [18-19]. Electric conductivity and total salt content measurement of soil sample is in accordance with Agricultural Chemical Analysis Method of Soil, pH measurement selects Rex pH meter. Organic matters content measurement uses potassium dichromate - concentrated sulfuric acid externally heating; available phosphorus is extracted with sodium bicarbonate, measured with UV/Vis spectrophotometer (700nm); available potassium is extracted with ammonium acetate, measured with flame photometer (M420); total nitrogen is measured with automatic Kjeldahl azotometer (UDK-129).

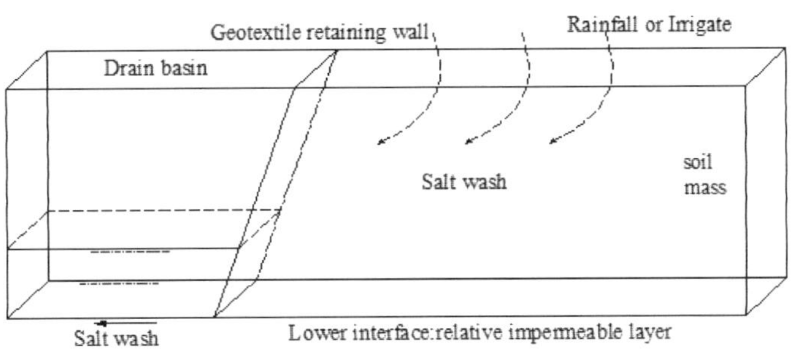

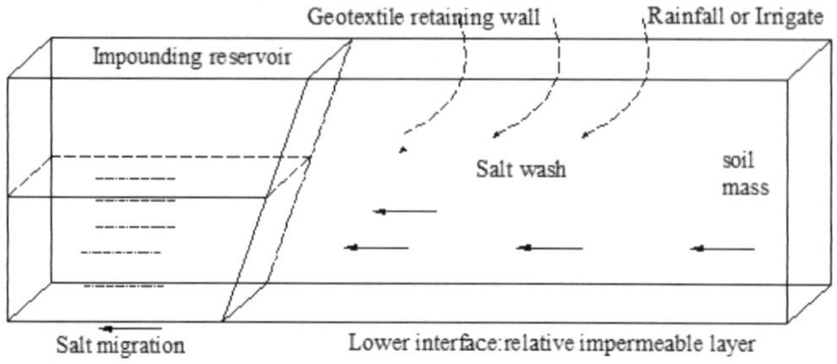

Figure 1. Experimental design sketch about water storage and waterdrainage treatments.

2.4. Statistical analyses

(1) Soil waterstorage calculation

$$SW=(M_1-M_2)/M_2\times 100\% \tag{1}$$

$$W=SW_i\times P_i\times H_i\times 10/100 \tag{2}$$

SW is soil moisture content, %; M_1 is Wet soil weight, g; M_2 is dry soil weight, g; W is the soil water storage, mm; SW_i is the ith layer soil moisture content, %; P_i is the layer of soil quality, g/cm^3, H_i is the ith layer soil thickness, cm.

Fractal dimension values were calculated using the equation below by Yang *et al.* [14]

$$\frac{M(r<\overline{T_n})}{M_t}=(\frac{\overline{T_n}}{T_{max}})^{3-D} \tag{3}$$

$\overline{Tn}$ means diameter of the aggregates is in a specific size, $M(r\le\overline{Tn})$ is the mass of aggregates in size less than $\overline{X_i}$, M_i is the total mass of aggregates (50 g), $Tmax$ is the maximum diameter of aggregates. D value can be obtained by linear regression using equation (3) through least square fitting method.

Using Origin 8 to process the data and figures, and then SPSS (PASW Statistics 18) was used to do the data statistical analysis and the multiple comparisons were done using duncan's new multiple range method (SSR).

3. Results and analysis

3.1. Characteristic parameter analysis of soil bulk density

Different treatment has significant influence on bulk density and porosity of 0-30 cm layer soil (Table 1). Soil porosity is negatively correlated with soil bulk density. Larger bulk density indicates that soil is more compact, so porosity decreases. At 0-30 cm layer, soil bulk density increases as soil depth increases, significantly different also under impounding and drainage, and significantly different between every soil layer ($p<0.05$). The soil bulk density of the water storage treatment in the 0-30 cm soil layer was 1.3%~4.2% lower than that of the drainage treatment. Soil porosity at every layer within 0-30 cm under impounding treatment is 3.1%, 4.7% and 1.5% higher than under drainage treatment. By comprehensively analyzing soil bulk density and porosity under impounding and drainage treatment, it is found that impounding treatment can effectively increase soil porosity at plough layer, increasing soil air and water permeability, conductive to maintaining soil bulk density at lower standard; and drainage treatment increases soil bulk density, intensifying soil hardness.

Table 1. Soil bulk density and soil porosity changes in 0-30 cm soil depth of different treatments.

Treatments	Soil depth/cm	Soil bulk density/g/cm^3	Soil porosity/%
	0-10	1.39±0.07c	48.18±2.56a
Storage	10-20	1.44±0.16ab	46.49±5.21b
	20-30	1.53±0.05a	43.48±1.73bc
	0-10	1.43±0.13ab	46.75±4.21b
Drainage	10-20	1.50±0.04b	44.41±1.20ab
	20-30	1.55±0.03a	42.84±0.98c

3.2. Spatial characteristics analysis of soil water

In December 2009, the winter wheat was irrigated under the impoundment and drainage treatment. The irrigation amount was the same. In the same year, the wheat harvest was determined to be 0-160 cm soil moisture in the same year, and the soil water storage under the storage and drainage treatment is shown in Figure 2. The soil moisture change trend of 0-160 cm soil layer under water storage and drainage treatment is consistent, in the 0-60 cm soil layer. The soil moisture decreased and the average soil moisture of the water storage and drainage treatments were 20.42% and 19.11%, respectively. In the 60-160 cm soil layer, the soil moisture increased and the average soil moisture was 24.65% and 23.86%, respectively. The average soil moisture of the 0-160 cm soil layer treated with water storage was 4.47% higher than that of the drainage treatment, and there was no significant difference between the treatments. The test results show that: the same amount of water, different treatments, after a long period of water transport, evaporation and utilization, the soil moisture under the water storage treatment is slightly higher than the drainage treatment, and the soil in the field is not easy to be knotted. Under impounding and drainage treatment water storage capacity of soil at 0-160cm layer varies as shown in Figure 3. Under impounding and drainage treatment water storage capacity of soil at 0-160cm layer shows identical trend, namely, at 0-60cm layer, soil water storage capacity both shows decreasing trend, under impounding and drainage treatment water storage capacity of soil at the layer decreases by 21.09% and 28.53% respectively, at 60-160cm layer, under impounding and drainage treatment water storage capacity of soil increase significantly by 26.16mm and 24.35mm respectively, and difference between different treatment is significant ($p<0.05$). At 0-160cm layer, under impounding and drainage treatment mean water storage capacity of soil is 65.36mm and 63.72mm respectively.

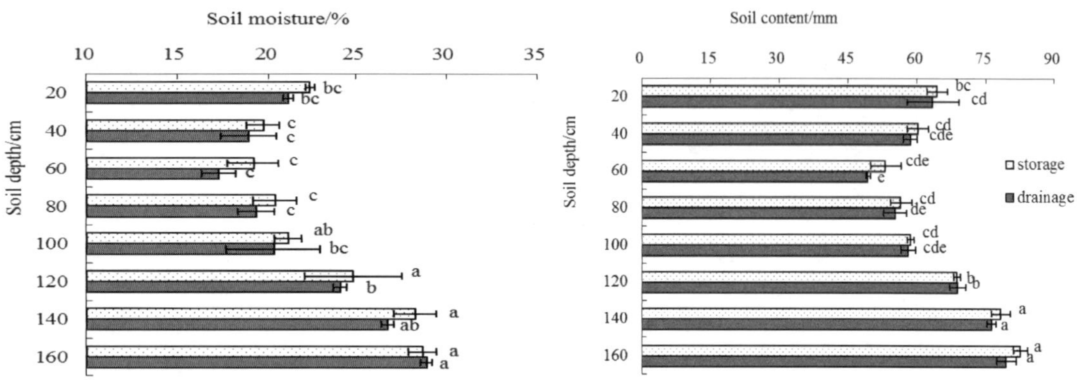

Figure 2. Soil water moisture changes and soil water content in 0-160cm soil profile of different models.

3.3. Spatial characteristic analysis of soil salt

The control of soil salinity plays a key role in the growth of crops. For the treatment of saline-alkali land, it is not only necessary to overcome its compaction, improve bulk density and increase soil porosity, and maintain its salt content in an appropriate range [20]. Effect of saline-alkali soil improvement under traditional drainage treatment and ecological impounding treatment is shown in Table 2.

Table 2. Discriptive characteristics of the soil pH, eletrical conductivity and salt content.

Treatments	Soil depth/cm	pH	Electrical conductivity/(s/cm)	Salt content /(g/kg)
	0-10	8.29±0.07bc	239.85±2.12a	1.77±0.12bc
Storage	10-20	8.38±0.05ab	205.45±3.74d	1.70±0.04bc
	20-30	8.48±0.05ab	203.91±1.44cd	1.40±0.03c
	0-10	8.35±0.02c	234.01±3.00ab	2.32±0.09a
Drainage	10-20	8.39±0.06ab	219.15±2.58bc	1.80±0.05b
	20-30	8.57±0.04a	212.05±2.53cd	1.53±0.06bc

At 0-30 cm layer, under impounding and drainage treatment soil pH, electric conductivity and total salt content shows identical trend, pH increases as soil depth increases; electric conductivity and total salt show identical trend, decreasing as soil depth increases, 3 indicators decrease obviously compared to before experiment, and difference is significant ($p<0.05$). In 0-30 cm soil depth, under impounding treatment soil means pH decreases by 0.64% than under drainage treatment, difference between treatments is insignificant. under impounding treatment 20-30 cm layer soil electric conductivity decreases by 35.95 us/cm than at 0-10 cm layer, drainage treatment decreases by 21.95 us/cm; at 0-30 cm layer, under impounding treatment mean electric conductivity decreases by 2.46% than under drainage treatment. impounding and drainage treatment significantly decrease soil total salt content than before experiment, through 3 year plantation experiment and salt control, at 0-30 cm layer, under impounding and drainage treatment mean total salt content is 1.62g/kg and 1.88g/kg, and impounding treatment decreases salt content more significantly than drainage, difference between treatment is significant ($p<0.05$). In general, impounding treatment can overall enhance soil quality, decrease soil salt content and pH integrated indicator more than drainage treatment, being a more suitable mode of saline-alkali soil improvement; drainage treatment can also decrease certain saline-alkali degree, less than impounding treatment, and easy to produce soil hardness, difficult to provide appropriate seedbed and growth environment to crops.

3.4. Spatial characteristic analysis of soil nutrient

Soil nutrient is indispensably important factor for sprouting and normal growth of crops, and also one of important indicators for evaluating soil fertility [21]. Saline-alkali soil improvement through drainage mode not only effectively improves soil structure, salt content, but also influences chemical property, namely nutrient of soil greatly, but also significantly improving soil nutrient content to meet fertility demand of crop growth (Table 3). Influence of impounding and drainage treatment on organic matters, total nitrogen, available potassium and available phosphorus content of soil at 0-30 cm layer is shown in Table 3, at 0-30 cm layer, under impounding, drainage treatment various nutrient content shows identical trend, namely decreases as soil depth increases. Under impounding, drainage treatment 0-10 cm soil layer organic matters content is 2.03 g/kg and 2.63g/kg higher than 20-30 cm soil layer respectively; total nitrogen content is 10.79% and 8.93% higher respectively; available potassium content 27.65% and 19.84% respectively, effective phosphorus content 46.52 mg/kg and 42.95mg/kg respectively. at 0-30 cm layer, mean content of organic matters, total nitrogen, available potassium and effective phosphorus under impounding treatment is all better than under drainage treatment, 18.96%, 4.76%, 10.67% and 9.35% higher respectively, difference between treatment is significant ($p<0.05$). In a word, under impounding, drainage treatment soil nutrient content increases obviously compared to before experiment, decreasing as soil depth increases. Effect of impounding treatment on improvement of soil nutrient content is more significant than drainage treatment, mainly because of different treatment mode. The drainage treatment and salt discharge method mainly adopts flood irrigation, which causes nutrient leaching and loss, and after the straw is returned to the field, the soil structure of the water storage treatment is better, and the water content is high, which is conducive to the decomposition and utilization of the straw. The drainage treatment of the soil is severe, which is not conducive to the decomposition of the straw.

Table 3. Discriptive characteristics of the soil nutrients of 0-30 cm soil depth.

Treatments	Soil depth/cm	SOM/(g/kg)	T.N/(g/kg)	A.K/(mg/kg)	A.P/(mg/kg)
storage	0-10	10.48±0.35a	0.70±0.05ab	242.74±1.48a	58.41±0.80a
	10-20	8.71±0.36bc	0.62±0.05ab	185.16±2.68c	22.63±0.33c
	20-30	8.45±0.50bc	0.62±0.01b	175.62±0.34cd	11.89±0.21d
drainage	0-10	9.24±0.05ab	0.65±0.04a	206.95±1.73b	53.58±0.70b
	10-20	7.38±0.31bc	0.61±0.03ab	172.49±1.29de	20.78±0.60c
	20-30	6.61±0.21c	0.59±0.02ab	165.88±0.46e	10.63±0.12d

3.5. Stability analysis of soil

Soil fractal dimension is an important parameter indicating soil structure and geometric shape, the smaller the soil coacervate fractal dimension, more stable the soil structure [22]. Using equation (3) to Fit calculation of agglomerate dry and wet sieve data, determine soil fractal dimension value (D), at 0-40cm layer, under different treatment fractal dimension of soil coacervate is shown in Table 4. In the 0-30cm soil layer, the fractal dimension of soil aggregates treated by dry sieve method decreases with the increase of soil depth, and the performance trend is consistent. The fractal dimension of soil aggregates is 1.65-2.15, and each Significant difference between treatments ($P<0.05$) Stability of impounding treatment at every soil layer is 7.02%~11.80% better than drainage treatment; mean stability of impounding treatment at 0-30 cm soil layer is 9.1% than drainage.at 0-30cm soil layer, results of wet screened soil coacervate fractal dimension under every treatment show a trend opposite to that of dry screened, namely every treatment shows it increases as soil depth increases, soil coacervate fractal dimension ranging 2.93-2.98.

Table 4. Fractal dimension of soil aggregates under different models

Treatments	Soil depth/cm	D (Dry sieving)	D (Water sieving)
	0-10	1.97±0.01ab	2.93±0.21c
storage	10-20	1.89±0.29ab	2.94±0.11bc
	20-30	1.64±0.06b	2.95±0.05ab
	0-10	2.15±0.09a	2.94±0.07bc
drainage	10-20	2.39±0.05a	2.96±0.04b
	20-30	1.83±0.04ab	2.97±0.01a

3.6. Correlation analysis of soil stability and chemical parameters

Correlation analysis between soil physical and chemical index data and soil stability, ie soil fractal dimension under wet sieve method (Figure 3), showed that there is a good linear relationship between soil stability and soil chemical characteristics. The effects of body stability on soil chemical characteristics were significant ($P<0.05$). Soil bulk density and soil fractal dimension are positively correlative, soil chemical indicators and fractal dimension are negatively correlative. The soil bulk density increases with the increase of fractal dimension, and the correlation coefficient between them is $R^2=0.8604$, which is significant.. as soil stability indicator fractal dimension is larger, various soil chemical indicator namely content of organic matters, total nitrogen, available potassium and available phosphorus shows decreasing trend, correlation coefficient of fractal dimension and various soil chemical property indicators is $R^2=0.9499$, $R^2=0.8472$, $R^2=0.8286$ and $R^2=0.8006$. The effect of soil stability on soil physical and chemical properties is mainly the impact of different treatments on soil structure, such as increasing soil compaction, moisture content, etc., causing changes in soil bulk density and nutrients, resulting in differences in soil water storage and moisture conservation, affecting soil. Quality and crop emergence, water and fertilizer absorption, and ultimately affect crop growth and yield formation.

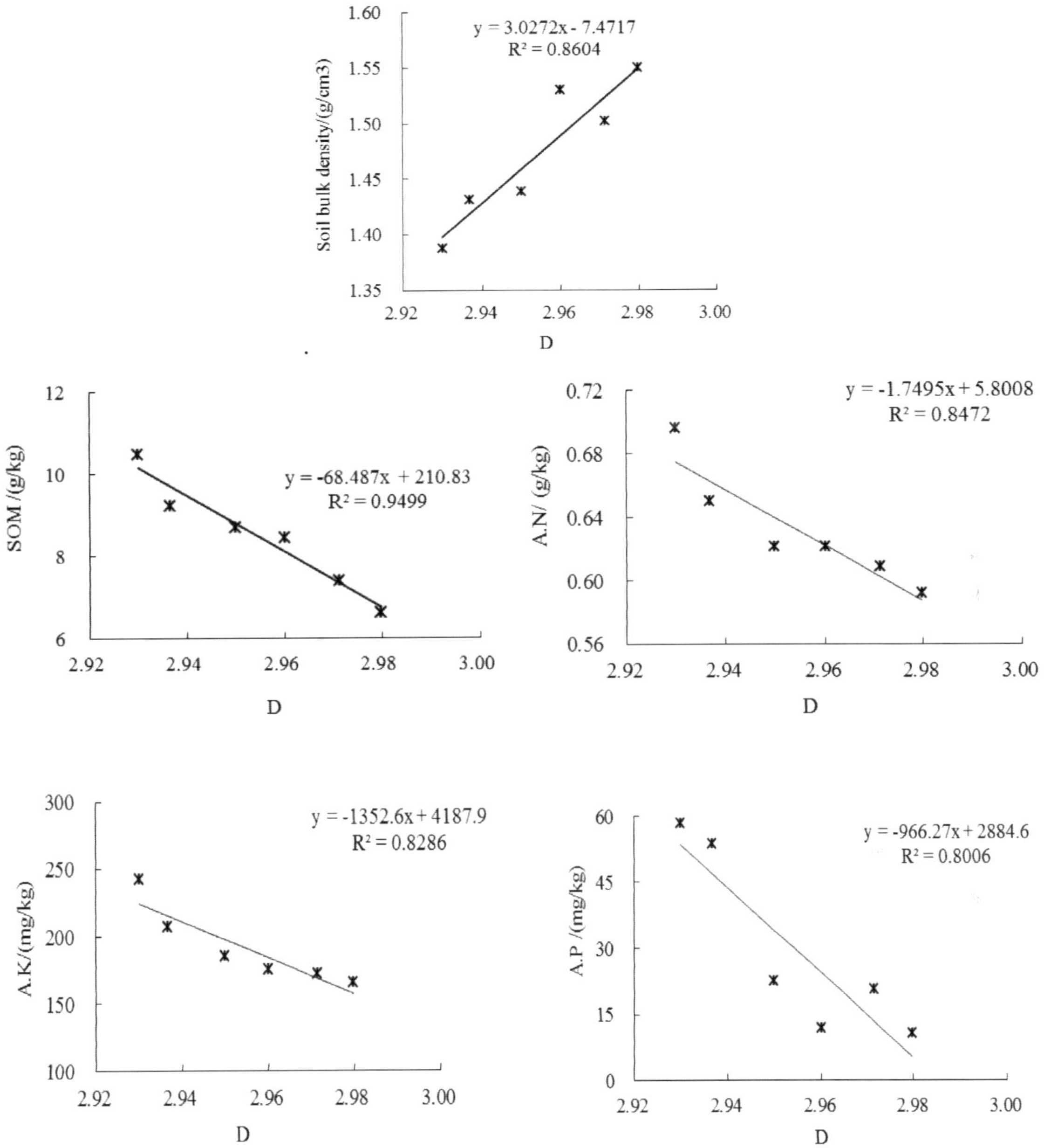

Figure 3. Correlations between Fractal dimension of soil aggregates and soil bulk density, SOM and nutrients.

4. Discussion

4.1. Influence on soil bulk density

Soil bulk density is one of important indicators assessing soil quality [23]. At present, the research on saline-alkali land mainly focuses on the selection of salt-tolerant varieties, the selection of irrigation methods, and the design of saline-alkali land treatment projects, etc., The use of circulating pressure salt and its implementation can change the spatial distribution of salt in the vertical direction and gradually press down the salt, but little research has been reported on the changes in soil physical characteristics [24-26]. Impounding treatment can effectively decrease salt compared to drainage treatment, effectively suppressing soil hardness, maintaining soil bulk density within appropriate range. In aid of straw recycling treatment, it can obviously improve soil structure. Impounding treatment effectively improves soil bulk density, significantly increasing field capacity and improving porosity distribution of soil, Combined with the return of straw, the distribution of solid, gas, and liquid accelerates decomposition of straw and improves soil structure to some extent. Irrigation under drainage treatment intensifies soil hardness and destruction of soil structure, easily forming soil runoff channel, intensifying nutrient loss and reducing water utilization. The research finds that impounding treatment significantly increases soil quality compared to drainage treatment.

4.2. Influence on soil impounding and moisture conservation

Utilization of effective water in soil and conservation of water in drought season are key factors for crop growth [27]. The experiment shows that under impounding, at 0-160 cm soil layer, soil humidity is obviously higher than drainage treatment, and soil water storage capacity and soil humidity show identical trend. At 0-60 cm soil layer and 60-160 cm soil, both show the trend is decreasing first and increasing then. Soil moisture conservation and increase can prevent from soil hardness, conductive to demand of crop growth.

4.3. Influence of soil salt and nutrient`

The salt-alkali salt washing process is the physical process of solute migration in the solute-migrating soil in the soil, including: convection, solute molecular diffusion, mechanical dispersion process, ion exchange adsorption at the interface between soil particles and soil solution. Traditional salt wash by drainage realizes rapid dissolution of various salt into water by effectively irrigating fresh water in aid of combination of wash and convection, reducing salt; core technology of ecological impounding mode is to use dynamic variation of impounding pool level, salt concentration and interaction of water body and adjacent soil to create dynamic balance, realizing circulation to suppress increase of salt. Overall objective factor and natural condition are influenced by variation of irrigated groundwater level, rainfall and water recession [28]. The ecological storage salt is based on the water level and salinity of the reservoir. The concentration continues to change and the water body and the adjacent soil salt reach a dynamic balance, which aims to achieve accelerated circulation of salt and alkali soil. Using “impounding without drainage” to improve saline-alkali soil effectively reduces labor, material and financial resources invested in other methods. Most importantly, it effectively avoids and reduces secondary pollution of and harm to natural resource, with broad application prospect and great significance [29-30]. 2 kinds of treatment significantly decreases soil pH and reduces soil content compared to before experiment, and at 0-30 cm layer, effect of impounding treatment reducing soil is better than drainage treatment.

4.4. Soil nutrients are necessary nutrients for crop growth in soil

The treatment of saline-alkali soil improvement not only effectively improves soil structure, soil salt content, but also has a greater impact on soil chemical characteristics, ie soil nutrients, it can significantly improve soil nutrient content, meeting fertility demand necessary for crop growth. The research shows that ecological impounding mode increases soil fertility better than traditional drainage treatment, agreeing with research result of Sun Bo *et al* [28]. In experiment by Han *et al* [29], it is also found that under impounding treatment, organic matter content in soil is improved, soil quality improves obviously.

4.5. Influence on soil stability and its relativity variation

The fractal dimension of soil can effectively reflect the stability of soil. In the 0-30cm soil layer, the fractal dimension of soil aggregates treated by dry sieve method decreases with the increase of soil depth, and the performance trend is consistent. Under impounding treatment various layer soil is more stable than under drainage treatment. This is mainly because the ecological water storage treatment effectively improves the soil structure, and the water storage mode effectively increases the soil aggregate due to the presence and maintenance of water. Water stability; water straw decomposition accelerates the input of organic matter, which plays a positive role in improving the structure. [30].

The improvement of soil stability is beneficial to increase soil nutrient. The correlation between the average fractal dimension of soil treated with different treatments in 0-30 cm soil layers and soil physical and chemical characteristics under wet sieve method is analyzed. Comprehensive analysis shows that soil mean fractal dimension and bulk density are positively correlative, and R^2=0.8604, difference is significant. Soil mean fractal dimension and mean content of soil organic matter, total nitrogen, available potassium and available phosphorus are negatively correlative, and correlation coefficient ranges 0.80~0.95, difference is significant ($P<0.05$) [31].

5. Conclusion

(1) Ecological impounding and traditional drainage treatment can effectively decrease soil bulk density, at 0-30 cm layer, it increases with the increase of soil layer, and impounding treatment improves soil bulk density and porosity better than drainage treatment, the reduction is 1.3%~4.2% and 1.5%~4.7%.

(2) Under the same amount of water, at 0-160 cm soil layer, under impounding treatment and drainage treatment soil humidity and water storage capacity shows identical trend, and under impounding and drainage treatment soil mean water storage capacity is 65.36mm and 63.72mm respectively, impounding treatment is helpful to water conservation.

(3) The storage and drainage treatment effectively reduced soil salinity and soil nutrient content compared to before experiment, and the effect of ecological water storage treatment on reducing salt and increasing nutrient content was better than drainage treatment.,

(4) The soil stability under ecological water storage treatment is higher than that of traditional drainage treatment. Under the wet sieve method, the average fractal dimension of soil in 0-30 cm soil layer and soil physical and chemical properties have good linear characteristics.

Acknowledgments

This research was supported by Shaanxi Provincial Key Technology Innovation Team Project Funding (2016KCT-23) and the Scientific Research Item of Shaanxi Provincial Land Engineering Construction Group (DJNY2018-14). We are grateful to researchers at the Fuping pilot base of Shaanxi Provincial Land Engineering Construction Group, for their help with the field experiments.

References

[1] Gorji T, Sertel E and Tanik A 2017 Monitoring soil salinity via remote sensing technology under data scarce conditions: A case study from Turkey *Ecol. Indic.* **74** 384-91.

[2] Manandhar R and Odeh I 2014 Interrelationships of land use/cover change and topography with soil acidity and salinity as indicators of land degradation *Land* **3** 282-99.

[3] Wu J, Li P, Qian H and Fang Y 2014 Assessment of soil salinization based on a low-cost method and its influencing factors in a semi-arid agricultural area, northwest China *Environ. Earth Sci.* **71** 3465-75.

[4] Li P, Qian H, Howard K and Wu J, 2015 Building a new and sustainable "Silk Road economic belt" *Environ. Earth Sci.* **74** 7267-70.

[5] Li P, Wu J and Qian H 2016 Regulation of secondary soil salinization in semi-arid regions: A simulation research in the Nanshantaizi area along the Silk Road, northwest China *Environ. Earth Sci.* **75** 1-12.

[6] Yang Z 1997 Land degradation and its control strategies in China *China Environ. Sci.* **17** 108-12.

[7] Li P, Tian R, Xue C and Wu J 2017 Progress, opportunities and key fields for groundwater quality research under the impacts of human activities in China with a special focus on western China *Environ. Sci. Poll. Res.*

[8] Li P, Qian H and Zhou W 2017 Finding harmony between the environment and humanity: an introduction to the thematic issue of the Silk Road *Environ. Earth Sci.* **76** 105.

[9] Wu J and Sun Z 2016 Evaluation of shallow groundwater contamination and associated human health risk in an alluvial plain impacted by agricultural and industrial activities, mid-west China *Expo. Health* **8** 311–29.

[10] Li P, Qian H and Wu J 2014 Origin and assessment of groundwater pollution and associated health risk: a case study in an industrial park, northwest China *Environ. Geochem. Health* **36** 693-712.

[11] Li P, Li X, Meng X, Li M and Zhang Y 2016 Appraising groundwater quality and health risks from contamination in a semiarid region of northwest China *Expo. Health* **8** 361-79.

[12] Liu J, Luo Z, Jia Z, Fang S and Wang N 2005 Feasibility study of adopting controlled drainage in Yinnan irrigation district based on salt and water balance. *T. Chin. Society Agr. Eng.* **21** 43-5.

[13] Yue W, Yang J, Tong X and Gao H 2008 Transfer and balance of water and salt in irrigation district of arid region *J. Hyd. Eng.* **39** 623-28.

[14] Yang G 1999 A study on water salt of Solonchak in Hetao plan, Inner Mongolia *Sci. Silvae Sinicae* **35** 107-10

[15] Lv D, Wang Q, Wang W and Shao M 2000 Soil water and salt transport features under one dimension infiltration *J. Soil Water Conserv.* **14** 91-4

[16] Wang Q and Wang W 2000 Water and salt transport features for salt effected soil through drip irrigation under film *Transactions Chin. Society Agr. Eng.* **16** 54-7.

[17] Khaledian Y, Kiani F, Ebrahimi S, Brevik E and Aitkenhead-Peterson J 2016 Assessment and monitoring of soil degradation during land use change using multivariate analysis *Land Degrad. Dev.* **28** 128-41.

[18] Lado M, Paz A and Ben-Hur M 2004 Organic matter and aggregate-size interactions in saturated hydraulic conductivity *Soil Sci. Soc. Am. J.* **68** 234.

[19] Taboada M, Barbosa O, Rodríguez M and Cosentino D 2004 Mechanisms of aggregation in a silty loam under different simulated management regimes *Geoderma* **123** 233–44.

[20] Rodrigo Comino J, Ruiz Sinoga J, Senciales González J, Guerra-Merchán A, Seeger M and Ries J 2016 High variability of soil erosion and hydrological processes in Mediterranean hillslope vineyards (Montes de Málaga, Spain) *Catena* **145** 274–84.

[21] Lu J and Li Z 2002 Advance in soil aggregate study *Res. Soil Water Conserv.* **9** 81-5.

[22] Yuan X, Jiao J, Zhu L, Liu M, Li H and Hu F 2011 Effects of earthworm activity on soil aggregates' stability and organic carbon distribution under different manipulations of corn straw *Soils* **43** 968-74

[23] Wei Y, Ma T, Wei X, Wang C, Hao M and Zhang M 2016 Effects of cropping systems on distribution of water-stable aggregates and organic carbon and nitrogen in soils in semiarid farmland of the Loess Plateau *J. Agro-Environ. Sci.* **35** 305-13.

[24] Sun G, Chen F, Li L, Wu F, Xiao X and Zhang H 2007 Effects of tillage on the carbon pool of paddy soil with long-term no-tillage *J. China Agr. U.* **12** 45-9.

[25] Liang A, Zhang X, Yang X and Drury C 2006 Short-term effects of tillage on soil organic carbon storage in the plow layer of black soil in northeast China *Sci. Agr. Sinica* **3** 1287-93.

[26] Vita D, Paolo E, Fecondo G, Fonzo N and Pisante M 2007 No-tillage and conventional tillage effects on durum wheat yield, grain quality and soil moisture content in southern Italy *Soil Till. Res.* **92** 69-78.

[27] Martinez E, Fuentes J, Silva P, Valle S and Acevedo E 2008 Soil physical properties and wheat root growth as affected by no-tillage and conventional tillage systems in a Mediterranean environment of Chile *Soil Till. Res.* **99** 232-44.

[28] Sun G, Chen F, Xiao X, Wu F and Zhang H 2007 Preliminary study on effects of rotational tillage on soil physical properties and rice yield *T. China. Soc. Agr. Eng.* **23** 109-13.

[29] Han J, Xie J, Zhu J and Wang T 2009 Comprehensive method for treatment of Saline lands *J. Hydraul. Eng.* **40** 372-77.

[30] Limon O, Govaerts B, Deckers J and Sayre K 2006 Soil aggregate and microbial biomass in a permanent bed wheat-maize planting system after 12 years *Field Crops Res.* **97** 302-09.

[31] The Chinese academy of sciences institute of Nanjing, soil soil physical and chemical analysis 1983 Shang Hai: Shanghai Sci. Press: 62-126.

IOP Conf. Series: Earth and Environmental Science **185** (2018) 012007 doi:10.1088/1755-1315/185/1/012007

Metropolitan residents' willingness to payment and factors affecting low-carbon agricultural products: an empirical analysis on low-carbon vegetables in Shanghai

J Ma, L Y Wang, X Y Zhang[1], Y Ma, J N Zhang and S Zhou

Shanghai Academy of Agricultural Sciences, Shanghai, China

E-mail: 18918162313@163.com

Abstract. In this paper, we investigated the willingness of metropolitan residents to pay for low-carbon agricultural products. By developing low-carbon agricultural products, carbon emission would be reduced significantly. Using the contingent valuation method, this paper explored Shanghai residents' willingness to pay for vegetables with carbon labeling. The results of multiple regression model and logistic regression model are as follows: while 91% of respondents were willing to buy vegetables with carbon labels, they were only willing to pay an additional fee of 6% for low-carbon vegetables. Secondly, gender, education level, awareness of carbon labeling, habits to buy vegetables in a traditional market, large supermarket chains, and community supermarket would significantly affect the respondent's willingness to buy low-carbon vegetables. Thirdly, family structure, household disposable income, awareness of carbon labeling and habits to buy vegetable in large supermarket chains would significantly affect the respondent's willingness for an additional pay. It is necessary to increase publicity for promoting the construction of low-carbon agricultural products market. Appropriate channels must be selected for the sale of low-carbon agricultural products because of the effect of vegetation buying habits to "willingness to buy" and "willingness to an additional pay". Due to the low level of additional payment of low-carbon agricultural products, producers need to reduce production costs of low-carbon agricultural products by developing circular agriculture.

1. Introduction

Food is a daily necessity of human life; a certain amount of greenhouse gases are emitted during every step, namely, production, processing, transportation, selling, etc. For example, carbon dioxide (CO_2) is directly emitted during the production of agricultural machinery, chemical fertilizer, pesticides, and farm equipment (plastic greenhouse). To increase agricultural production, deforestation is being carried out extensively. This indirectly increases CO_2 emissions. According to a study conducted by Vermeulen *et al* [1], 19–29% of global greenhouse gas emissions come from various processes of food production and consumption. Approximately 47–61% of food-related carbon emissions are associated with agricultural production in China. About 9–18% of greenhouse gas emissions are associated with agricultural production. According to a study conducted by Hu *et al* [2], the amount of CO_2 emissions rose at an annual average speed of 1.52% from the year 2000 to 2015 in China thanks to increasing agricultural production. According to a paper presented by China's Climate Change National Independent Contribution (INDC) Program, CO_2 emission intensity would be reduced by 60–65% in China [3] by the year 2030 compared to that in 2005. This paper was formulated and submitted to the Secretariat of the United Nations' Framework Convention on Climate Change (UNFCCC) by the Chinese Government in 2015. To reduce carbon emissions associated with agricultural production, low-carbon agriculture must be practiced with definite guidelines. With this strategy, reduction targets can be defined for energy-saving emission.

Presently, 525 types of products with carbon labels have been defined in the carbon labeling program of the British government in 2007. According to a corporate responsibility report issued by Tesco in 2011, 59% of respondents purchased low-carbon products [4]. Climate change is a reality today and various campaigns have been held to increase awareness about this issue among the public; a carbon labeling system that reflects carbon footprint of products has been popularized among the public. In a study conducted by the Gallup organization [5], 72% of EU citizens support the carbon labeling system and believe that carbon labels should be mandatory in the future. According to the Information and Product Labeling Economic theory proposed by Cohen and other researchers [6], the main approach of carbon labeling would be to transform "certification" to "search". With this approach, consumers can make comparisons and more decisions to achieve maximum personal utility. To correct market failures of publicly listed commodities, the product labeling system has to reduce information asymmetry through information disclosure. Similar researches and case studies have been conducted in China and all over the world. These studies have shown that some consumers are willing to pay through the product labeling system to reduce carbon emissions: in the view of product labeling, the Energy Star Labeling Scheme [7] would reduce carbon emissions by 22.1 million tonnes in 2006. According to studies conducted by Ward et al [8], respondents are willing to pay an additional $ 249.82–349.30 for refrigerators that have received the Energy STAR label. Factors such as price, product characteristics, gender, educational background, and income, would affect respondents' willingness to buy these products. In Xiao Jianhong's and other research studies [9], ecological labeling of marine ecological products has been conducted. These studies point out that gender, age, educational level, and income would significantly affect the willingness of respondents to pay for "Eco-labeled" marine ecological products. Gadema's and other researchers carried out a survey of 428 supermarket shoppers in the UK: [10] 72% of people preferred to buy carbon-labeled goods. On the other hand, 89% of people have a relatively low cognitive degree of carbon labeling. Respondents' preferences for the purchase of low-carbon products are based on the following factors: age, educational level, attitudes towards environmental protection, taste, price, and quality of low-carbon products. Brécard and other researchers [11] performed a study to understand "eco-labeling parameters" of fish in Europe. Gender, age, education, income, and living areas (urban or rural) would have a significant impact on the demand for Eco-labeling. Ying Ruiyao and other researchers [12] performed a survey to know the willingness and motivation of urban residents to pay for low-carbon agricultural products. The contingent valuation method and comparison method were performed on the data collected from Tianjin, Shandong, and Jiangsu provinces in China. They found that the average willingness of urban consumers in China to pay for pork with low-carbon properties was 7.74 yuan/kg. Consumers would buy food with low-carbon properties based on two motives: altruism motive and self-interest motive. Age, occupation, and income levels would on the whole affect the willingness of urban residents to consume low-carbon agricultural products. Zhou Yingheng and other researchers [13] performed a study in Nanjing province to determine whether urban consumers were willing to purchase low-carbon pork. In this research study, the average consumers were willing to pay 3.95 yuan for low-carbon pork. Consumers' willingness to pay for products would depend on the following factors: low-carbon pork prices, awareness of low-carbon agricultural products, household income, family population, and level of education. In relevant research studies, questionnaire surveys have been conducted and the data has been analyzed using contingent valuation method and scenario experiment method. This is because carbon labeling or ecological labeling is still not popular in different areas and different regions. In fact, the market of low-carbon products is still in nascent stages of development. Related research studies mainly focus on foods consumed by residents, such as pork, seafood (including fish), and other agricultural products. Very few studies have been conducted on low-carbon vegetables. Most related research studies mainly consider "willingness to buy" as the explained variable, while "willingness to pay" was considered as the explanatory variable. Very few research studies [14] consider both the variables at the same time. This sort of relational-comparative analysis would provide more useful inspirations for market construction.

The willingness to buy (WTB) and the willingness to pay (WTP) for low-carbon vegetable (LCV) were more likely to be influenced by personal characteristics, cognitive degree of carbon labeling, vegetable consumption habits, and other factors. By analyzing the influence of these factors, we can effectively develop LCV market. By comparing the differences between the influencing factors of WTB and WTP, we put forward some suggestions for the establishment of LCV market.

2. Methods

It has been pointed out that per capita energy consumption of urban residents was 3.4-4 times that of rural residents. Moreover, urban residents were well aware of low-carbon products. On the other hand, rural residents were not quite aware of low-carbon products [15]. Therefore, it is more significant to determine low carbon consumption among urban residents. Besides, rural residents mainly purchase vegetables for self-sufficiency; therefore, they purchase few vegetables. It is difficult to build a hypothetical market of low-carbon vegetables. Urban residents usually purchase vegetables, so it is possible to build a hypothetical market of low-carbon vegetables. Therefore, we performed our research study on urban residents.

Shanghai is located in the estuary of Yangtze River (N 31° 14 ', E 121 ° 29 ') in China. Owing to a subtropical monsoon climate summer here is characterized with high temperature and lots of rain. On the other hand, winter has moderate temperature and little rain. The permanent population of Shanghai city was 24,152,700. Shanghai residents have a preference for vegetables, especially green leafy vegetables. According to statistics in Shanghai Statistical Yearbook, per capita vegetable consumption had an annual growth of 0.38% from 2010 to 2015. Per capita fresh vegetable consumption was 236.83 kg in 2015. Shanghai is a city with high population density, high per capita income, and high consumption of vegetables; it is the biggest metropolitan city in China.

In this study, WTB and WTP for LCV were considered as explained variables. Gender, age, educational qualifications, family structure, cognitive degree of carbon labeling, and personal vegetable consumption habits of respondents were regarded as explanatory variables. To obtain data from research study, we used a questionnaire based on the contingent valuation method. Binary logistic regression model was used to analyze factors influencing residents' WTB LCV. Multiple regression model was used to analyze factors influencing residents' WTP for LCV. Then, we compared the connections and differences of the two factors, namely, WTB and WTP for LCV.

The research data mainly takes the questionnaire network as its carrier. The instant messaging software, Wechat, was used to deliver the questionnaire to subjects and obtain requisite results. The sample size was usually 500–1000 for regional surveys. For correlation or causality surveys, sample sizes were usually greater than 30. Our survey was conducted between April 28 and July 4, 2017. During this period, 771 questionnaires were issued but only 759 questionnaires were received; the effective ratio was 98.44%. The questionnaire mainly consisted of four parts: basic information of respondents, cognition of low-carbon agricultural products, willingness to pay for low-carbon agricultural products, preferences, and vegetable consumption habits of residents. To determine residents' WTP for LCV, we investigated whether they were willing to buy vegetables with carbon labels (WTB) and how much they were willing to pay (WTP) for carbon-labeled vegetables. Table 1 presents basic characteristics of residents included in this survey.

Table 1. Basic characteristics of residents included in this survey.

Basic characteristics	**Classification indicators**	**Percent‖%**
Gender	Male	43.74
	Female	56.26
Age	18-25 years old	12.12
	26-35 years old	38.34
	36-40 years old	18.84
	41-55 years old	24.11
	56-60 years old	2.64
	61-65 years old	3.16
	More than 66 - year - old	0.79
Educational level	Junior high and below	1.71
	Senior high school	5.40
	Junior college	9.88
	Undergraduate	41.63
	Graduate and above	41.37
Whether the family has minor children	Yes	45.32
	No	54.68

Family population	1 person	10.41
	2 people	15.81
	3-4 people	54.41
	More than 5 people	19.37
Annual disposable income of the family	Less than 50,000 yuan	12.25
	50,000-100,000 yuan	21.48
	100,000-200,000 yuan	29.12
	200,000-300,000 yuan	20.69
	300,000-400,000 yuan	8.43
	400,000-500,000 yuan	3.43
	More than 500,000 yuan	4.60
Whether it is for purchasing decision makers	Yes	39.92
	No	60.08
Purchase frequency	Everyday	32.81
	Two to three times a week	49.01
	Once a week	13.31
	Less than once a week	4.87
Purchase place	The vegetable market or morning market	69.57
	Convenience stores	2.24
	Large chain-supermarkets	17.79
	Small community supermarkets	5.14
	Online or telephone purchase	2.37
	The farmer	1.05
	Other places	1.84

3. Results and discussion

3.1. Shanghai residents' WTP for LCV

Although 91% of respondents had the WTB LCV, their WTP for LCV was limited. Nobody had the WTP more than 50% of extra fees for LCV; more than 85% of respondents were not WTP more than 50% of extra fees for LCV; 67% of residents were WTP 10% of surcharge to buy LCV. The results indicate that the average WTP for LCV was 1.06 standard units, i.e. 6% of the additional WTP.

3.2. Factors influencing WTP based on binary logistic regression model

In this study, regression calculation was carried out with STATA software (Cabit Information Technology Co. Ltd, Shanghai, China). The WTB (whether people are willing to buy LCV) was considered as explained variable and the three influencing factors, i.e. the respondents' personal characteristics, cognitive degree of LCV, and their vegetable-buying habits were considered as independent variables. Based on the regression model, we conclude that several factors have a remarkable effect on the WTB for LCV, including gender, levels of education, and cognition of buying LCV. The survey was conducted in vegetable markets, large chain-supermarkets, and small community supermarkets.

More women were WTB LCV than men. Influenced by the traditional social concept of "Men outside, women inside", women played a very important role in household consumption and family life. Compared with men, women were more eager to know about the vegetable production. Compared to men, woman collected information more systematically. Thus, women have stronger WTP for LCV than men. Higher the education level, higher would be WTB for LCV. With an improvement in education levels, people have gained a more profound knowledge of carbon emission and the harm of greenhouse effect. They also have a clearer understanding of how the living environment can be improved immensely with the consumption of LCV. All these factors play a very active role in environmental protection. In general, highly educated people prefer to consume LCV. Moreover, LCV consumption was less in lower-educated people. In terms of cognitive situation, residents who know LCV were obviously more willing to buy it as compared to people who didn't know about LCV; however, the respondents' WTP for LCV does not really depend upon their level of understanding low carbon levels. To improve residents' WTB for LCV, it is more important to let more respondents know about vegetables with low carbon label. It is not really necessary to know the exact meaning of carbon labels. Respondents who purchased vegetables from morning markets, large chain-supermarkets, and small community-supermarkets had higher WTB for LCV. This implies that the market value of LCV was higher in these locations. Channels as online and telephone purchase as well as vegetables directly bought from farmers' hands were less suitable for selling LCV. We conclude that the emerging online to offline (O2O) model and the customer to customer (C2C) model did not suit the selling condition of LCV. In the current phase of economic and urban development, the business to customer (B2C) model was more suitable for the consumption of LCV.

3.3. Factors influencing the willingness to pay extra fees based on the multiple regression model

The price (LOP for LCV) was fixed as an explained variable and other influencing factors, including respondents' personal characteristics, cognitive degree of LCV, vegetable-buying habits, etc. were considered as independent variables in this study; multiple regression calculation was carried out with the help of STATA software. Regression model results show that several factors have a significant impact on the LOP for LCV. These factors are as follows: minor children, family population, annual disposable income of the family, cognition of LCV, and the habit of purchasing vegetables in large chain-supermarkets.

From the perspective of family structure, families with minor children have lower LOP for LCV consumption. In a family with minor children, education is a major expense. Since higher education is expensive, price allocation for food consumption is limited. Education of minor children is the biggest expense in Shanghai. In some families, direct spending on education accounts for more than half of the total child-rearing expenditure. Besides, families also spend a sizeable amount of money in renting property near the school, which can be considered as indirect costs (immovable property) of education. Thus, the consumption structure of such families also changes dramatically. In case of hypothesized household income, an increasing number of direct and indirect costs of education is bound to affect food consumption choices. Larger the family population, more willing they will be to pay for a higher level of LCV. This is mainly influenced by the scale of family economy—larger the family size, lower would be the proportion of food consumption in the entire family. Therefore, they are more suitable to raise LOP for LCV. From the perspective of annual disposable income of families, each increase of 10,000 yuan would increase the residents' LOP for LCV by 0.41%. By increasing the families' annual disposable income, LOP can be improved for LCV. Families with high income pay for precious commodities. In terms of cognitive situation of low carbon vegetables, residents knowing about LCV obviously have higher LOP than the ones who do not know about it. However, the extent of understanding of low-carbon labels does not affect the respondents' LOP for LCV. From the vegetable consumption places of residents included in this survey, residents who choose large supermarket chains as their main channel of vegetable consumption have higher LOP for LCV. This implies that if the price of LCV is slightly higher than general vegetables, there would be better sales for low-carbon labeled vegetables at large supermarket chains.

3.4. Empirical results discussion

The results indicate that although the degree of cognition is a mutual factor that affects both WTB and LOP, there are three differences between factors that affect WTB and LOP for LCV.

Among residents in Shanghai, gender and education levels significantly affected WTB for LCV, but they have no significant impact on the LOP of Shanghai residents. This implies that women and residents with high-level of education are more likely to buy LCV, yet the levels of payment (LOP) may not be defined.

Secondly, family situations would greatly affect Shanghai residents' LOP for LCV, but they have no remarkable effect on their WTB for LCV. This indicates that three factors, namely, families with no minor children, families with a larger population, and families with higher level of annual income, have a higher LOP for LCV. On the other hand, WTB would not be affected by these factors. The characteristics of respondents were divided into two types: personal characteristics and family characteristics; the former ones have an overt influence on WTB, while the latter ones tend to affect LOP to a much greater extent.

Residents who are used to buying vegetables at large supermarket chains have both higher WTB and LOP for LCV. This implies that large-scale chain supermarket is a primary location to promote LCV. Conversely, people who are used to buying vegetables in vegetable markets, morning markets, and small community supermarkets have higher WTB but that of LOP are not as strong as their WTB. The habit of buying vegetables in other sites of the remaining population doesn't have a remarkable effect on their LOP and WTB. The possible reasons are as follows: the business to customer (B2C) model is more suitable for LCV consumption; the investigated sample of this model is relatively smaller, which may result in a relatively larger deviation from results.

4. Conclusions

4.1. Main conclusions

By surveying Shanghai residents' consumption of LCV, this study explores Shanghai residents' WTB and LOP for LCV. Then, the influencing factors of their WTB and LOP were analyzed using the binary logistic model and the multiple regression model. The main conclusions are as follows:

Shanghai residents have a certain degree of WTB for LCV, but their LOP is not high. In this study, 91% of respondents are willing to buy vegetables with carbon labels, but their average level of willingness to pay is only 1.06 standard units. In other words, they are willing to pay 6% additional fees for LCV.

Many factors influence Shanghai residents' WTB and LOP for LCV. There are both connections and differences between these factors:

(1) The degree of cognition affects their WTB and LOP simultaneously, and it has the same direction of function; (2) personal characteristics, such as gender and levels of education, etc. have a significant influence on their WTB but have no effect on their LOP. Conversely, family characteristics, such as family population, whether the families have minor children or not, and the amount of household disposable income, have a significant influence on their LOP, but they have no effect on their WTB; (3) people who are used to consuming in large supermarket chains have overt WTB for LCV and higher LOP, while those who consume in vegetable markets or morning markets as well as small community supermarkets have higher WTB.

4.2. *Future works*

Low-carbon lifestyle was popularized and carbon emission was reduced by promoting the consumption of LCV. Related market guidance and policies were drafted with respect to the following three aspects: the first step was to gain good publicity for low-carbon agricultural products, including LCV in an effort to increase the cognitive degree of low carbon agricultural products among residents, which will continue to improve their WTP and LOP. The second one is to take the large supermarket chains as primary and principal channel to promote the consumption of LCV. At the same time, conventional markets, morning markets and small community supermarkets, etc., should be considered as important channels that promote its consumption by enlarging the coverage of consumers. To realize the production of LCV, strive to develop recycling agriculture and other modes of agricultural production. All of these will, in turn, take an effective control of its cost, thereby changing the production means that only depend on chemical fertilizers and pesticides and avoiding the increase of cost associated with the reduction of output.

In this study, e-questionnaire was adopted in the questionnaire survey process. Limited by the transmission route of the e-questionnaire, the sample's regional distribution is uneven. Hence, further research can be completed by performing a more targeted supplemental study on the spatial equilibrium distribution of sample collection.

References

[1] Vermeulen S J and Campbell B M 2012 Ingram J S I. climate change and food systems *Annual Revi. .Envir. Res..* **37**

[2] Hu J B and Guo F 2017 Study on dynamic change of carbon efficiency of main crops in China *World Agri.* **6**:134-141

[3] Chinese government network. Li Keqiang announced the Chinese government will submit a national contribution to climate change [EB/OL]. (2015-06-30) [2017-07-28]. http://www.gov.cn/guowuyuan/2015-06/30/content_2887287.htm.

[4] Tesco. Corporate Responsibility Report 2011 [EB/OL]. (2011-06-30) [2017-07-28].

https://www.tescoplc.com/media/1460/tesco_cr_report_2011.pdf.

[5] The Gallup Organisation 2009 Europeans' attitudes towards the issue of sustainable consumption and production *Flash Eurobarometer*, 256

[6] Cohen M A and Vandenbergh M P 2012 The potential role of carbon labeling in a green economy *Energy Econ.*. **34**:53-63

[7] Sanchez M, Webber C A, Brown R E, *et al.* 2007 Status report: savings estimates for the ENERGY STAR (R) Voluntary Labeling Program *Lawrence Berkeley National Laboratory.*

[8] David O W, Christopher D C, Kimberly J, Steven T Y and Clifford S R 2011 Factors influencing willingness-to-pay for the ENERGY STAR® label *Energ. Poli.* **39**: 1450-8

[9] Xiao J H, Wang M, Liu J and Zhang R 2016 Study on the regional difference of eco-compensation standards for the marine ecological products based on marine fisheries' eco-label schemes *J. Natur. Resour.*. **31**:402-12

[10] Gadema Z and Oglethorpe D 2011 The use and usefulness of carbon labelling food: a policy perspective from a survey of UK supermarket shoppers *Food Poli.* **36**:815-22

[11] Brécard D, Hlaimi B, Lucas S, Perraudeau Y and Salladarré F 2009 Determinants of demand for green products: An application to eco-label demand for fish in Europe *Ecol. Econ.* **69**: 115-25

[12] Ying R Y, Xu B and Hu H 2012 China urban resident's motives of willingness to pay for low-carbon agricultural products *China Popul. Resour. Envir.* **22**:165-71

[13] Zhou Y H and Wu L F 2012 A study on the willingness of payment of low - carbon agricultural products by urban consumers *J. Agric. Sci. Tech.*. **8**:4-12

[14] Liu X L, Wu L H and Xu L L 2015 Study of factors affecting consumer willingness to pay for certified traceable tea in China *China Popul. Resour. Envir.* **25**:170-6

[15] Wang X D and Jing F J 2012 Model research on residents' Low-carbon purchasing behavior based on the Survey Data in Five Cities *China Popul. Resour. Envir.* **22**:47-55

IOP Conf. Series: Earth and Environmental Science **185** (2018) 012008 doi:10.1088/1755-1315/185/1/012008

Evaluation of nutritional quality in different varieties of green and purple leaf lettuces

J H Hao[1a], L Wang[1a], C J Liu[1], X D Zhang[1], Z Y Qi[1], B Y Li[1], S Xiao[1] and S X Fan[1,2]

[1] Beijing Key Laboratory for Agricultural Application and New Technique, National Demonstration Center for Experimental Plant Production Education, College of Plant Science and Technology, Beijing University of Agriculture. No. 7 Beinong Road, Changping District, Beijing, P. R. of China

[a] J H Hao and L Wang are co-first authors.

E-mail: fsx20@163.com

Abstract. To determine the quality difference in different varieties of leaf lettuces, we investigated the characteristics of 74 varieties of leaf lettuces (55 green and 19 purple varieties). Both domestic and foreign varieties of lettuce were included in this study. We determined six nutritional indicators of these varieties: such as total soluble sugar, starch, cellulose, proteins, and vitamin C. The grey correlation analysis was used to analyze the 74 varieties of leaf lettuces. Thus, the merits of these traits were compared. The results indicate that the purple leaf lettuce P-S23 had grey comprehensive evaluation value of about 0.8, which was the highest among all varieties. For green leaf lettuces G-S68 and G-S10, grey comprehensive evaluation values were less than 0.5. This indicates that nutritional quality and nutritional value of the purple leaf varieties were higher than that of the green leaf varieties. Furthermore, the quality of green varieties of G-S68 and G-S10 were of poor quality, whereas the purple variety P-S23 was most suitable for commercial production.

1. Introduction

Lettuce (*Lactuca sativa* L.) has high nutritive value and quality, so it has received considerable attention. Lettuce has been considered as a model system for plant biology; moreover, it forms an integral component of diet thanks to its health benefits [1][2]. Lettuce is an important class of green leafy vegetables that contain nutrient-rich protein [3], carbohydrates, vitamins, and minerals. In addition, it offers several health benefits: it helps in preventing anemia; it possesses anti-cancer and anti-aging qualities; it lowers blood pressure and prevents heart rhythm disorders [4][5]. Sugar is one of the major metabolites of higher plants. Sugar content is high in higher plants; moreover, different isomers of sugar can be isolated from these plants. Sugar plays an important role in many physiological processes of plant growth and development as it is either a metabolic intermediate or final product in these processes [6]. Lettuce contains the following vitamins: vitamin B9, vitamin C, and vitamin E [7]; however, vitamin C content in lettuce is lower than that of other leafy vegetables, such as spinach and kale. Vitamin C (ascorbic acid) is a water-soluble antioxidant, and several studies have proved that vitamin C plays an important role as a plant growth regulator [8].

Vegetables stimulate appetite in humans by improving digestion: they regulate pH of stomach and promote intestinal peristalsis. Recently, domestic varieties of vegetables, such as spinach [9], carrots [10], tomatoes [11], and asparagus [12], have been assessed by nutritional quality analysis and evaluation studies. By implementing the fuzzy membership function method, Min Q *et al.* [13] determined seven quality indicators in 34 types of spinach. Thus, they conducted a comprehensive analysis of different spinach types in terms of nutritional quality; they determined nitrate, tannin, and acid contents of these species.

Grey correlation analysis is a method of treating and analyzing data according to grey system theory. This method can be performed on relatively small amounts of sampling data. With this method, scientists can easily analyze relationships of many factors [14]. Currently, grey correlation analysis is widely used in agricultural science studies. In a previous study, Ma *et al.* [15] performed grey relational analysis to evaluate 13 main agronomic characters of 10 sugarcane lines. The 13 agronomic characterized were as follows: sprouting rate, seeding emergence rate, tillering rate, stalk diameter, effective stalks, single stalk weight, sugar content, stalk yield, sugar yield, fiber content, reducing sugar, rate of internal blight rate, and dry top rot. Grey relational analysis (GRA) was used to classify dried roselle (*Hibiscus sabdariffa* L.). All the roselle calyces (undried and dried) were correctly identified and classified according to GRA grades [16].

Loose leaf lettuce is extremely rich in nutrients, especially in vitamins A and C and calcium; however, different cultivars have significantly different proportion of nutrients. Therefore, this study analyzed nutritional quality differences of different varieties of loose leaf lettuce. Then, we screened these lettuce varieties to identify the ones with higher nutritive value and quality. The results served as a theoretical basis for germplasm improvements and quality breeding in future studies.

2. Materials and methods

2.1. Plant material and growth conditions

The experiment was conducted in greenhouses of Beijing University of Agriculture, Beijing, China. In this experiment, we included the following varieties of lettuce: G-S1, 2, 3, 5, 6, 7, 8, 9, 10, 12, 13, 18, 19, 20, 21, 22, 23, 24, 25, 26, 30, 31, 33, 34, 37, 39, 41, 42, 43, 45, 47, 48, 49, 50, 51, 52, 53, 54, 55, 56, 57, 58, 59, 60, 63, 64, 65, 66, 67, 68, 70, 71, 72, 73, and 75 and P-S1, 2, 3, 4, 5, 6, 9, 12, 16, 20, 21, 22, 23, 28, 29, 30, 32, 33, and 34 (These were assigned by our laboratory. Green leaf lettuces were marked as G-S, while purple leaf lettuces were marked as P-S.). We used the following product seeds: G-S1 ('Jinwangjie'), G-S3 ('Aosaidi'), G-S6 ('Ruibite'), G-S19 ('Xiangshengcai'), G-S42 ('Huangshengcai'), G-S45 ('Jieqiushengcai'), G-S49 ('Jiguanshengcai'), G-S50 ('Sanyeshengcai'), G-S53 ('Bolishengcai No.93'), G-S54 ('Dongshanshengcai'), G-S59 ('Aijiaoshengcai'), G-S59 ('Aijiaoshengcai'), G-S63 ('Luosheng No.3'), G-S68 ('Yingzi'), P-S20 ('Meiguoziyeshengcai'), P-S21 ('Zishengcai'), P-S22 ('Hongsheng No.1') and P-S34 ('Dxiangsheng No.1').

In a green house, germination of seeds was carried out in cell trays filled with soil. A variety of seeds were sown in a cell tray with 50 cells, each cell containing one seed. The daytime temperature of the green house was fixed at 25 °C, while the night time temperature was maintained at 20 °C. Loose leaf lettuce was sampled 50 days after sowing, and samples the same parts of each lettuce. After removing main veins, we measured the weight of shredded leaves on an analytical balance (BSA224S, Sartorius, Germany); then, they were frozen in liquid nitrogen. The samples were stored at –80°C in an ultra-low temperature freezer for further analysis.

2.2. Method of determination

In this experiment, we measured five nutrition indicators of vegetables, namely, total soluble sugar, starch, cellulose, proteins, and vitamin C. A titrimetric method was used to measure the content of vitamin C; the indicator 2,6-dichlrophenolindopehnol was used in titration. Anthrone colorimetry was used to measure soluble sugar, crude fiber, and starch content. Coomassie brilliant G-250 method was used to calculate protein content in samples. The above tests were carried out using UV-visible spectrophotometer (UV-5200, METASH Shanghai, China).

Grey correlation analysis was performed to assess the quality of loose leaf lettuces. The specific method of analysis can be summarized as follows:

A. Determine the comparison sequence $X_i=\{x_i(k)|k=1,2,\ldots,n\}(i=1,2,\ldots,n)$ and the reference sequence $x_0=\{x_0(k)|k=1,2,\cdots,n\}$. The maximum value of each nutrient index is selected as the reference sequence.

B. Dimensionless treatment:

$$X_i = \frac{X_i(k) - \min X_i(k)}{\max X_i(k) - \min X_i(k)}$$

$i = 0,1,2 \ldots m;\; k = 0,1,2 \ldots, n$

C. Coefficient of correlation:

$$\zeta_i(k) = \frac{\min \Delta_i(k) + \rho \max \Delta_i(k)}{\Delta_i(k) + \rho \max \Delta_i(k)},$$

$$\Delta_i(k) = |X_0(k) - X_i(k)|$$

the distinguish coefficient(ρ) is 0.5 generally.

D. Relationship of degree

$$\gamma_i = \frac{1}{n}\sum_{X=1}^{n} W_k \zeta_i(k)$$

W_k is the weight value of the kth quality index. In this study, W_k is 0.2.

E. Ranking of related degrees

2.3. Statistics and data analysis

The experiment was arranged in a randomized block design and was performed in three replicates. Data were analyzed with OriginPro software (OriginLab Corporation, Northampton, USA) and SPSS 10.0 software (International Business Machine, Chicago, IL, USA).

3. Results

3.1. Comparison of soluble sugar content of different varieties of leaf lettuces

As shown in Figure 1, total soluble sugar content (0.97±0.03%.FW) was highest in the variety G-S66. On the other hand, total sugar content (0.03±0.01%.FW) was lowest in the variety G-S12. There was a difference of 0.94% between the maximum and minimum values of total sugar content. In the purple leaf lettuces, the highest total soluble sugar content was P-S20 (0.72±0.18%.FW), and the lowest soluble total sugar content was P-S23 (0.09±0.01%.FW).

In 31 varieties (41.89%), total soluble sugar content was in the range 0–0.3%.FW, these include 10 purple leaf lettuce varieties. In 27 varieties (36.49%), total sugar content was in the range 0.30-0.60%.FW, these include 5 purple leaf lettuce varieties. In fifteen varieties (20.27%), total sugar content was in the range 0.60–0.90%.FW, these include 4 purple leaf lettuce varieties. Total sugar content of the variety G-S66 was the highest, ranging from 0.90 to 1.00%.FW. In most varieties, total soluble sugar content was in the range 0–0.30%.FW; therefore, these varieties were more abundant than the varieties containing sugar contents in the range 0.90–1.00%.FW.

3.2. Comparison of starch content of different varieties of leaf lettuces

As shown in Figure 2, the variety G-S68 had the highest starch content at 0.98±0.04%.FW. On the other hand, the variety G-S12 had the lowest starch content at 0.04±0.01%.FW. There was a difference of 0.94%FW between the maximum and minimum values of starch content. In the purple leaf lettuces, the highest starch content was P-S32, 0.91±0.02%.FW, and the lowest starch content was P-S12, 0.08±0.01%.FW.

Among the 74 lettuce varieties, the starch content of 44 varieties (59.46%) was in the range 0–0.30%.FW, these include 6 purple leaf lettuce varieties. Starch content of 22 varieties (28.73%) was in the range 0.30–0.60%.FW, these include 7 purple leaf lettuce varieties. Starch content of seven varieties (9.46%) was in the range 0.60–0.90%.FW, these include 4 purple leaf lettuce varieties. Only the variety G-S66 had starch content in the range 0.90–1.20%.FW. In most varieties, starch content was in the range 0–0.60%.FW; therefore, these varieties were more abundant than the ones whose starch contents ranged from 0.60 to 1.20%.FW.

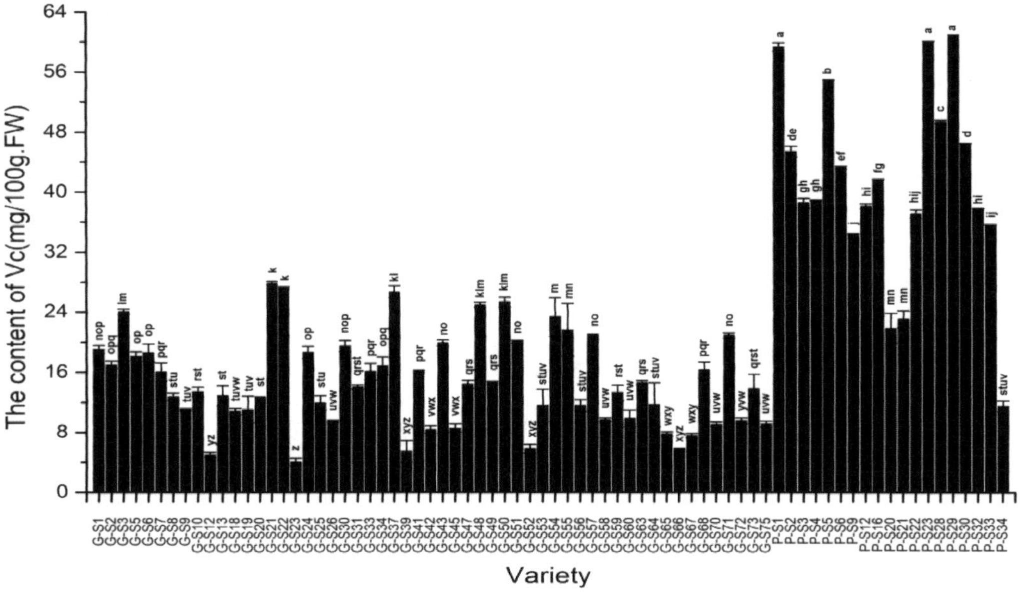

Figure 1. Comparison of total soluble sugar content in different varieties of green and purple lettuces (bars, Standard Error; the difference is not significant when the miniscule is same, whereas different minuscules imply significant difference (P<0.01)).

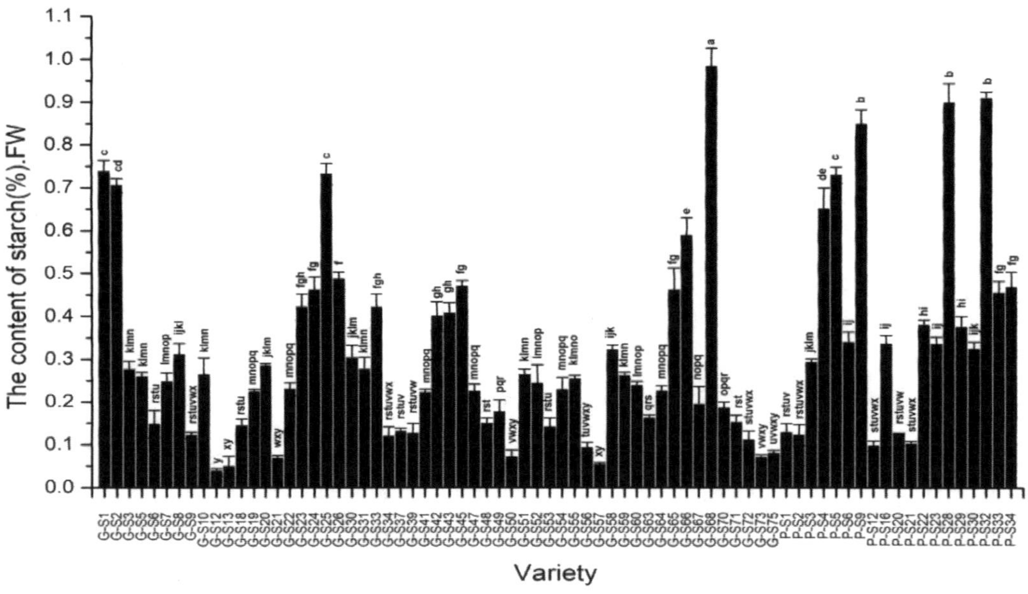

Figure 2. Comparison of starch content in different varieties of green and purple lettuces (bars, Standard Error; the same minuscule indicates that the difference is not significant, whereas different minuscules imply significant difference. (P<0.01))

3.3. Comparison of cellulose content of different varieties of leaf lettuces

As shown in Figure 3, the variety G-S9 had the highest cellulose content at 0.46±0.01%.FW. On the other hand, the variety G-S70 had the lowest cellulose content of 0.02±0.01%.FW. There was a difference of 0.44%.FW between the maximum and minimum values of cellulose content. In the purple leaf lettuces, the highest cellulose content was P-S30, 0.43±0.02%.FW, and the lowest cellulose content was P-S3, 0.04±0.01%.FW.

Among the 74 lettuce varieties, cellulose content of 18 varieties (24.32%) was in the range 0–0.10%.FW, these include 5 purple leaf lettuce varieties. Cellulose content of 25 varieties (33.78%) was in the range 0.10–0.20%.FW, these include 5 purple leaf lettuce varieties. Cellulose content of 22 varieties (29.73%) was in the range 0.20–0.30%.FW, these include 3 purple leaf lettuce varieties. Cellulose content of seven varieties (9.46%) was in the range 0.30–0.40%.FW, these include 3 purple leaf lettuce varieties. The varieties G-S9 and G-S30 represented 2.70% of total samples: cellulose content was in the range 0.40–0.50%.FW for the varieties G–S9 and G-S30. Thus, most varieties had cellulose content in the range 0–0.30%.FW; these varieties were more abundant than the ones whose cellulose contents were in the range 0.30–0.50%.FW.

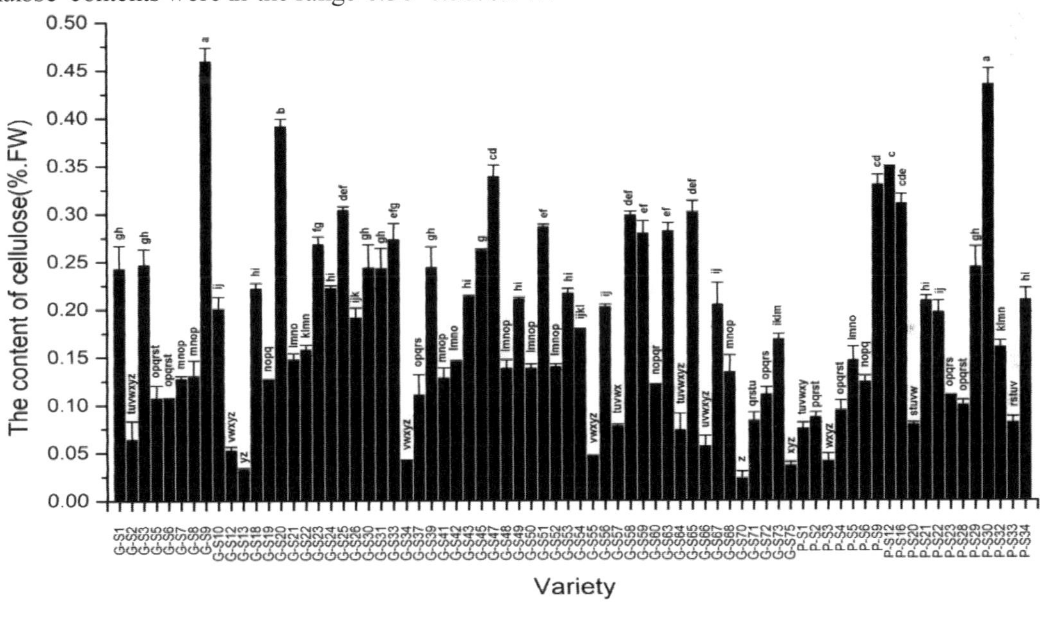

Figure 3. Comparison of cellulose content in different varieties of green and purple lettuces (bars, Standard Error; the difference is not significant if miniscules are same, while there is significant difference when miniscules are different (P<0.01).).

3.4. Comparison of protein content of different varieties of leaf lettuces

As shown in Figure 4, variety G-S42 had the highest protein content at 27.62±0.34 mg/g.FW. On the other hand, variety G-S10 had the lowest protein content of 8.42±0.01 mg/g.FW. There was a difference of 19.20 mg/g.FW between the maximum and minimum values of protein content. In the purple leaf lettuces, the protein content was P-S5, 22.12±019mg/g.FW, and the lowest protein content was P-S6, 12.14±0.11mg/g.FW.

Among the various lettuce varieties, only G-S10 had protein content in the range 0–10 mg/g.FW; the variety G-S10 represented 1.35% of all the samples. Protein content of 62 varieties (83.78%) was in the range 10–20 mg/g.FW, these include 16 purple leaf lettuce varieties. Protein content of ten varieties (13.51%) was in the range 20–25 mg/g.FW, these include 3 purple leaf lettuce varieties. Protein content of variety G-S42 was 27.62±0.34 mg/g.FW; only this variety had protein content greater than 25 mg/g.FW. Thus, a majority of varieties had protein content in the range 10–20 mg/g.FW; these varieties were more abundant and evenly distributed than the varieties with protein contents in the ranges 0–10 mg/g.FW and 25–30 mg/g.FW.

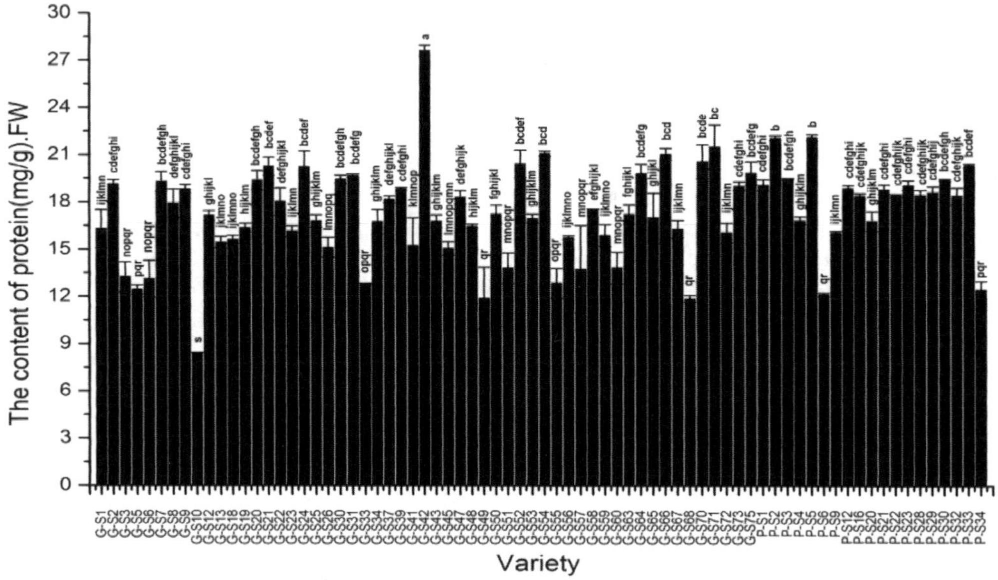

Figure 4. A comparison of protein content in different varieties of green and purple lettuces (bars, Standard Error; the same minuscules indicate that the difference is not significant, and different minuscules mean significant difference (P<0.01)).

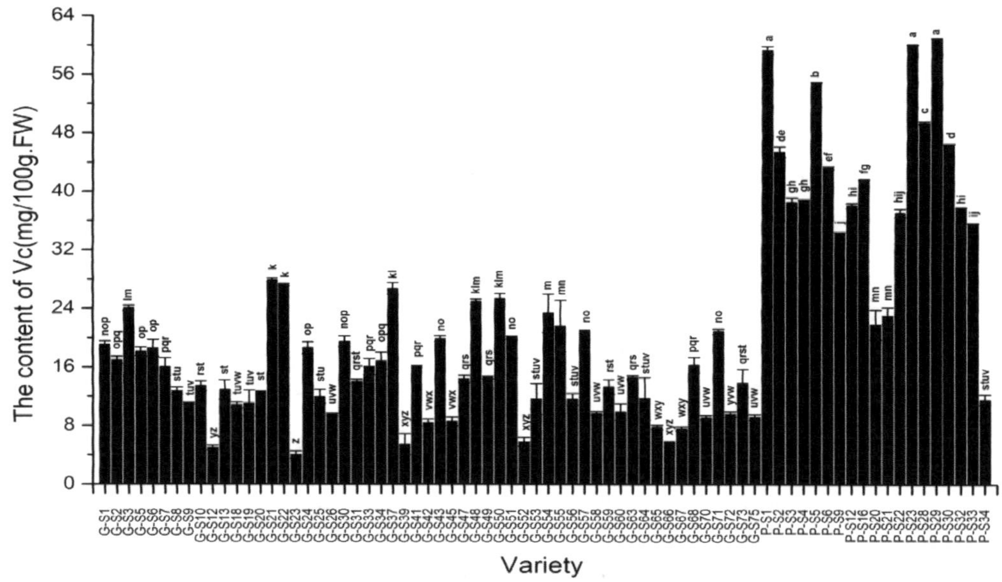

Figure 5. A comparison of V_c content in different varieties of green and purple lettuces (bars, Standard Error; the difference is not significant when the miniscule is same, whereas different minuscules imply significant difference (P<0.01)).

3.5. Comparison of V_c content of different varieties of leaf lettuces

As shown in Figure 5, variety P-S29 had the highest V_c content of 60.99±0.01 mg/100 g.FW. In contrast, variety G-S23 showed the lowest V_c content of 3.35±0.51 mg/100 g.FW. There was a difference of 57.64 mg/100 g.FW between the maximum and minimum values of V_c. In the purple leaf lettuces, the lowest Vc content is P-S34, 11.56±0.07mg/100 g.FW. In the green leaf lettuces, the highest Vc content is G-S21, 27.91±0.24mg/100 g.FW.

Among the 74 lettuce varieties, V_c content of 17 varieties (22.97%) was in the range 0–10.00 mg/100 g.FW. Moreover, V_c content of 30 varieties (40.54%) was in the range 10.00–20.00 mg/100 g.FW. Furthermore, V_c content of 18 varieties (24.32%) was in the range 20.00–40.00 mg/100 g.FW, these include 9 purple leaf lettuce varieties. Finally, V_c content of nine varieties (12.20%) was greater than 40.00 mg/100 g.FW, these 9 varieties are purple leaf lettuces. Thus, a majority of varieties had V_c content in the range 10.00–20.00 mg/100 g.FW; these varieties were more abundant than the ones whose Vc content was greater than 40.00 mg/100g.FW. Vitamin C content of purple leaf lettuces was generally greater than that of green leaf lettuces.

3.6. Grey fuzzy relational analysis of the nutritional quality of different varieties of leaf lettuces

When relational degree was between 0 and 1, we observed the following trends: higher the relational degree, better would be the quality of the lettuce. As shown in Table 1, grey relations were highest in the following varieties: P-S23, 6, 5, 32, 3, and 4 and G-S60, 55, and 75; grey relation degree was greater than 0.7 for all these varieties. Grey relations were lowest for varieties G-S68 and 10; the relational degree was less than 0.5 for both these varieties. The mean of grey relations in purple lettuces is 0.67±0.05, the mean of grey relations in green lettuces is 0.62±0.05. The mean of grey relations in purple lettuces was higher than that of green lettuces.

Table 1. Grey-fuzzy relational analysis of the nutritional quality of different varieties of leaf lettuces.

Variety	Relational degree	Variety	Relational degree
P-S23	0.7948	G-S19	0.6285
P-S6	0.7339	G-S71	0.628
P-S5	0.7281	G-S50	0.6251
G-S60	0.7251	P-S20	0.6215
G-S55	0.7198	P-S9	0.6211
P-S32	0.7197	G-S56	0.6204
P-S3	0.7197	G-S7	0.6203
P-S4	0.7023	G-S47	0.6189
G-S75	0.7008	G-S26	0.6187
P-S33	0.6935	G-S30	0.6179
G-S67	0.689	G-S3	0.6136
P-S22	0.6821	G-S73	0.6134
P-S28	0.682	G-S1	0.6127
G-S2	0.6727	P-S30	0.6119

G-S66	0.6696	G-S43	0.6045
G-S70	0.6678	G-S54	0.6034
P-S29	0.6667	P-S12	0.603
P-S34	0.6582	G-S53	0.6011
P-S2	0.6557	G-S65	0.5994
G-S72	0.6487	G-S33	0.5991
G-S6	0.646	G-S20	0.598
G-S57	0.6456	G-S48	0.5974
G-S25	0.6419	P-S21	0.5951
G-S63	0.6417	G-S45	0.593
G-S42	0.6415	G-S31	0.5925
G-S37	0.6395	G-S12	0.5853
G-S22	0.6393	G-S59	0.5843
G-S8	0.6392	G-S5	0.577
P-S16	0.6337	G-S9	0.5757
G-S13	0.6331	G-S21	0.564
G-S24	0.6308	G-S58	0.5639
G-S41	0.6306	G-S51	0.5621
P-S1	0.6303	G-S39	0.5574
G-S18	0.6301	G-S23	0.5322
G-S34	0.629	G-S49	0.5252
G-S52	0.6287	G-S68	0.497
G-S64	0.6286	G-S10	0.4842

4. Discussion

Among modern consumers, leaf lettuce has become very popular because it is a refreshing green leafy vegetable with tremendous health benefits. However, the quality of domestic variety is poor. Although many foreign varieties are available in China, their growth conditions are quite inconsistent. Therefore, scientists need to urgently select and cultivate varieties, which suite the local climate and growth conditions in China. Leaf lettuce varieties were judged in terms of their nutritional quality and adaptability to the following environmental conditions: lighting, temperature, humidity, soil, etc. Based on the results, we selected the varieties suitable for cultivation on a commercial scale.

Soluble sugar is an important parameter for determining crop quality[17]. The levels of soluble sugar, protein, and cellulose were high in all the included varieties, with maximum values being

0.97±0.03% (G-S66 variety), 27.62±0.34 mg/g (G-S42 variety), and 0.46±0.01% (G-S9 variety), respectively. In a study conducted by Yuan *et al*, the content of soluble sugar, crude protein, and crude fiber were much lower at 1.23%, 11.26 mg/g, and 0.64%, respectively[18]. Compared to our current study, Yuan *et al.* obtained different results in their experimental study as they investigated different varieties of leaf lettuces.

Energy is produced from starch in the human body; therefore, starch content is one of the criteria for evaluating the quality of vegetables. Variety G-S68 had the highest starch content at 0.98%; however, this value did not agree with the results of a previous study conducted by Garmendia and Idoia [19]. This is because Garmendia and Idoia used romaine lettuce for investigation, while other varieties of leaf lettuce were included in our experiment. In this study, variety P-S29 had the highest vitamin C content of 60.99 ±0.01mg/100 g, while variety G-S23 had the lowest level of Vitamin C of 3.35 mg/100 g. Vitamin C content of 30 varieties was in the range 10.00–20.00 mg/100 g; these varieties represented 40.54% of the total samples. Ahmet Turhan [20] found that vitamin C content of lettuce cultivar 'Funly' was 16.75 mg/100 g, which agreed with our results.

Grey correlation analysis was first proposed by Ju long Deng [14]. Thereafter, it has been widely used in the fields of agriculture and social economics [21]-[22]. By performing grey correlation analysis, we determined the quality of optimal varieties and the weight value of each quality in the market; moreover, we determined the correlation degree between different varieties of leaf lettuce and the ideal variety. Larger the correlation degree, closer would be the ideal situation for identification: a variety that is suitable for commercialization as it completely adapts with the market and production conditions. In this study, highest grey relations were observed between the following varieties: P-S3, 4, 5, 6, 23, and 32 and G-S55, 60, and 75. This indicates that the quality of content in these varieties was good enough to be considered as a final choice or reference variety. On the other hand, varieties G-S68 and 10 had the lowest grey relations. This implies that the quality of these varieties was poor.

In summary, different varieties showed differences in their quality: total soluble sugar, starch, cellulose, protein, and vitamin C content were highest in the varieties G-S66, G-S68, G-S9, G-S42, and P-S29, respectively. But, P-S23 had the highest grey relations. Four varieties of green leaf lettuces had the best quality; however, five varieties of green leaf lettuces were of worst quality. At the same time, the mean of grey relations in purple lettuces was higher.

Nutritional properties of plants are affected by phenolic compounds [23]. Phenolic acids and flavonoids are the most common phenolic compounds in lettuce [24]. In purple lettuce, the main phenolic compounds are phenolic acids and flavonoids, whereas green lettuce usually contains phenolic acids in significant amounts. In purple lettuce, the content of phenolic compounds is higher than that in green lettuce [24]. Interestingly, several studies have shown that phenolic compounds interact with proteins and polysaccharides. The functional properties of proteins are affected when they interact with polyphenols. Chung *et al.* found that tannins react with proteins to form compounds that reduce the nutritional value of food [26]. In some polysaccharides (including starch), biological macromolecules exhibit specific biological activity towards polyphenol; these reactions can be combined with static electricity, hydrogen bonding, and hydrophobic forces. On the other hand, polysaccharides can be combined with phenolics through the following chemical pathways: methylation, acyl interaction with sweets, esterification, etc [26]. In this study, the overall quality of purple leaf lettuce was higher than that of green leaf lettuce. The difference in quality was probably due to the difference in the phenolic content in the different varieties of purple lettuce and green lettuce.

In this study, we included only 74 varieties of leaf lettuce. For germplasm improvement and quality breeding of lettuce, future studies have to include more varieties and determine other nutritional indicators, such as anthocyanin and calcium.

5. Conclusion

In this study, five nutritional indicators (total soluble sugar, starch, cellulose, proteins, and vitamin C) of 74 lettuce varieties were examined. Grey correlation analysis was performed to appraise the quality of leaf lettuces. The relational degree of P-S23 was 0.7948, which was the highest among all varieties. Therefore, purple leaf lettuce P-S23 is the preferred variety for commercial production. The overall quality of purple leaf lettuces was better than that of green leaf lettuces.

Acknowledgments
This work was financially supported by the Fund of the National Natural Science Foundation of China (31372057), the 2017 Agriculture Education Teaching Reform Project of Beijing University of Agriculture (BUA2017JG030), 2017 Research Fund for Academic Degree & Graduate Education of Beijing University of Agriculture, and the Innovation Team Construction of Leafy Vegetables of Beijing (blvt-02).

References
[1] Almaskri A, Alkharusi L and Almiqbali H 2010 Effects of salinity stress on growth of lettuce (*Lactuca sativa*) under closed-recycle nutrientilm technique *Int. J. Agric. Biol.* **12** 377-80
[2] Yuan H R, Huang X, Li X J, Tian L B, Zhu C and Zou D X 2014 Effect of three-dimension fertilizing mode using CO_2 and liquid/solid fractions of the digesate from anaerobic digester on yield and quality of lettuce *J. Agric. Sci.* **6** 212-20
[3] Santamaria P 2006 Nitrate in vegetables: toxicity, content, intake and EC regulation *J. Sci. Food Agric.* **86** 10–17
[4] Liu R S, Qiu Y L and Jian-Wen L I 2005 Study on varietal difference of mineral nutrient characteristics in *Lactuca sativa Life Sci. Res.* **9** 258-62
[5] Krężel J and Kołota E 2014 Source of nitrogen affects the yield and quality of spinach cultivars grown for autumn harvest *Acta Agric. Scand.* **64** 583-9
[6] Garg O P and Kapoor V 1972 Retardation of leaf senescence by ascorbic acid. *J. Exp. Bot* **23** 699-703
[7] Moo Jung Kim, Youyoun Moona, Janet C, *et al.* 2016 Nutritional value, bioactive compounds and health benefits of lettuce (*Lactuca sativa* L.) *J. Food Compos. Anal.* **49** 19-34
[8] Conklin P L 2001 Recent advances in the role and biosynthesis of ascorbic acid in plants *Plant, Cell Environ.* **24** 383-94
[9] Ito A, Shimizu H, Hiroki R, Nakashima H, Miyasaka J and Ohdoi K 2013 Effect of acclimation to low temperature on nutritional quality of spinach *IFAC Proceedings* **46** 29-32
[10] Cefola M, Pace B, Renna M, Santamaria P, Signore A and Serio F 2012 Compositional analysis and antioxidant profile of yellow,orange and purple polignano carrots *Ital. J. Food Sci.* **24** 284-91
[11] Samaila A, Amans E B, Abubakar I U and Babaji B A 2011 Nutritional quality of tomato (*Lycopersicon esculentum* Mill) as influenced by mulching, nitrogen and irrigation interval *J. Agric. Sci.* **3** 266-70
[12] Amaro-López M A, Zurera-Cosano G and Moreno-Rojas R 1998 Trends and nutritional significance of mineral content in fresh white asparagus spears *Int. J. Food Sci. Nutr.* **49** 353
[13] Qi M, Chen H L, Tang X W, *et al.* 2009 Analysis and assessm on nutritional quality of spinach(Spinaciao leraces L) varieties from different sources *China Vegetables* **10**:20-7
[14] Deng J L 1989 Introduction to grey system *J.Grey.systl.* **1** 1-24
[15] Ma W Q, Guo Q, Wei C Z, Li H R and Peng C 2014 Comprehensive evaluation of new self-cultivated sugarcane lines by grey relational analysis *Asian Agric. Res.* **6** 85-8
[16] Chen H H, Tsai P J, Chen S H, Su YM and Chung C C 2005 Grey relational analysis of dried roselle (*Hibiscus sabdariffa* L.) *J. Food Process. Preserv.* **29** 28-245
[17] Moretti C L, Sargent S A, Huber D J, Calbo A G and Puschmann R 1998 Chemical Composition and physical properties of pericarp, locule and placental tissues of tomatoes with internal bruising, *J. Am. Soc. Hortic. Sci.* **123** 656-60
[18] Yuan H R, Huang X and Li X J 2014 Effect of three-dimension fertilizing mode using CO_2 and liquid/solid fractions of the digesate from anaerobic digester on yield and quality of lettuce *J.*

Agric. Sci. **6** 212-20

[19] Garmendia I and Mangas V J 2014 Comparative study of substrate-based and commercial formulations of arbuscular mycorrhizal fungi in romaine lettuce subjected to salt stress *J. Plant Nutr.* **37** 1717-31

[20] Turhan A, Kuscu H and Ozmen N 2014 Effect of different concentrations of diluted seawater on yield and quality of lettuce *Chilean J. Agric. Res.* **74** 111-6

[21] Trivedi H V and Singh J K 2005 Application of grey system theory in the development of a runoff prediction model *Biosyst. Eng.* **92** 521-6

[22] Jiang Z H, Weng B Q, Lei J G, Xiao S X and Wang Y X 2012 Grey relational grade analysis on the relationship of amino acid contents and Cd content in agaricus brasiliensis under interaction of selenium and lanthanium *J. Agro-Environ. Sci.* **31** 1066-9

[23] Karakaya S 2004 Bioavailability of phenolic compounds *Crit. Rev. Food Sci. Nutr.* **44** (6) 453-64

[24] Kim M J, Moon Y and Tou J C 2016 Nutritional value, bioactive compounds and health benefits of lettuce (*Lactuca sativa* L.) J. *Food Compos. Anal.* **49** 19-34

[25] Llorach R, Martínez-Sánchez A, Tomás-Barberán F A, Gil M I and Ferreres F 2008 Characterisation of polyphenols and antioxidant properties varieties and escarole *Food Chem.* **108** 1028-38

[26] Chung K T, Lu Z and Chou M W 1998 Mechanism of inhibition of trannic acid and related compounds on the growth of intestinal bacteria *Food Chem. Toxicol.* **36** 1053-60

[27] Wu L X, Hu R K, Chen Y X, *et al.* 2017 Research progress in the interaction between polysaccharides and phenolics *Sci. Technol. Food Ind.* **12** 1-7

IOP Conf. Series: Earth and Environmental Science **185** (2018) 012009 doi:10.1088/1755-1315/185/1/012009

Effects of grafting on free fatty acid contents and related synthetic enzyme activities in peel and flesh tissues of oriental sweet melon during the different development period

J H Hao[1a], Z Y Qi[1a], J R Li[2], C J Liu[1], L Wang[1], W T Jin[1], X Li[3] and H Y Qi[4,5]

[1] Beijing Key Laboratory for Agricultural Application and New Technique, College of Plant Science and Technology, Beijing University of Agriculture, No. 7 Beinong Road, Changping District, Beijing, P. R. of China

[2] Administration Office for public service in Benxi city of Liaoning Province, No. 101 Jiefang North Road, Mingshan District, Benxi, P. R. of China

[3] Agricultural Bureau of Jing Country of Hebei Province, No.36 Jinghua Road, Jing Country, Hebei, P. R. of China

[4] Key Laboratory of Protected Horticulture of Education Ministry and Liaoning Province, College of Horticulture, Shenyang Agricultural University, No. 120 Dong Lin Road, Shenhe District, Shenyang, P. R. of China

[a] J H Hao and Z Y Qi are co-first authors.

E-mail: hyqiaaa@126.com

Abstract. In this experiment, white seed pumpkin 'Shengzhen No.1' was used as rootstock for the oriental sweet melon (*Cucumis melo L.*)'Yumeiren'. The fruits of own-rooted melons were used as control. The objective of this experiment was to study the effects of grafting on the following parameters: content of free fatty acids and activities of synthetic enzymes (LOX, ADH, and AAT). These parameters were measured in the peel and flesh tissues of oriental sweet melon at different time-points. Due to grafting, the content of oleic acid and linolenic acid decreased in peel tissues of melons at early stages of maturity. Therefore, total contents of free fatty acids were lower than those of control. However, there was no significant difference in the content of three free fatty acids in flesh tissue; this content was lower than that in peel tissue. Due to grafting, enzymatic activities of ADH and AAT decreased in peel tissue of melons; however, these enzymes showed different activities in flesh tissues of melons subjected to grafting. Nevertheless, there was no significant difference in enzymatic activities of flesh tissues; however, these activities were lower than that of peel tissue. These results indicate that free fatty acid content and related synthetic enzyme activities decreased when grafting was included in the cultivation of oriental sweet melons. Consequently, the aroma of these melons decreased due to grafting.

1. Introduction

Melon (*Cucumis melo L.*) is one of the most popular fruits in the world, and it is known for its unique flavor. Aromatic components are the most valuable fraction of volatile compounds in melon fruits, because they possess high sensory values and characteristic physiological values. Therefore, aromatic components have become one of the most important indices that define the quality of melon fruits [1-3]. Aroma is a quality characteristic that attracts a lot of attention from fruit growers, consumers, and quality breeders. In our previous study [4], we investigated oriental sweet melons and detected 40 types of aromatic compounds, including 32 different esters. Peak area of esters accounted for 70.14–95.77% of the total peak area, indicating that esters are the main aromatic components of melon fruits.

Fatty acids are the precursors of most straight-chain esters in aromatic volatile compounds [5-6]. In this study, we found that large amounts of free fatty acids accumulate in apples even before aromatic compounds develop in the fruit. Six-carbon aldehydes and alcohols were detected by crushing the leaf and fruit of tomato (Lycopersicon esculentum Mill.); these compounds result from the degradation of lipids and free fatty acids [7]. Collectively, these C-6 volatile compounds are considered as the most important aromatic compounds. As fruit aromas become stronger, the content of free fatty acids decreases rapidly [8]. In a study conducted by Francisc V Dulf, the dominating fatty acids of berry pulp/peel oils were arranged in a descending order: oleic (20–40%), palmitic (17–27%), palmitoleic (10–22%), linoleic (10–20%), and á-linolenic (4–9%) acids [9]. In addition, palmitoleic acid (0.1–0.5%) was found in trace amounts in berry seed oils. The proportions of oleic (13–21%) and linoleic (33–43%) acids showed relatively high deviations in different varieties of berry seed oils. Unlike pulp oils, seed oils contain higher amounts of polyunsaturated fatty acids (PUFAs) (65–72%) but relatively lower proportions of monounsaturated fatty acids (MUFAs) (16–21.5%) and saturated fatty acids (SFAs) (11–16%). The study was conducted by Schaffer *et al.* [10]. They showed that alcohol dehydrogenase (ADH) and alcohol acyltransferase (AAT) play a pivotal role in the synthesis of ester from amino acids. Moreover, only a handful of genes were associated with the production of esters in apple, including alcohol acyltransferase (AAT) and alcohol dehydrogenase (ADH). As the fruit develops, ADH activity gradually decreases but AAT activity steadily increases in the fruit [11-12].

Grafting is the effective method to prevent blight and successive cropping obstacle in protected cultivation [13-15].However, grafting leads to a decline in quality and fragrance of melon fruits. Krumbein and Schwarz [16] reported that total aromatic compounds and special esters decreased when grafting was implemented in tomato cultivation; there was a decline in the content of aromatic substances, which primarily attract people with their beautiful fragrance. Therefore, it is very important to determine the influence of grafting on the metabolism of volatile substances in melon fruits. With this strategy, we could improve the metabolism of volatile substances in melon fruits. However, very few studies have investigated how grafting affects the metabolism of aromatic substances in melon fruits. Therefore, we analyzed the content of free fatty acids and related activities of synthetic enzymes (LOX, ADH, AAT) in peel and flesh tissues of sweet melons at different development periods; both own-rotten and grafted varieties of oriental sweet melons were included in this study. Our main aim was to investigate the influence of grafting on aromatic substances and to elucidate the underlying physiological mechanism. This study served as the theoretical basis for regulating the quality of melons in the future.

2. Materials and methods

2.1. Plant material and growth conditions

Yumeiren is a cultivar of oriental sweet melon (*Cucumis melo L.*); it was used as scion with white-seeded pumpkin (*Cucurbita moschata Duch.*). Shengzhen1 was used as rootstock, and conventional methods were used to cultivate seedlings. The first two true leaves were produced on melon plants; these plants and a terminal bud were directly grafted onto white-seeded pumpkins by grafting (inarching) method. The own-rooted melon plants were used as controls (CK). Melons that underwent grafting treatment (G) were planted along with the controls in cultivation barrels (Meidolin agricultural equipment co. LTD), which were incubated in a greenhouse. Each barrel had a height of

30 cm and a diameter of 30 cm. These barrels were filled with a fermented potting substrate (soil: peat: chicken manure = 3:2:1). The bottoms of barrels were covered with a layer of perlite, which was 2 cm in thickness. Within-row plant spacing was 60 cm, whereas between-row spacing was 80 cm. Melon plants were pruned to single stems during the process of cultivation. We preserved all the melons that grew beyond the 10th node of the main stem, so only 2–3 melons were excluded from each plant. We waited for the day when the female flowers bloomed completely. Then, growth regulator Fengchanji 2 (Shenyi Biology Co., Ltd.) was sprayed on these freshly bloomed flowers in the morning. Thereafter, they were pollinated and tagged for further analyses. After the blooming of grafted and own-rooted plants, we collected samples from the flesh and peels of melons. Following the blooming of plants, samples were collected on the following days: 20, 25, 30, and 35. These samples were quickly frozen in liquid nitrogen and then stored at –80 °C. These samples were analyzed to determine free fatty acid content and the activity of enzymes (LOX, ADH, and AAH), which participated in the synthesis of esters.

2.2. Fatty acid determination

Fresh samples (5 g each) were ground into a homogeneous fine powder. Then, 10 mL was taken out from a solution mixture of petroleum and ether (4/3, v/v), and it was added to the fine powder and mixed thoroughly. This suspension was extracted for 24 h at 0–4 °C. Then, 10 mL of 0.4 mol/L KOH methanol solution was added to the extract. Finally, methyl esterification was carried out for 2 h. Redistilled water was added to separate out the organic phase. Then, organic phase was concentrated or evaporated to dryness. Finally, it was resolubilized and filtered through an acetate filter (0.45 μm pore size, Nalgene).

Gas chromatography (VARIAN CP-3800, USA) was performed to characterize fatty acids in this solution. To perform analysis, CP 7485 column was used along with splitless injection in the gas chromatography instrument. Injection pore temperature was maintained at 240 °C, while detector temperature was maintained at 250 °C. Temperature programming was set at 150 °C for 1 min, and then it was raised to 230 °C at a rate of 10 °C/min; this temperature was again maintained for 1 min. Finally, it was raised to 240 °C at a rate of 3 °C/min. This temperature was maintained for 10 min. The flow rate was 10 mL/min. Experimental related drugs were purchased from Sigma-Aldrich (St. Louis City, MO, USA).

2.3. Analysis of lipoxygenase (LOX) activity

Fresh samples (3 g each) were ground into a homogeneous fine powder. To prepare crude enzyme extracts, this powder was mixed thoroughly with 8 mL of 50 mmol/L phosphate buffer (pH 7.0). Then, it was precooled to 4 °C. This mixture was centrifuged (Thermo Scientific™, Sorvall™ ST 16R, USA) at 15000 g and 4°C for 15 min. The supernatant was used to determine the activity of LOX enzyme.

Analysis of enzyme activity: The reaction system (total volume 3 mL) contained 25 μl of sodium linoleate stock solution (substrate), 2.775 mL of 100 mmol/L acetate buffer (pH 5.5), and 0.2 mL of crude enzyme extracts. The reaction temperature was fixed at 30 °C. Lysyl oxidase (LOX) activity was measured with a spectrophotometer at 234 nm. After adding crude enzyme solution, we first measured optical density (OD) value at 15 s. Then, OD values were continuously monitored and recorded for 1 min. Thus, changes in OD values were observed for 1 min. Enzyme activity was expressed as U/mg protein. Enzyme activity assay was repeated thrice.

2.4. Analysis of alcohol dehydrogenase (ADH) activity

Preparation of crude enzyme extracts: Fresh samples (3 g each) were placed in liquid nitrogen medium and ground thoroughly into a homogeneous fine powder. Six milliliters of enzyme extraction buffer was precooled to 4 °C. Then, it was added to the powder and mixed thoroughly. The composition of enzyme extraction buffer was as follows: 100 mM MES-tris buffer solution (pH 6.5), 2 mM dithiothreitol (DTT), and 1% polyvinylpolypyrrolidone (PVPP). After incubation, the mixture was centrifuged at 15000 g and 4 °C for 30 min. The supernatant was used to determine ADH activity.

Analysis of enzyme activity: The reaction system (total volume 3 mL) consisted of the following reagents: 2.4 mL of 100 mM MES-Tris buffer (pH 6.5), 0.15 mL of 1.6 mM nicotinamide adenine dinucleotide (reduced form) (NADH), 0.15 mL of 80 mM acetaldehyde, and 0.3 mL of crude enzyme extracts. The temperature of the reaction was fixed at 30 °C. The OD values were measured with a spectrophotometer at 340 nm. After adding crude enzyme extracts, we first measured OD values at 15 s. Then, changes in OD values were continuously monitored and recorded for 1 min. The enzyme activity assay was repeated three times.

2.5. Analysis of alcohol acyltransferase (AAH) activity

Preparation of crude enzyme extracts: Fresh samples (3 g each) were placed in liquid nitrogen medium and ground thoroughly into a homogeneous fine powder. Then, 1 mg of polyvinylpolypyrrolidone (PVPP) and 6 ml 0.1M Tris-HCl (pH 8.0) were added to the powder and mixed thoroughly. After incubation, the mixture was centrifuged at 16000 g and 4 °C for 30 min. The supernatant was used to determine AAT activity.

Analysis of enzyme activity: The reaction system consisted of the following reagents: 0.5 M Tris-HCl buffer (pH 8.0), 5 mM $MgCl_2$, 5 mM acetyl coenzyme A, 200 mM butanol, and 0.6 mL of crude enzyme extracts. The temperature of reaction was fixed at 35 °C for 10 min. Then, 150 μl of 20 mM DTNB was added and mixed thoroughly with the reaction mixture. This mixture was allowed to stand at room temperature for 10 min. The OD values were measured at 412 nm. The enzyme activity assay was repeated three times.

2.6. Statistics and data analysis

The experiment was arranged in a randomized block design and performed in three replicates. Data were analyzed with DPS software (Refine Information Tech, Hangzhou, Zhejiang, China). Statistical significance of the difference was estimated by least significant difference (LSD) test at the 5% or 1% level [17]. Figures representing physiological parameters were drawn using OriginPro software (OriginLab Corporation, Northampton, MA, USA).

3. Results

3.1. The effects of grafting on the oleic acid, linoleic acid and linolenic acid contents in the peel and flesh tissues of melons at different stages of development

In the peel and flesh of oriental sweet melons, Oleic acid, linoleic acid, and linolenic acid content reached a peak on 20th day after full bloom. The content of three free fatty acids decreased steadily with the gradual development of melon fruits. On 35th day after blooming, there was significant increase in the content of three free fatty acids, except for the oleic acid in G-flesh and the oleic acid content in G-peel (Figure 1). In melon fruit, linolenic acid content was the highest among the three free fatty acids, followed by the linoleic acid. Oleic acid content was the lowest in melon fruits. Compared to the flesh of melon fruit, the content of three free fatty acids was significantly higher in the peel. At the early stage of maturity, grafting reduced the content of oleic acid and linolenic acid in the peel of melon fruit; however, grafting exerted little effect on the 30th day after blooming. Grafting reduced linoleic acid content in the melon peel on 20th and 35th days after blooming. In the flesh of grafted and own-rooted melons, no significant differences were observed in the content of three free fatty acids.

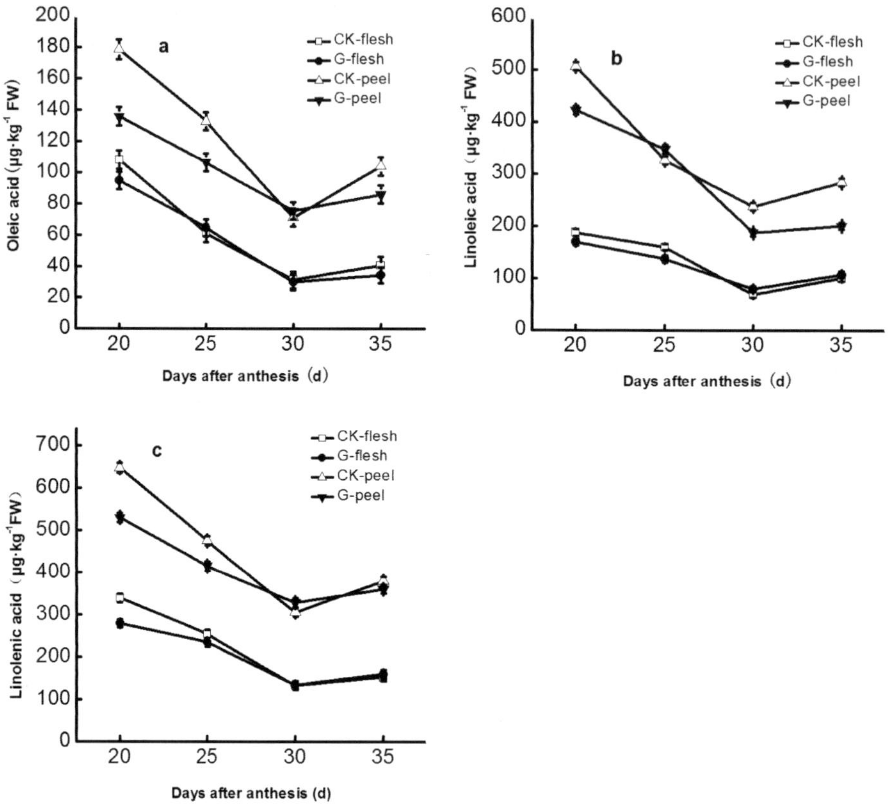

Figure 1. Effects of grafting on the content of three free fatty acids in the peel and flesh tissues of oriental sweet melon during fruit development(a, Content of oleic acid; b, Content of linoleic acid; c, Content of linolenic acid; bars, SE).

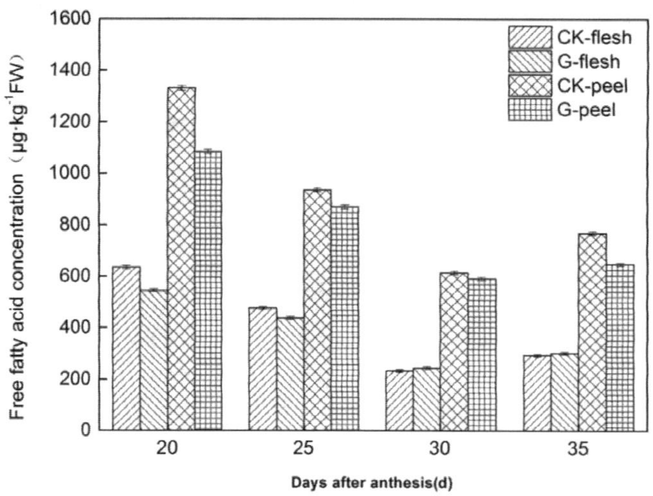

Figure 2. Effect of grafting on the total content of free fatty acid of oriental sweet melon during fruit development (bars, SE).

3.2. The effect of grafting on the total content of the three free fatty acids (TFFA) at different stages of development

It is now believed that oleic acid, linoleic acid, and linolenic acid are precursors in the synthesis of straight-chain fatty alcohols, fatty aldehydes, and fatty esters; the synthesis reaction proceeds through LOX pathway [6]. Figure 2 illustrates that changes in the total content of TFFA were essentially similar to those of three free fatty acids. The total content of TFFA reached a maximum on 20th day after blooming. The total content of TFFA decreased gradually with the development of melon fruits. The total content of TFFA was significantly higher in mature fruits. Compared to the flesh of melon fruit, the peel contained a significantly higher content of TFFA. Compared with own-rooted melons, the peel of grafted melons contained considerably lower content of TFFA at all stages of development. On 20th and 25th days after blooming, the flesh of grafted melons had sharply lower levels of free fatty acids than in own-rooted melons. On 30th and 35th day after blooming, no significant differences were observed in the total content of TFFA in the flesh of grafted melons and own-rooted melons. The results indicated that grafting reduced the total content of TFFA in the peel of oriental sweet melons. In the flesh of melons, grafting showed various effects on the total content of TFFA; these effects depended on the different stages of development.

3.3. The effect of grafting on the LOX, ADH and AAT activity levels in the peel of oriental sweet melon

In the peel of grafted and own-rooted melons, LOX activity decreased initially; however, it increased subsequently and decreased again (Figure 3). Interestingly, LOX activity reached a peak on 20th day after blooming. Moreover, LOX activity reached its peak before a large number of volatile esters developed in the peel of oriental sweet melons. The results indicated that LOX activity occurred prior to the production of volatile esters on a large scale. Furthermore, LOX activity gradually decreased with the development of melon fruits. At 25–30 days after blooming, LOX activity increased. After the fruit became fully mature, LOX activity decreased slightly. Grafting reduced LOX activity in the melon peel on 20th and 35th days after blooming; however, no significant changes were observed in the enzymatic activity levels on 25th and 30th days after blooming.

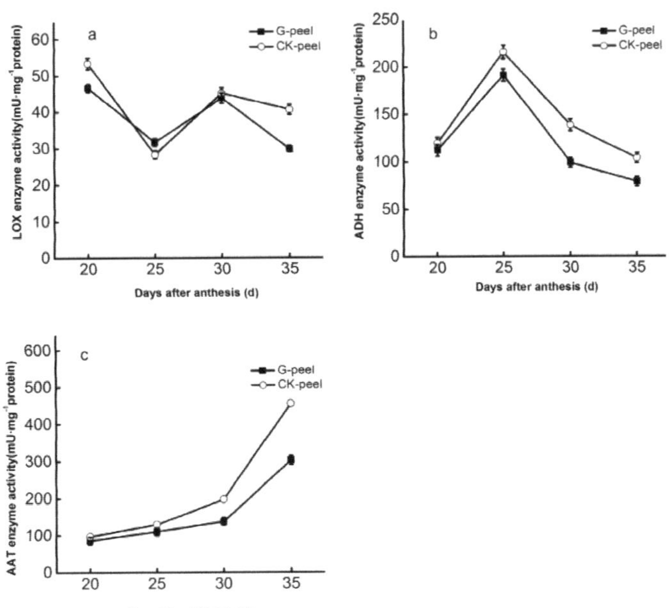

Figure 3. Effects of grafting on LOX, ADH, and AAT activities in peel tissue of oriental sweet melon during fruit development (a, activity of LOX enzyme; b, activity of ADH enzyme; c, activity of AAT enzyme; bars, SE).

As shown in Figure 3, ADH activity first increased and then decreased as the fruit matured gradually. Moreover, ADH activity reached its peak on 25th day after blooming. Consequently, a large number of intermediate products were available for esterification. However, ADH activity gradually declined with the development of melon fruits. Due to grafting, ADH activity decreased in melon peels at 25–35 days after bloom.

In the peel of oriental sweet melons, AAT activity increased steadily with the gradual maturation of fruit. Moreover, ATT activity increased slowly at 20–30 days after bloom; then, it increased rapidly at 30–35 days after bloom. Compared to controls, ATT activity was significantly reduced in grafted melons at 25–35 days after blooming (Figure 3).

3.4. The effect of grafting on the LOX, ADH and AAT activity levels in the flesh of oriental sweet melons

The activity of LOX, ADH, and AAT was lower in the flesh than that in peel of oriental sweet melons. In the flesh of melons, the activity of LOX, ADH and AAT enzymes underwent changes, which were similar to those observed in the peel of melons (Figure 4). Due to grafting, the LOX activity decreased in melon flesh on 20th and 35th days after bloom; however, no significant changes were observed in LOX activity at any other time point. Due to grafting, ADH activity decreased on 20th day after bloom; however, it hardly showed any significant changes at other time points. No significant differences were observed in ATT activity of grafted melons and own-rooted melons.

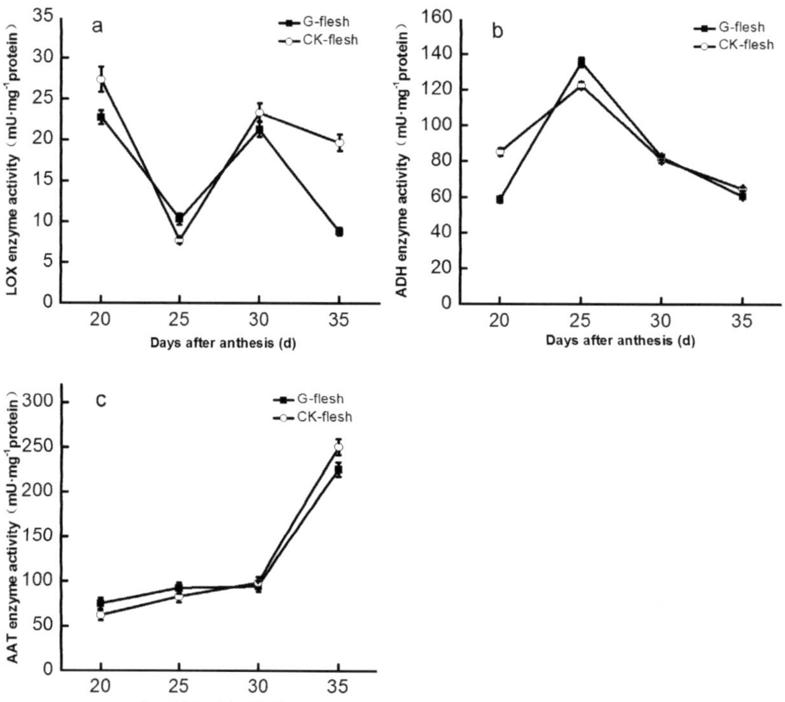

Figure 4. Effects of grafting on activities of LOX, ADH, and AAT enzymes in flesh tissues of oriental sweet melon during fruit development (a, activity of LOX enzyme; b, activity of ADH enzyme; c, activity of AAT enzyme; bars, SE).

4. Discussion

Fatty acids are precursors of most straight chain esters with aromas [5, 6]. Fatty acids are undergoing α-oxidation reaction to produce acetic acid, butyric acid, and hexanoic acid: these carboxylic acids are further reduced to their corresponding alcohols [18]. In the presence of acyl-CoA, alcohols are

IOP Conf. Series: Earth and Environmental Science **185** (2018) 012009 doi:10.1088/1755-1315/185/1/012009

esterified by AAT [19-21]. Certain alcohols are used in the synthesis of straight-chain esters; these alcohols are derived from LOX pathway. Due to the enzymatic activity of LOX, fatty acids get converted into aldehydes; these compounds are further reduced to alcohols with the help of ADH [22]. In the synthesis of esters, the following fatty acids act as precursors: oleic acid, linoleic acid, and linolenic acid.

In this study, a large amount of free fatty acids (oleic acid, linoleic acid, and linolenic acid) accumulate in melon fruits; they accumulate before aromatic volatiles are formed in melons. Fruit aromas become stronger as the fruit matures gradually. At the same time, the content of free fatty acids also decreases rapidly. These results comply with the findings of an earlier study on grapes; this study was conducted by Agudelo-Romero P *et al* [23]. They showed that immature apple fruits contain rather low levels of free fatty acids; these fruits produce little aroma as they are immature. As these fruits mature gradually, fatty acid content increases steadily. In particular, the content of unsaturated fatty acids (oleic acid, linoleic acid, and linolenic acid) increases sharply. Consequently, the aroma becomes stronger with the maturation of fruits [24]. α-linoleic acid and linolenic acid were the two essential fatty acids in banana peels; the content of α-linoleic acid and linolenic acid was 20.75% and 19.56%, respectively [25]. Khan et al. [26] analyzed and compared the ratio of linoleic acid to linolenic acid in 10 different varieties of cucumbers. They found a strong positive correlation between the flavor of cucumbers and the ratio of linolenic acid to linoleic acid. In addition, exogenous linoleic and linolenic acids were added as supplements during the cultivation of cucumbers. Consequently, contents of two flavor compounds increased sharply with these supplements. These compounds were characteristic of fresh cucumbers: trans-2, cis-6-nonadienal and trans-2-nonenal. Duan *et al.* [27] investigated the correlation between peach fruit firmness and fatty acid content. They found that the content of linolenic acid increased significantly with fruit ripening, but only oleic acid was positively correlated with fruit firmness. In addition, linoleic acid content increased initially but then decreased during fruit ripening [27].

In this study, the activity of LOX enzyme reached its peak value on 20th day after blooming. Moreover, LOX activity reaches its peak before the formation of several volatile esters in oriental sweet melons. The results indicate that LOX activity occurs prior to the production of volatile esters on a large scale. Moreover, ADH activity reached its peak value on 25th day after bloom. Owing to ADH activity, a large number of intermediate compounds participated in ester synthesis. In addition, ADH activity gradually decreased with the development of fruit melons. In contrast, AAT activity increased steadily with the maturation of fruits. Compared to flesh tissues, the three enzymes showed higher activity in melon peel. Owing to grafting, there was a sharp decrease in the activity of ADH and AAT enzymes in the peel. This adversely affected the conversion of aldehydes and biosynthesis of esters. However, future studies must be conducted to know whether grafting affects the expression of key enzyme genes. The investigation has to be carried out at transcriptional level of genes.

Previous studies have investigated the activity of AAT gene in melon fruits. The expression of AAT genes has been determined at the molecular level in these studies. Therefore, these studies are valuable references, which can be used to determine the expression of key enzyme genes. Yahyaoui *et al.* [28] and Islam *et al.* [29] cloned four genes from cantaloupe melon fruit: CM-AAT1, CM-AAT2, CM-AAT3, and CM-AAT4. All four genes are members of acyltransferase gene family of plants. These genes encode AAT proteins. Except for CM-AAT2, the three genes are expressed in the fruit. They encode proteins that display a strong preference for substrates. It is speculated that ATT activity decreases after grafting due to the expression of these four genes. Islam *et al.* [29] and Manri'quez *et al.* [30] studied the expression patterns of genes that encoded AAT enzymes in melon fruits. They also analyzed functional characteristics of Cm-AAT1, Cm-AAT2, Cm-AAT3, and Cm-AAT4 enzymes, which played a pivotal role in ester synthesis. To restore the activity of enzymes, alanine residue was mutated into threonine residues of AAT. The results indicate that threonine residues are essential for AAT activity, which governs the process of ester biosynthesis. The present study showed that AAT activity decreased in melon peel when grafting was conducted during cultivation.

Previous studies have also determined the expression of ADH gene in melon fruits. Manri'quez *et al.* [30] investigated the effects of CM-ADH1 and CM-ADH2, the two ADH genes cloned from melon fruit, on the aromas of the fruit [30]. In melon fruits, they found that ADH plays an important role in aroma development. This implies that CM-ADH1 and CM-ADH2 are highly specific ADH genes in melon fruits; these genes are expressed specifically in mature melons and their expression is regulated by ethylene. The enzyme aldehyde reductase was strongly preferred by proteins, which were encoded by the two genes. In our current study, ADH activity was strongly reduced in melon peel by including grafting technique during cultivation. However, further studies must be conducted to confirm whether grafting affects the expression of the ADH gene.

In summary, the peel of free fatty acids (oleic acid, linolenic and acid linoleic acid) content and related synthetic enzyme (LOX, ADH and AAT) activities are higher than those in flesh tissue of melons. By grafting method, free fatty acid (oleic acid, linolenic and acid linoleic acid) contents and related activities of synthetic enzymes were reduced in peel tissue of melons. Consequently, the aroma of melon fruits declined due to grafting. To elucidate the mechanism through which grafting brings about these changes, we have to further analyze the expression of genes and proteins during various stages of fruit development. These analyses must include both own-rooted and grafted melons. Moreover, the interaction between proteins and genes must be further elucidated in future studies.

Acknowledgements
This work was financially supported by the Fund of the National Natural Science Foundation of China (31372057), the 2017 Agriculture Education Teaching Reform Project of Beijing University of Agriculture (BUA2017JG030), 2017 Research Fund for Academic Degree & Graduate Education of Beijing University of Agriculture (5056516005/033), and the Innovation Team Construction of Leafy Vegetables of Beijing (blvt-02).

References

[1] Condurso C, Verzera A, Dima G, Tripodi G, Crinò P, Paratore A and Romano D 2012 Effects of different rootstocks on aroma volatile compounds and carotenoid content of melon fruits *Sci. Hortic.* **148**: 9-16.

[2] Dos-Santos N, Bueso M C and Fernández-Trujillo J P 2013 Aroma volatiles as biomarkers of textural differences at harvest in non-climacteric near-isogenic lines of melon *Food Res. Int.* **54**: 1801-12.

[3] Lucchetta L, Manriquez D, El-Sharkawy I, Flores F B, Sanchez-Bel P, Zouine M, Ginies C, Bouzayen M, Rombaldi C, Pech J C and Latché A 2007 Biochemical and catalytic properties of three recombinant alcohol acyltransferases of melon. Sulfur-containing ester formation, regulatory, role of CoA-SH activity, and sequence elements conferring substrate preference *J. Agric. Food Chem.* **55**: 5213-20.

[4] Qi H, Wei M and Liu Y 2008 Elementary studies on influences of grafting on aromatic compounds in muskmelon *China Vecetables* **2008**(2):21-24.

[5] Mekni M, Flamini G, Garrab M, Hmida R B, Cheraief I, Mastouri M and Hammami M 2013 Aroma volatile components, fatty acids and antibacterial activity of four Tunisian Punica granatum L. flower cultivars *Ind. Crops Prod.* **48**: 111-7.

[6] Rowan D D, Allen J M, Fielder A S and Hunt M B 1999 Biosynthesis of straight-chain ester volatiles in red delicious and granny smith apples using deuterium-labeled precursors *J. Agric. Food Chem.* **47**: 2553-62.

[7] Tandon K S, Baldwin E A and Shewfelt R L 2000 Aroma perception of individual volatile compounds in fresh tomatoes as affected by the medium of evaluation *Postharvest Biol. Technol.* **20**:261-268.

[8] Canoles M A, Beaudry R M, Li C and Howe G 2006 Deficiency of linolenic acid in Lefad 7 mutant tomato changes the volatile profile and sensory perception of disrupted leaf and fruit tissue *J. Amer. Soci. Hortic. Sci.* **131**: 284-9.

[9] Francisc V D 2012 Fatty acids in berry lipids of six sea buckthorn (Hippophae rhamnoides L., subspecies carpatica) cultivars grown in Romania *Chem. Cent. J.* **6**:106.

[10] Schaffer R J, Friel E N, Souleyre E J, Bolitho K, Thodey K, Ledger S, Bowen J H, Ma J H, Nain B and Cohen D et al. 2007 A Genomics Approach Reveals That Aroma Production in Apple Is Controlled by Ethylene Predominantly at the Final Step in Each Biosynthetic Pathway *Plant Physiol.* **144** : 1899-912.

[11] Defilippi B G, Kader A A and Dandekar A M 2005 Apple aroma: alcohol acyltransferase, a rate limiting step for ester biosynthesis, is regulated by ethylene *Plant Sci.* **168**:1199-210.

[12] Souleyre E J F, Greenwood D R, Friel E N, Karunairetnam S and Newcomb R D 2005 An alcohol acyl transferase from apple (cv. Royal Gala), MpAAT1, produces esters involved in apple fruit flavor *FEBS J.* **272**: 3132–44.

[13] Bautista A S, Calatayud A, Nebauer S G, Pascual B, Maroto J V and López-Galarza S 2011 Effects of simple and double grafting melon plants on mineral absorption, photosynthesis, biomass and yield *Sci. Hortic.* **130**: 575-80.

[14] Martinez-Rodriguez M M, Estañ M T, Moyano E, Garcia-Abellan J O, Flores F B, Campos J F, Al-Azzawi M J, Flowers T J and Al-Azzawi M J 2008 The effectiveness of grafting to improve salt tolerance in tomato when an 'excluder' genotype is used as scion Environ. *Exp. Bot.* **63**: 392-401.

[15] Sakata Y, Ohara T and Sugiyama M 2008 The history of melon and cucumber grafting in Japan *Acta Hortic.* **767**:217–28.

[16] Krumbein A and Schwarz D 2013 Grafting: A possibility to enhance health-promoting and flavour compounds in tomato fruits of shaded plants *Sci. Hortic.* **149**(1):97-107.

[17] Reid and Julian R W 2017 Experimental Design and Data Analysis for Biologists *Austral Ecol.* **28**(2):781–3.

[18] Paillard N M M 1990 Flavour of apples, pears and quinces *Dev. Food Sci.* **3**:1-41

[19] Günther C S, Chervin C, Marsh K B, Newcomb R D and Souleyre E J 2011 Characterisation of two alcohol acyltransferases from kiwifruit (Actinidia spp.) reveals distinct substrate preferences *Phytochemistry* **72**: 700-10.

[20] Khanom M M and Ueda Y 2008 Bioconversion of aliphatic and aromatic alcohols to their corresponding esters in melons (Cucumis melo L. cv. Prince melon and cv. Earl's favorite melon) *Postharvest Biol. Technol.* **50**: 18-24.

[21] Knee M and Hatfield S G S 1981 The metabolism of alcohols by apple fruit tissue *J. Sci. Food Agric.* **32**: 593-600.

[22] Moummou H, Tonfack L B, Chervin C, Benichou M, Youmbi E, Ginies C, Latché A and Pech J C 2012 Functional characterization of SlscADH1, a fruit-ripening-associated short-chain alcohol dehydrogenase of tomato *J. Plant Physiol.* **169**: 1435-44.

[23] Agudelo-Romero P, Erban A, Sousa L, Pais M S and Kopka J 2013 Search for Transcriptional and metabolic markers of grape pre-ripening and ripening and insights into specific aroma development in three portuguese cultivars *PLoS One* **8**(4): e60422.

[24] Dixon J and Hewett E W 2000 Factors affecting apple aroma/flavour volatile concention:review *N. Z. J. Exp. Agric.* **8**: 55-173.

[25] Guo X X, Song W D, Zhu S W, Meng Y Y, Huang J J and Chen Y F 2008 Withdrawal and GC-MS analysis of fatty acids from banana peel *J. Fruit Sci.* **25**(4): 604-6

[26] Khan A L, Hamayun M, Radhakrishnan R, Waqas M, Kang S M, Kim Y H, Shin J H, Choo Y S, Kim J G and Lee I J 2011 Mutualistic association of paecilomyces formosus LHL10 offers thermotolerance to cucumis sativus *Antonie van Leeuwenhoek* **101**:267-79.

[27] Duan Y X, Dong X Y, Liu B H and Li P H 2013 Relationship of changes in the fatty acid compositions and fruit softening in peach (Prunus persica L. Batsch) *Acta Physiol. Plant.* **35**(3):0137-5881.

[28] Yahyaoui F E L, Wongs-Aree C, Latche A, Hackett R, Grierson D and Pech J C 2002 Molecular and biochemical characteristics of a gene encoding an alcohol acyl-transferase

involved in the generation aroma volatile esters during melon ripening *Eur. J. Biochem.* **269**: 2359-66.

[29] Islam E S, Manríquez D, Flores F B, Regad F, Bouzayen M, Latché A and Pech J C 2005 Functional characterization of a melon alcohol acyl-transferase gene family involved in the biosynthesis of ester volatiles *Plant Mol. Biol.* **59**: 345-62.

[30] Manríquez D, El-Sharkawy I, Flores F B, El-Yahyaoui F, Regad F, Bouzayen M, Latché A and Pech J C 2006 Two highly divergent alcohol dehydrogenases of melon exhibit fruit ripening-specific expression and distinct biochemical characteristics *Plant Mol. Biol.* **61**: 675-85.

IOP Conf. Series: Earth and Environmental Science **185** (2018) 012010 doi:10.1088/1755-1315/185/1/012010

Effects of different types of polyamine on growth, physiological and biochemical nature of lettuce under drought stress

C J Liu[#], H R Wang[#], L Wang, Y Y Han, J H Hao and S X Fan[1]

College of Plant Science and Technology, Beijing University of Agriculture/Beijing Key Laboratory for Agricultural Application and New Technique/National Demonstration Center for Experimental Plant Production Education, Beijing 102206, China

#These authors contributed equally to this work

E-mail: fsx20@163.com

Abstract. This study explored the effects of drought stress on scattered leaf lettuce seedlings and the effects of different types of polyamines (PAs) on the morphological indexes and antioxidant enzyme activities of scattered leaf lettuce. The drought-sensitive variety "Victoria crispy" was used as the plant material, and 10% Polyethylene Glycol (PEG) 6000 was used to simulate drought stress. Foliar applications of 0.1 millimoles per litre of three polyamines [putrescine (Put), spermidine (Spd) and spermine (Spm)] were performed at the seedling stage (the control group was sprayed with water). The morphological indices and antioxidant enzyme activity were recorded on the eighth day after treatment. The results showed that drought stress significantly decreased the morphological indices of scattered leaf lettuce, and spraying different types of PAs had a mitigating effect. Drought stress significantly reduced the ascorbic acid peroxidase (APX), catalase (CAT), superoxide dismutase (SOD), and peroxidase (POD) activity, increased the content of malondialdehyde (MDA), and exogenous polyamine spraying significantly affected the antioxidant enzyme activity. Among the polyamines, Put showed the most noticeable effect of mitigation. Drought stress had a significant effect on the morphological and antioxidant enzyme activities of scattered leaves lettuce, and Put was the most effective PA to increase drought tolerance.

1. Introduction

Lettuce (*Lactuca sativa* L.) belonging to the family Compositae, is a one or two-year-old herb; its common name is lettuce [1]. Lettuce is a low-sugar, low-fat vegetable that is not only rich in vitamin [2] , vitamin E [2-3], calcium, phosphorus, iron and other minerals and vitamins but also contains malic acid, tartaric acid, lactic acid, mannitol and the active thyroid hormone [4-5]; thus, it has a high nutritional value [6]. The stem and leaf extract of lettuce can enhance gastric juices, stimulate digestion, increase appetite, and has a hypnotic analgesic effect [2, 7]. At the same time, the extract of leaf lettuce can help treat neurasthenia and effectively lower cholesterol, in addition to other effects, which can reduce the risk of chronic diseases such as cancer, cardiovascular disease and age-related decline [8-10]. Thus, lettuce has gradually become a popular vegetable, and its productivity increases every year [11]. Lettuce has shallow roots, large leaf area characteristics, and does not tolerate drought

[12]. It has high water content, and arid environments significantly reduce its quality and production, causing significant economic losses.

Drought stress poses an immense challenge to global agricultural production. Due to its action, the balance between the generation and elimination of reactive oxygen species (ROS) disappears [13], causing plants to reduce the true potential for stress resistance. China is a country with a shortage of freshwater resources, and its per capita share is only one-fourth of the world average; there are drought and semi-arid phenomena in many areas. Water is the most important environmental factor in plant growth and development, and water deficits may be produced due to soil conditions and other factors, even if there is sufficient water. Photosynthesis, material transport, protein synthesis and other physiological processes are severely affected [14]. Drought stress affects plant growth and development in many ways, including reactive oxygen species, osmolytes, photosynthesis, and water use efficiency [13, 15-16]. To ensure their continuous growth and development under drought stress, plants act via a series of responses to stress, such as antioxidant regulation, stomatal regulation and osmotic adjustment [17-19]. Drought stress and subsequent plant responses have become a hot topic in plant physiology, and it is vital to cultivate drought-resistant varieties by studying plant drought resistance.

Polyamines (PAs) are ubiquitous low molecular aliphatic amines generated during the metabolic processes of an organism. They are the second messengers of signaling pathways, and have recently been added to plant growth regulators [20-22]. Putrescine (Put), Spermine (Spm), and Spermidine (Spd) play pivotal roles in plant growth and development [23]. Polyamines affect DNA, RNA and protein biosynthesis, promote plant growth and development, delay ageing, and improve disease resistance in plants [24]. Since PA can clean free radicals in plants, they also protect the membrane to some extent from oxidative damage [25]. And the most effective PA for alleviating drought stress in rice is Spm [26]. Exogenously applied PA also increased plant tolerance under several abiotic stresses [27-28]. Nayyar *et al.* found that drought tolerance in soybean can be increased with exogenous application of Put and Spd [29]. To our knowledge, previous studies on polyamines mostly involve studies on the drought resistance of crops; however, to date, no research has been found on the potential role, basis and influence of exogenous PA on the dryness resistance of vegetables such as leaf lettuce. In this study, we investigated the effects of different types of PAs on drought stress, determining its physiological indicators and screening out the most effective polyamines, which alleviate drought stress. This study aimed to provide a theoretical basis for screening more efficient polyamine types for leafy vegetables.

2. Materials and Methods

2.1. Experiment material

The lettuce varieties were screened based on their polyamine content. The variety Victoria crispy, (origin: Italy; manufacturer: Tianjin Priority Seed Co., Ltd.) contains polyamine species Spm, Spd and Put.

2.2. Experimental method

The seeds were germinated on a qualitative filter paper tile, wetted with distilled water, in a 9-cm-diameter dish. The seeds were selected such that they were the same size, whole, and showed no disease. The germinated seedlings were planted on a nursery sponge in water culture; half of the total seedlings were watered with the nutrient solution (the formula is shown in Table 1). At the four-leaf stage, the growing seedlings were placed in a hydroponic set-up to add the nutrient solution.

After slow seedling growth for 2-3 days, 10% polyethylene glycol (PEG 6000) was added for drought simulation. Different polyamines were sprayed after 6 pm every day (the control group was sprayed with water) until the leaves began to drip solution, and on the 8th day, the morphological index was measured. The test was carried out in a growth chamber; the temperature period was 20°C

(14 h, 7: 00 ~ 21: 00) / 15°C (10 h, 22: 00 ~ 6: 00) and the photoperiod was 14 h (7: 00 ~ 21: 00) / night 10 h (22: 00 ~ 6: 00), with a light intensity 1200 Lux and 60% relative humidity.

Five test groups were set up for the experiment. Treatment I (CK-W) was set up as a control and treatment II (PEG-W) for drought stress. Two controls were maintained, one of them was maintained under drought conditions and the other was maintained under full watering conditions; both received no PA treatment as a foliar application or seed treatment, Treatment III (PEG-Spd) included drought treatment + 0.1 mmol / L of the spermine aqueous (drought treatment 2-3 days after spraying 0.1 mmol / L of the spermine aqueous solution); treatment IV included (PEG-Spm) for drought + 0.1 mmol / L aqueous spermidine; treatment V (PEG-Put) included drought + putrescine (drought treatment 2-3 days after spraying 0.1 mmol / L aqueous solution of Put). The three groups were treated as described above to observe the reversal and alleviation effects of the three PAs on drought stress.

Table 1. Nutrient solution.

Name	Consumption (1 L)
Calcium nitrate tetrahydrate ($Ca(NO_3)_2 \cdot 4H_20$)	0.589 g
Potassium nitrate (KNO_3)	0.887 g
Ammonium nitrate (NH_4NO_3)	0.0571 g
Magnesium sulfate heptahydrate ($MgSO_4 \cdot 7H_2O$)	0.183 g
Potassium sulphate (K_2SO_4)	0.054 g
Phosphoric acid (H_3PO_4)	0.223 mL
Ethylenediaminetetraacetic Acid Ferric Sodium Salt (EDTA-FeNa)	0.020 g
Perboric acid (H_3BO_4)	0.00286 g
Manganese sulphate ($MnSO_4$)	0.002 g
Zinc sulphate ($ZnSO_4$)	0.00022 g
Copper sulfate ($CuSO_4$)	0.00008 g
Ammonium heptamolybdate tetrahydrate ($(NH_4)_6Mo_7O_{24} \cdot 4H_2O$)	0.00002 g

2.3. Determination of indicators and methods

Fresh and dry weights were recorded using a balance. Water content and relative moisture content of the samples were calculated. To determine the relative water content (RWC), fresh leaves (0.5 g) (Wf) were weighed to obtain a fresh weight. The leaves were then floated in water for 4 hours and the saturated weight (Wt) was measured. The same leaves were dried at 85 °C for 24 hours and the dry weight (Wd) was measured.

RWC (%) were calculated as: $RWC= (W_f - W_d)/(W_t - W_d)*100\%$

Plant height and root length were measured using a ruler, and the root / shoot ratio was calculated.

To analyze the activity of antioxidant enzymes, 0.2 g of fresh leaves were arbitrarily taken from each reservoir at each sampling time, immediately frozen in liquid nitrogen and stored at -80℃ until use. For the extraction, frozen samples were crushed on ice with 4 ml of 50 mM cold phosphate buffer (pH 7.8) containing 1%(w/v) Polyvinylpyrrolidone. The homogenate was centrifuged at 12,000 g for 30 minutes at 4℃. The supernatant was used to determine the activity of the antioxidant enzyme and the MDA content, which was measured as the degree of lipid peroxidation. The activity of SOD enzyme is mainly determined by measuring the rate of decrease in absorbance of nitroblue tetrazolium chloride at 560 nm [30]. The activity information of CAT, POD and APX is determined by analyzing the changes in the absorbance values of CAT, POD and APX at 240, 470 and 290 nm [31-32].

2.4. Data processing and analysis

Data analysis was performed using SPSS 19.0 for determining variance and significance analysis.

3. Results and analysis

3.1. Morphological indicators

3.1.1. Effects of different polyamines treatments on fresh and dry weight of lettuce under drought stress. When the lettuce leaves are under drought conditions, their growth is affected by severe stress, but the exogenous application of polyamines reduces the damage caused by drought. Figure 1 shows the effect of different treatments on fresh weight and dry weight on the 8th day. In the control plants, the fresh weight of the shoots was significantly higher. Under drought stress treatment, exogenous application of Spm and Spd can restore the fresh weight to a certain extent, basically reaching the level of control treatment, but the difference is not obvious. However, spraying Put resulted in significantly higher fresh weight compared with that of other treatments, even more than that in the control treatment. The fresh weight of the roots was significantly higher in the control compared with the drought treatment. Under drought stress, the spraying of these three PAs could significantly increase the fresh weight of the roots, and there were significant differences among spraying Put Spm, and Spd, although the effect of Put was the most obvious.

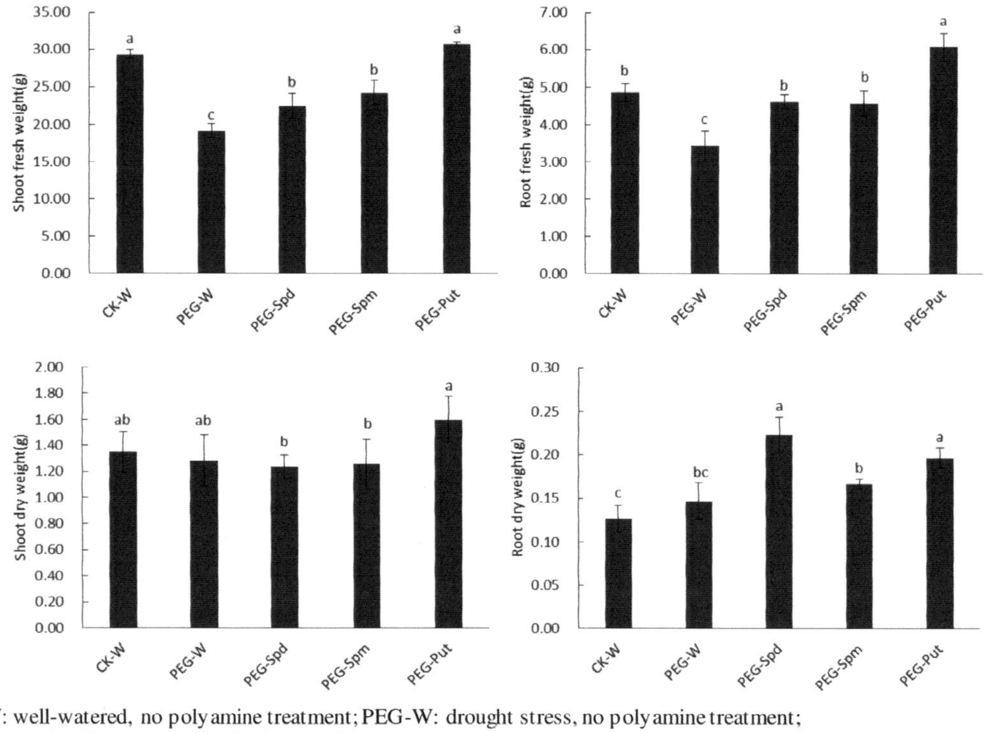

CK-W: well-watered, no polyamine treatment; PEG-W: drought stress, no polyamine treatment;
Put: putrescine; Spm: spermine; Spd: spermidine
Note: Different letters showed significant differences ($P<0.05$)

Figure 1. Effects of different polyamine treatments on lettuce fresh weight and dry weight.

There was a small difference in the dry weight of the shoots treated with different treatments, and there were no significant differences among spraying Spm, Spd and water in drought treatment. The dry weight of plants treated with Put was slightly higher than the other treatments. Root dry weight differed slightly among different treatments; the dry weights were significantly different with Spd, Put and spraying water. However, Spm spraying did not result in significantly different dry weights compared with the control. It can be seen from the test conclusion that drought stress significantly reduces the dry weight and fresh weight of leaf lettuce.

3.1.2. Effects of different polyamines treatments on plant height, root length and root / shoot ratio of lettuce in drought stress. Figure 2 shows the significance of plant height, root length and root/shoot ratio at eight days after different treatments. Drought stress significantly reduced plant height. During drought treatment, spraying PA had a certain effect on plant height, but it was not apparent. Drought stress reduced the root length; spraying polyamine can increase root length, but Spm and Spd treatments did not show significant differences. The application of Put drought treatment showed increased root length, which was similar to the control group. Drought stress had no significant effect on the root/shoot ratio of seedlings. In the drought treatment, the effect of spraying Put on root/shoot ratio was the highest, but there was no significant difference between spraying and PA.

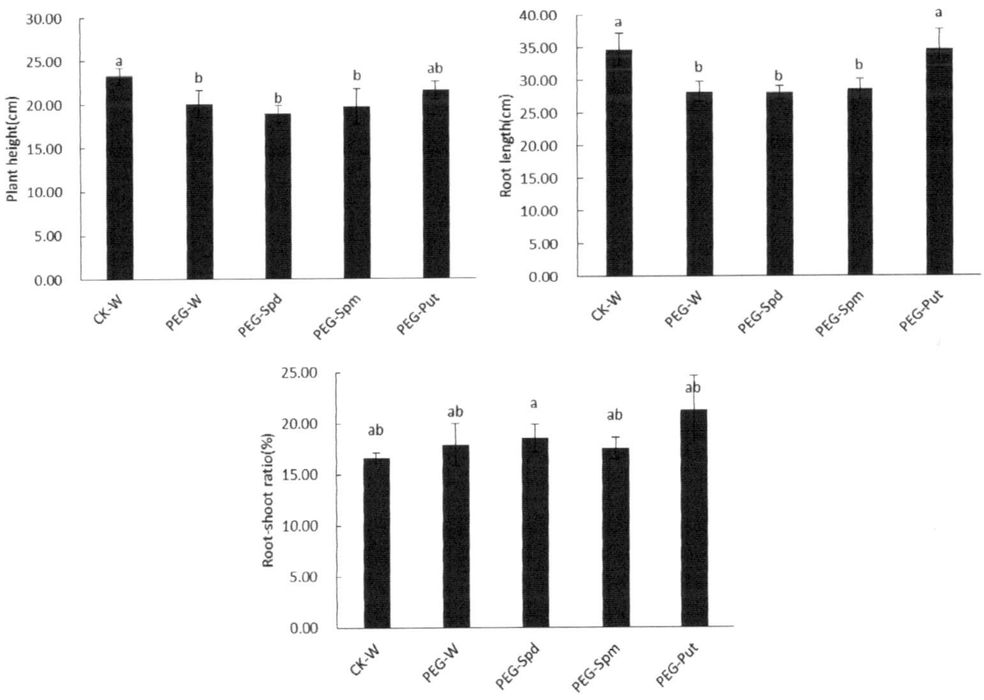

CK-W: well-watered, no polyamine treatment; PEG-W: drought stress, no polyamine treatment;
Put: putrescine; Spm: spermine; Spd: spermidine
Note: Different letters showed significant differences ($P<0.05$)

Figure 2. Effects of different polyamines treatments on plant height, root length and ratio of root to leaf lettuce.

3.1.3. Effects of different polyamine treatments on water content and relative water content of lettuce in drought stress. Figure 3 shows the water content and relative water content on the 8th day after the different treatments; drought caused a decline in water content and relative moisture content in lettuce. Spraying polyamines, particularly putrescine, can significantly increase the water content of seedlings. Moreover, spraying PAs has little effect on plant relative water content.

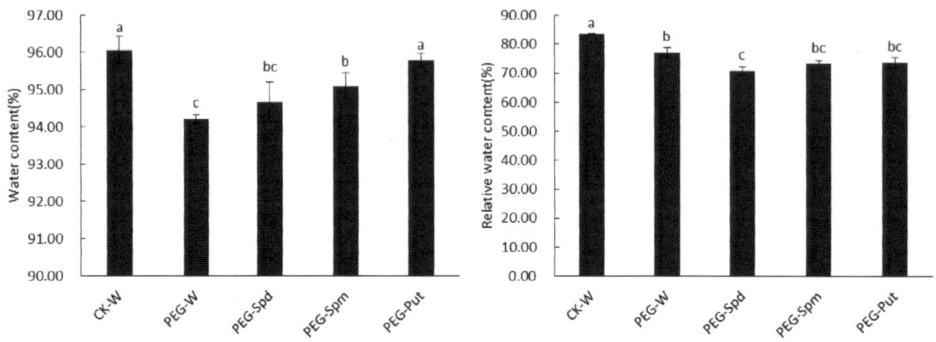

CK-W: well-watered, no polyamine treatment; PEG-W: drought stress, no polyamine treatment;
Put: putrescine; Spm: spermine; Spd: spermidine
Note: Different letters showed significant differences ($P<0.05$)

Figure 3. Effects of different polyamine treatments on lettuce water content, relative water content.

3.2. *Effects of polyamines application on the antioxidants under drought stress in lettuce*

Figure 4 shows a significant analysis of the antioxidant enzyme activity on day 8 after different treatments. APX content decreased due to drought stress, but the application of PA under stress conditions improved its content. Exogenous spraying of PAs under drought stress could significantly increase the activity of APX. The effect of Put was the most obvious, and the activity of APX was higher than that of the control. Compared with other PAs, there was a significant difference with Put treatment. Drought stress caused a significant increase in the MDA content in lettuce. Exogenous PAs application inhibited this increase such that there were no significant differences between the polyamine treatments and the control, and the MDA content was slightly reduced, which was significantly ameliorated by Spm and Put treatments, with Put effect being the most obvious. In the determination of CAT content, it can be found that the maximum amount can be obtained from fully watered lettuce, but it is significantly reduced under drought stress, and the application of exogenous PA shows a small increase in enzyme activity and is not obvious.

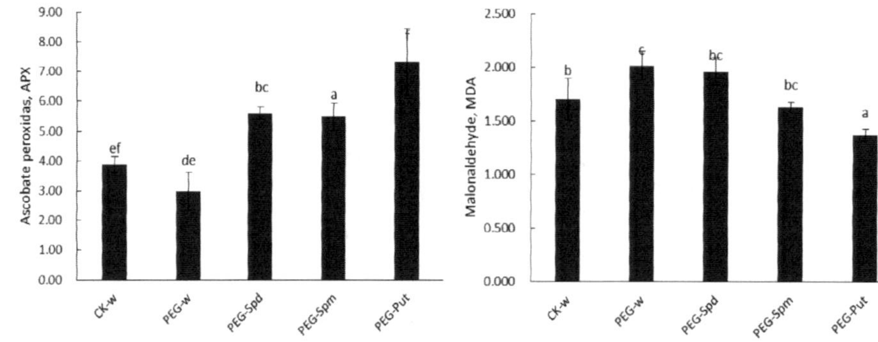

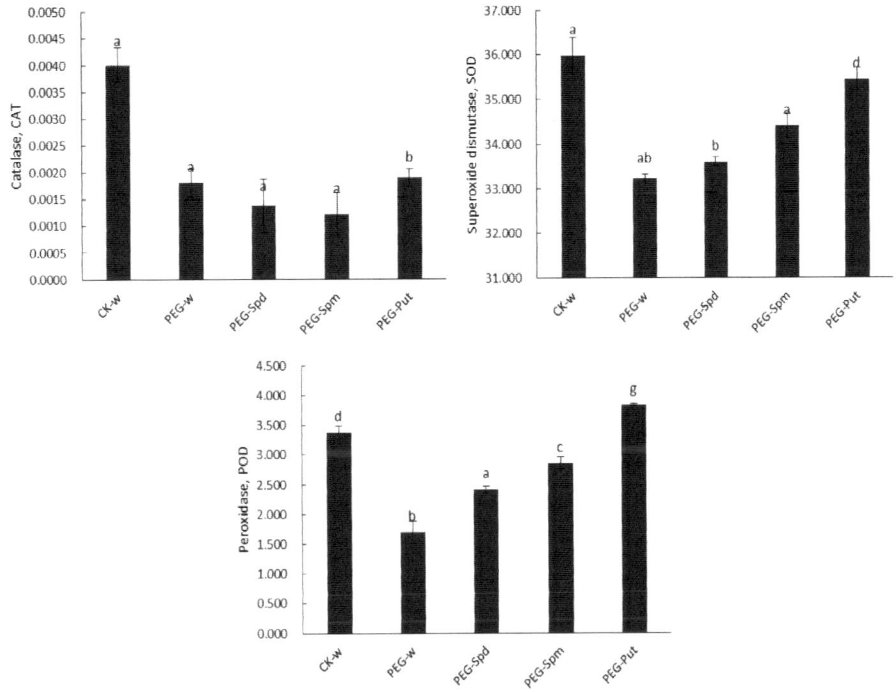

CK-W: well-watered, no polyamine treatment; PEG-W: drought stress, no polyamine treatment;
Put: putrescine; Spm: spermine; Spd: spermidine
Note: Different letters showed significant differences ($P<0.05$)

Figure 4. Effects of different polyamine treatments on lettuce antioxidant enzyme activity.

Although the content of SOD was reduced due to drought stress, this was significantly improved by the application of PA. The Spd treatment and the control group showed almost no difference, the rest were significantly different, and the most obvious effect was shown by Put. Similarly, the activity of POD was significantly increased by exogenous polyamines, and the effect of putrescine was the most obvious. The results showed that APX, CAT, SOD and POD activities were significantly reduced due to drought stress treatment, increased MDA content, and produced different types of PA. In general, the activity of antioxidant enzymes is most obvious under exogenous application under drought stress.

4. Discussion

Polyamines are now considered as plant growth regulators and secondary messengers in signalling pathways [20, 22-33]. Studies have shown that PAs play an important role in the growth and development of plants under drought stress, the scavenging of reactive oxygen species and maintaining a stable membrane structure [34-35]. Therefore, it is important to study the use of PAs to alleviate the harm caused by drought stress and help reduce economic losses. The exogenous spraying of Spd could alleviate the increase in chili physiological indicators, and the initial injection of Spd in drought stress can make the mitigation effect more obvious [36]. And exogenous spermidine could significantly increase dry weight, relative water content, root number and the chlorophyll content of maize seedlings under drought stress; the cell membrane permeability of maize seedling leaves was significantly decreased, and the effect on the varieties with weak drought resistance was more apparent [37]. In this study, the morphological indexes on the 8th day after different treatments were assessed, and the results showed that drought stress caused a significant decrease in fresh weight, plant height, root length, water content and relative water content of lettuce seedlings but had no significant effect on dry weight and root/shoot ratio. In general, the exogenous spraying of polyamines has a certain mitigation effect on the morphological indicators of drought stress in lettuce seedlings. A comprehensive analysis of the morphological indicators showed that the exogenous spraying of putrescine had the most significant mitigation effect. The choice of polyamines in this experiment was different from that of predecessors, but the effect of polyamine on drought stress was similar. In addition, in plants, in addition to biotic and abiotic reactions, PA also involves several other physiological processes including embryogenesis, organogenesis, flower germination and development, fruit development, maturation and leaf senescence [38-39].

Plants produce reactive oxygen free radicals (ROS) in normal and adverse environments; the ROS react with proteins, lipids and DNA, thus impairing normal cellular functions [40]. Under non-stress conditions, plants maintain a balance between producing and scavenging ROS through a well-coordinated and rapidly responsive antioxidant system. Under these circumstances, the antioxidant enzymes CAT, SOD, POD and others coordinate to promote low levels of free radicals to prevent cell damage [41-43]. Water stress obliterates the equilibrium state in the production and removal of reactive oxygen species in plants, thus causing the accumulation of free radicals of active oxygen, which accelerates the oxidation of many vital substances on the cell membrane, resulting in the membrane lipid peroxidation product malondialdehyde (MDA) [26]. The accumulation of MDA can be used as a measure of the degree of cell damage and as an indicator of drought. Previous studies showed that the accumulation of MDA increased with the increase in the time of stress; the degree of membrane lipid peroxidation increased, but there were differences among the cultivars [44]. Under drought stress, the degree of membrane lipid peroxidation was lower compared with that in non-drought resistant varieties. Under drought stress, the activities of malondialdehyde (MDA), superoxide dismutase (SOD), peroxidase (POD) and catalase (CAT) in cucumber leaves increased significantly [45]. Moreover, MDA content in tomato seedlings increased first and then decreased, and the content of MDA also decreased after exogenous Spd treatment. The content of MDA was lower, and the alleviation effect was more obvious in drought-treated plants. This indicated that drought stress in tomato seedlings could accumulate reactive oxygen species (ROS) in plant cells and attack the cell membrane, which would increase the MDA content in membrane lipid peroxidation products. However, exogenous spermidine can reduce the MDA content, enhance membrane stability, and relieve drought damage in plants [46].

SOD is the first line of defence that constitutes the elimination of reactive oxygen species in plants. The high expression of these enzymes can reduce the damage of reactive oxygen species to plants under stress conditions. Under drought stress, the activity of SOD in drought-resistant cultivars was higher than that in susceptible cultivars, which indicated that drought-tolerant varieties had a higher ability to scavenge superoxide free radicals, and similar results were obtained in other crop studies [47]. CAT and POD can decompose H_2O_2 [48]. The activity of SOD and POD in drought stress increased during a specified period, probably because of the increase in active oxygen free radicals, such as O_2^- and other active oxygen in drought stress, and the accumulation of reactive oxygen species in the plant. Adaptability regulation reduces the increase in reactive oxygen free radicals caused by cell damage; however, prolonged accumulation of reactive oxygen free radicals and the resulting accumulation of MDA in the cells leads to decreased enzyme activity, which was significantly lower in drought-resistant varieties compared with drought-sensitive varieties. Under drought stress, CAT, SOD and POD activities increased first and then decreased. Exogenous Spd treatment significantly increased the protective enzyme activity of seedlings under drought stress, and the effect was more evident in drought-sensitive cultivars. Under drought stress, the APX activity of tomato seedlings continued to increase, and exogenous Spd treatment significantly improved its APX activity [46]. Moreover, under drought stress, exogenous Spd helps improve the antioxidant enzyme activity of tomato seedlings, thus enhancing the ability of ROS removal and improving the resistance of seedlings.

In this study, drought stress could decrease the activity of APX, CAT, SOD and POD, and increase the content of MDA in the leaves of lettuce seedlings. The exogenous spraying of different polyamines could alleviate the activities of antioxidant enzymes in leaves under drought stress.

Exogenous spraying of Put showed the best effect in alleviating drought effects. The results are similar to those of previous studies, but there are some differences. According to the study of Muhammad Farooq·Abdul Wahid·Dong-Jin Lee, 10 μM Spm, Spd and Put can significantly alleviate the net photosynthetic rate and water use efficiency of the leaves after drought stress [26]. Moreover, among the three types of PAs, the effect of Spm was the most apparent. The main reasons for these differences are as follows: First, there are differences in the tested varieties. The drought resistance of rice and lettuce is different, and the physiological structure and the number of polyamines required are different. The concentration of the three polyamines was different. In this experiment, the concentration of polyamines used was 0.1 mM, which is 10 times the concentration described by Muhammad Farooq *et al.*, so there was a difference. Finally, foliar spraying did not result in seed irrigation, but it may be a reason for the differences in test results. There are many studies on the relationship between protective enzyme systems and plant drought tolerance, but the results are different [49]. This may be related to the activity of protective enzyme systems in plants, different drought resistance in different plants and the content of nutrients, such as calcium ion distribution and antioxidant content, and other factors related to the root. Moreover, the drought stress response was also different in different ground parts, and this view can be obtained from the experimental results.

In summary, the effect of drought stress in lettuce lowers the morphological indicators and antioxidant enzyme activities, and exogenous polyamine sprays can effectively alleviate the situation; Putrescine showed the most significant effect in alleviating drought stress in lettuce seedlings.

Acknowledgment
This research was financially supported by the National Key Research and Development Program of China (2016YFD0201010) and the earmarked fund for Beijing Leafy Vegetables Innovation Team of Modern Agro-industry Technology Research System (blvt-02).

References

[1] Martí n-Diana A B, Rico D, Frí as J, Henehan G T M, Mulcahy J, Barat J M and Barry-Ryan C. 2006 Effect of calcium lactate and heat-shock on texture in fresh-cut lettuce during storage. *J. Food Engineering.* **77**(4): 1069-1077.

[2] Nicolle C, Carna A, Fraisse D, Lamaison J-L, Rock E, Michel H, Amouroux P and Remesy C. 2004 Characterisation and variationof antioxidant micronutrients in lettuce (Lactuca sativa folium). *J. Sci. Food Agri..* **84**(15): 2061-2069.

[3] Altunkaya A, Becker E M, Gökmen V and Skibsted L F. 2009 Antioxidant activity of lettuce extract (Lactuca sativa) and synergism with added phenolic antioxidants. *Food Chem..* **115**(1): 163-168.

[4] Shandong Agricultural University. 1980 *Cultivation theory of vegetables* (Beijing: Agriculture Press), Beijing, China.

[5] Liu X Y, Fan S X, Huang J, Wang Z Y and Liu X P. 2016 Nutritional quality analysis of different varieties of leaf lettuce. *Chin. Agri. Sci. Bulletin.* **32** (22): 57-62

[6] Pepe G, Sommella E, Manfra M, Nisco M D, Tenore G C, Scopa A, Sofo A, Marzocco S, Adesso S, Novellino T and Campigli P. 2015 Evaluation of anti - inflammatory activity an d fast UHPLC-DAD-IT-TOF profiling of polyphenolic compounds extracted from green lettuce (Lactuca sativa L. var. Maravilla de Verano) *Food Chem.* **167:** 153-161

[7] Chen Y. 2009 Seed vigor test after initiation of primed PEG *J. Shanxi Agri. Uni. (Natural Science Edition)* **29**(2): 123-125

[8] Hung H-C, Joshipura K, Jiang R, Hu F B, Hunter D, Smith-Warner S A, Colditz G A, Rosner B, Spiegelma D and Willett W C. 2004. Fruit and vegetable intake and risk of major chronic disease. *J. Natl. Cancer Inst.* **96** (21): 1577-1584.

[9] Morris M C, Evans D A, Tangney C C, Bienias J L and Wilson R, S. 2006. Associations of vegetable and fruit consumption with age-related cognitive change. *Neurology* **67**(8): 1370-1376.

[10] Pavia M, Pileggi C, Nobile C G A and Angelillo I F 2006. Association between fruit and vegetable consumption and oral cancer: a meta-analysis of observational studies. *The American J. Clinical Nutrition* **83** (5): 1126-1134.
[11] Mulabagala V, Ngouajiob M, Nair A, Zhang Y J, Gottumukkala A L and Nair M G 2010 In vitro evaluation of red and green lettuce (Lactuca sativa) for functional food properties. *Food Chem.* **118** 300-306
[12] Wang Y. 2005 Lettuce cultivation techniques. *New Agri.* (07):18-19
[13] Smirnoff N 1998 Plant resistance to environmental stress. *Curr Opin Biotech.* **9**:214–219.
[14] Baier M, Kandlbinder A, Golldack D and Dietz K-J. 2005 Oxidative stress and ozone: perception, signaling and response. *Plant cell Envir.* **28:** 1012-1020
[15] Garcia-Plazaola J I and Becerril J M 2000 Effects of drought on photoprotective mechanisms in European beech (Fagus sylvatica L) seedlings from different provenances. *Trees* (*Berl*) **14**:485-490.
[16] Close T J and Wahid A 1997 A commonality in the response of plants to dehydration and low temperature. *Physiologia Plantarum* **100:** 291-296
[17] Seki M, Umezawa T, Urano K and Shinozaki K 2007 Regulatory metabolic networks in drought stress responses. *Current Opinion in Plant Bio.* **10** (3): 296-302.
[18] Manavalan L P. Guttikonda S K, Tran P L-S and Nguyen H T. 2009 Physiological and molecular approaches to improve drought resistance in soybean. *Plant Cell Physio.* **50** (7): 1260-1276
[19] Wu Q, Xia R X, Zou Y N and Wang G Y 2007 Osmotic solute responses of mycorrhizal citrus (Poncirus trifoliata) seedlings todrought stress. *Acta Physiologiae Plantarum* **29**(6): 543-549)
[20] Davies P J 2010 The plant hormones: their nature, occurrence and function. *In: Davies P.J.* (*eds*) *Plant hormones*
[21] Liu J-H, Kitashiba H, Wang J, Ban Y and Moriguchi T 2007 Polyamines and their ability to provide environmental stress tolerance to plants. *Plant Biotech.* **24:** 117-126
[22] Kusano T, Berberich T, Tateda C and Takahashi Y 2008 Polyamines: essential factors for growth and survival. *Planta* **228:** 367-381. 7
[23] Gill S S and Tuteja N 2010 Polyamines and abiotic stress tolerance in plants. *Plant Signaling & Behavior* **5**:26-33.
[24] Kaur-Sawhney R , Tiburcio A F, Altabella T and Galston A W 2003 Polyamines in plants: an overview. *Ameri. J. Respirat. Cell Mole. Biol.*, **2:**1-12
[25] Besford R T, Richardson C M, Campos J L and Tiburcio A F 1993 Effects of polyamines on stabilization of molecular complexes in thylakoid membranes of osmotically stressed oat leaves. *Planta* **189:**201-206.
[26] Farooq M, Wahid A and Lee D J 2009 Exogenously applied polyamines increase drought tolerance of rice by improving leaf water status, photosynthesis and membrane properties. Acta Physiol Plant **31:**937 – 945
[27] Basra R K, Basra A S, Malik C P and Grover I S 1997 Are polyamines involved in the heat-shock protection of mung bean seedlings? *Bot Bull Acad Sin* **38:**165-169
[28] Nayyar, H. and Chander, S. 2004 Protective Effects of polyamines against oxidative stress induced by water and cold stress in chickpea. *J. Agronomy Crop Sci.*, **190**: 355-365.
[29] Nayyar H, Kaur S, Kumar SS, Singh K J and Dhir K K 2005 Involvement of polyamines in the contrasting sensitivity of chickpea (Cicer arietinum L.) and soybean (Glycine max (L.) Merrill.) to water deficit stress. *Bot Bull Acad Sin* **46:** 333-338
[30] Giannopolitis C N and Ries S K 1977 Superoxide dismutase. I. Occurrence in higher plants. *Plant Physiol* **59:**309-314
[31] Chance B and Maehly A C 1955 Assay of catalase and peroxidases. *Method Enzymol* **2:**764-775
[32] Nakano Y and Asada K 1981 Hydrogen peroxide is scavenged by ascorbate-specific peroxidase in spinach chloroplasts. *Plant Cell Physiol* **22(5):**867-880

[33] Liu J-H, Kitashiba H, Wang J, Ban Yand Moriguchi T 2007 Polyamines and their ability to provide environmental stress tolerance to plants. *Plant Biotech.* **24:**117-126

[34] Roberts D R, Dumbroff E B and Thompson J E 1986 Exogenous polyamines alter membrane fluidity in bean leaves-a basis for their potential misinterpretation of their true physiological role. *Planta* **167:**395-401.

[35] Besford R T, Richardson C M, Campos J L and Tiburcio A F 1993 Effects of polyamines on stabilization of molecular complexes in thylakoid membranes of osmotically stressed oat leaves. *Planta* **189:**201-206.

[36] Lu M H, Wang X N and Gong Z H 2015 Alleviochemical effect of exogenous spermidine on pepper seedling 's drought stress. *Pepper Magazine* (**04**):40-44

[37] Li L J, Gu W R, Zhang Q, Zhao D X, Zhang H, Yang D G, Wei S and Hao W P. 2015 Effects of exogenous spermidine on leaf physiology and root system parameters of maize seedlings under drought stress. *J. Pesticide Sci.* (**03**)**:**291-299

[38] Hussain S S, Ali M, Ahmad M and Siddique K H 2011 Polyamines: natural and engineered abiotic and biotic stress tolerance in plants. *Biotech. Advan.* **29**(**3**)**:**300-311

[39] Alca´zar R, Altabella T, Marco F, Bortolotti C, Reymond M, Koncz C, Carrasco P and Tiburcio A 2010 Polyamines: molecules with regulatory functions in plant abiotic stress tolerance. *Planta* **231:**1237-1249

[40] Apel K and Hirt H 2004 Reactive oxygen species metabolism, oxidative stress, and signaling transduction. *Annu Rev Plant Biol* **55:**373-399.

[41] Fath A, Bethke P C and Jones R L 2001 Enzymes that scavenge reactive oxygen species are down-regulated prior to gibberellic acid-induced programmed cell death in barley aleurone. *Plant Physio.* **126:**156-166

[42] Joo J H, Bae Y S and Lee J S 2001 Role of auxin-induced reactive oxygen species in root gravitropism. *Plant physio.* **126:**1055-1060

[43] Garg N and Manchanda G 2009 ROS generation in plants: Boon or bane? *Plant Biosyst.* **143:**81-96

[44] Yang T, Liu X, Zhang X Y and Wei X J 2006 Effects of osmotic stress on membrane lipid peroxidation and catalase activities in maize seedlings. *J. Kashgar Teachers College* **27**(3):63-65

[45] Zhang T 2011 Effect of Drought Stress on Physiological indexes of cucumber seedlings *J. Southern Agri.* **42**(12):1466-1468

[46] Zhang C M, Zou Z R, Zhang Z X and Huang Z 2009 Effects of exogenous spermidine on reactive oxygen species and antioxidant system in tomato seedlings under simulated drought stress. *J. Applied Environ. Bio.* (**03**)**:**301-307

[47] Wang W B, Kim Y H, Lee H S, Kim K Y, Deng X P and Kwak S S 2009 Analysis of antioxidant enzyme activity during germination of alfalfa under salt and drought stresses. *Plant Physio. Biochem.* **47**(**7**)**:**570-577.

[48] Asada K. 2006 Production and scavenging of reactive oxygen species in chloroplasts and their functions. *Plant Physio.* **141:**391-396.

[49] Lu S Y, Chen S P, Chen S M, Liang X and Guo Z F 2003 Changes of proline content and antioxidant enzyme activities in three warm-season turfgrasses under drought stress *Horticulture J.* (**03**)**:**303-306.

Evaluation of Nutrient Expert system in improving nutrient use efficiency and environmental benefits for winter wheat in the Hebei Plain

Y M Sun, Y M Yang, J F Yang, S H Huang, L L Jia[1], B W Han and M C Liu

Institute of Agro-Resources and Environment, Hebei Academy of Agricultural and Forestry Sciences, Shijiazhuang, P. R. China

Email: jiall990@126.com

Abstract. This research evaluates the nutrient management system of Nutrient Expert (NE) in winter wheat grain yield production and greenhouse gas (GHG) emission in the Hebei Plain. Field experiments were conducted to check the yield effects, nutrient use efficiency (NUE) and GHG under NE (nutrient management based on Nutrient Expert recommendation) and FP (nutrient management based on farmers' practices). The results showed that NE significantly reduced N fertilizer input while maintain higher yield and nutrient use efficiency than FP. The total N_2O emission and GHG emissions for NE were about 0.69 kg N ha^{-1} and 1686 kg CO_2 eq ha^{-1} respectively, significantly lower than for FP ($p<0.05$), which was about 1.85 kg N ha^{-1} and 3329 kg CO_2 eq ha^{-1}, respectively. The NE system showed great potential to be easily used in smallholder farmers in the Hebei Plain.

1. Introduction

Winter wheat is one of the most important crops in the Hebei Plain (HBP). The total cultivation area of winter wheat in HBP was about 2.3×10^6 hectares, and the yearly production was about 14.3×10^6 tons, accounts for more than 9.7% of wheat cultivated land area and 11.0% wheat production in China in 2015 [1], therefore it plays a very important role for food security and increasing the farmer's income in China. However, in order to get higher grain yield, excessive or imbalanced fertilization by smallholder farms in HBP has become very common [2-3]. Over fertilization can cause serious environment impacts, such as surface water eutrophication, ground water nitrate contamination and greenhouse gas emission [4-6]. The increasing environmental concern asked the optimizing nutrient management to improve grain yield and nutrients use efficiency (NUE) while maintaining the environment sound.

Multiple-point field experiments based on soil testing fertilization method or yield targets fertilization methods have been conducted and showed significant yield and NUE improvement in China [7]. However, the shortcomings for soil testing method be used in the HBP is too labor cost, expensive and time consuming. The soil testing fertilizer recommendation needs comprehensive field sampling and expensive laboratory analysis. Due to the great variations of fertilization history for different farmers, the soil testing fertilizer recommendation cannot be performed well in hundreds and millions of smallholder farmers' fields [8]. Therefore, an easy use interface friendly nutrient management decision support tools, which can give suggestions to reducing nutrient inputs while maintaining high grain yield, improving NUE and mitigating GHG emissions is important for the smallholder farmers in the HBP.

In recent years, a new winter wheat fertilizer recommendation method named Nutrient Expert [9], which is based on yield response and agronomy efficiencies, has been developed and used in the North China Plain [10-11]. This method provides the ability of maintaining the crop yields while reducing the N fertilizer application rates, but few evaluations of the Nutrient Expert system on improving the winter wheat NUE and environmental benefits in HBP were conducted. In this context, we conducted multi-points on-farm experiments to study the long-term sustainability of Nutrient Expert for Winter Wheat in improving NUE and reducing the GHG emissions in the HBP. We compared the amount of fertilizer nitrogen, phosphate and postassium input rates, crop yield, NUE and GHG emissions between the Nutrient Expert (NE) and Farmers' Practice (FP), and assessed the stability of NE fertilizer recommendation for grain yield increasing, NUE and reducing GHG emission intensity.

2. Materials and Methods

2.1. Experiment arrangement

These field trails were arranged in farmers' fields in the HBP from 2010-2012. There were totally 43 field trails were conducted in Xinji County (XJ) in 2 crop seasons, and 8 field trails were conducted in ZhengDing County (ZD) in 2011-2012 crop season. These fields in this research are the typical loamy soil, which can represent the typical soil types and the local farmers' agricultural management strategies. The warm temperate, sub-humid continental monsoon climate is the dominated climate type in this region. The main cropping system in this research is a wheat-maize rotation system, and winter wheat was planted in the beginning of October and was harvested in the beginning of June the next year. The wheat varieties were Jimai 22 and Shimai 15, which were popular in this region. All the plots received the same cultivation practices, such as seeding rate, irrigation rate, weed, pest and disease controls. All these cultivation practices were similar as the best local management practices as the local farmers.

Each experiment included 3 treatments of no any fertilizer input (CK, control), Nutrient Expert fertilizer recommendation (NE) and Farmers fertilization Practices (FP). Fertilization input rates of N, P_2O_5, K_2O for NE were based on Nutrient Expert; FP fertilization rate was come from a survey of farmer's fertilization rates in the last wheat season. The fertilizers used in this research were urea (N46%), Superphosphates (P_2O_5 46%), and potassium chloride (K_2O, 60). The basal fertilizers including all the P and K fertilizers and 1/3 N fertilizer were applied before wheat sowing in the early October, the 2/3 of N fertilizer as topdressing was fertilized at shooting stage in the beginning of April the next year. The detailed fertilization rates for NE and FP treatments were listed in Table 1.

Table 1. Fertilization rates by Nutrient Expert recommendation and farmers' practices.

Year	Location	Sites	NE Recommended （kg ha^{-1}）			Farmers' Practices （kg ha^{-1}）		
			N	P_2O_5	K_2O	N	P_2O_5	K_2O
2010-2011	XJ	31	135 (130-150)	52 (50-56)	60 (48-70)	278 (196-344)	42 (30-68)	24 (0-68)
2011-2012	XJ	12	182 (182)	72 (67-79)	85 (80-105)	346 (234-460)	33 (0-113)	47 (0-158)
2011-2012	ZD	8	182 (182)	76 (67-89)	76 (65-105)	247 (218-266)	47 (38-54)	53 (44-68)

*Notes: numbers in brackets are fertilization rates distribution range for NE and FP in different experiment sites.

2.2. Plant sampling and management

For each treatment, plants from three separated 3 m^2 area of wheat plants in each plot were harvested manually, dry weights of stems and grains were determined after seperation and oven-dried at 60℃. The partial factor productivity of applied N (PFPN), which is defined as the average productivity of nitrogen fertilizer application, was calculated to interpret the nutrient use efficiency. The equation of PFPN was as below:

$$\text{PFPN (kg kg}^{-1}\text{)} = \text{Yf / N input}$$

where Yf (kg ha^{-1}) is the grain yield in N application, and N input (kg ha^{-1}) is the fertilizer N input.

2.3. Estimation of GHG emission and emission intensity

In the whole life cycle of wheat production, the total GHG emissions include CO_2, CH_4 and N_2O (CH_4 emission was not calculated in this research). The detailed calculation method is listed as below [12-13]:

$$GHG_{Total} = (GHG_m + GHG_t) \times N\ \text{rate} + \text{total}\ N_2O \times 44/28 \times 298 + GHG_{others}$$

where GHG_{total} (kg CO_2 eq ha^{-1}) is the total greenhouse gas emissions in the whole life cycle of wheat production. The GHG_m is the GHG emission in the process of industry's energy source from the fossil mining to N products manufacturing. The GHG_m is 8.21 kg CO_2 eq ha^{-1} [14]. The GHG_t is the transportation emission factor for nitrogen fertilizer, which was calculated based on energy consumption of transport and the average transportation distance by train and truck. The GHG_t is 0.09 kg CO_2 eq ha^{-1} [15].

N rate is the nitrogen fertilization amount (kg N ha^{-1}). GHG_{others} are the greenhouse gas emissions for phosphorus and potassium fertilizers production and transportation, which for phosphorus fertilizer production was 0.73 kg CO_2 eq ha^{-1}, and for transportation was 0.06 kg CO_2 eq ha^{-1}, and for potassium fertilizer production was 0.5 kg CO_2 eq ha^{-1}, and for potassium fertilizer transportation was 0.05 kg CO_2 eq ha^{-1}. The GHG emission per unit area, expressed as kg CO_2 eq ha^{-1} and the GHG emission intensity, was expressed as kg CO_2 eq Mg^{-1} grain.

The N_2O emission that come from nitrogen fertilizer applications to fields occur by direct and indirect pathways. The indirect pathway includes the NH_3 volatilization and NO_3^- leaching; the direct pathway is the N_2O emission from the soil. The calculation equations of direct N_2O emission and indirect N_2O emission for wheat [16-17] are as below:

$$\text{Direct } N_2O \text{ emission: } y = 0.54\ exp\ (0.0063 \times N\ surplus)$$

Note: The N surplus was the N fertilizer application rate minus the wheat total nitrogen uptake.

Indirect N_2O emission was estimated as 1% of NH_3 volatilization and 0.75% of NO_3^- leaching.

$$NH_3 \text{ volatilization} = 13.59\ exp\ (0.009 \times N\ rate)$$

$$NO_3^- \text{ leaching} = 4.95 + 0.17 \times N\ rate$$

The total N_2O emission was calculated in units of CO_2 equivalents (CO_2 eq), and expressed as kg CO_2 eq ha^{-1}. The significance of GHG emissions between NE and FP at the 0.05 level of probability was compared based on least significant difference (LSD). And the correlations between PFPN with GHG emissions were simulated by using the Proc GLM and Proc NLIN procedures in SAS 9.2 software, respectively.

3. Results

3.1. Fertilizer application rates

Of all sites and years average (Table 2), the fertilizer surveys for farmer's practices in this research showed that the fertilizer N application for FP was ranged from 245.6 to 301.6 kg ha^{-1}, the averaged N fertilization rate was 276.2 kg ha^{-1}. This data is similar to the reported farmers' fertilizer survey of 250-300 kg N ha^{-1} for winter wheat in this region [2, 17-18]. The NE fertilizer recommendation ranged from 149.3 to 182.0 kg ha^{-1}, the averaged N fertilization rate was 161.5 kg ha^{-1}, significantly lower than the FP.

Fertilizer P application in the FP ranged from 34.9 to 52.8 kg ha^{-1}, the averaged P fertilizer application rate was 45.0 kg ha^{-1}. The NE treatment P fertilization rate ranged from 58.2 to 68.2 kg ha^{-1}, the averaged P application rate was 62.0 kg ha^{-1}. The K fertilizer input for FP was from 25.5 to 59.2 kg ha^{-1}, and the average application rate was 43.9 kg ha^{-1}. The NE recommended K input was from 66.3 to 71.2 kg ha^{-1}, the average K application rate was 67.8 kg ha^{-1}. The NE system significantly reduced the nitrogen application rate while increased the phosphate and potassium fertilizer application rates to meet the wheat production need.

Table 2. Comparison of fertilization rates in wheat for NE and FP (kg ha^{-1}).

Sites and Years	N application rate[1]		P_2O_5 application rate		K_2O application rate	
	NE	FP	NE	FP	NE	FP
XJ-2011	149.3b[2]	301.6a	58.2a	34.9b	66.3a	25.5b
XJ-2012	153.1b	281.3a	59.5a	52.8a	71.2a	47.0b
ZD-2012	182.0b	245.6a	68.2a	47.2b	66.0a	59.2a
All sites & years	161.5b	276.2a	62.0a	45.0b	67.8a	43.9b

[1] For each sites and years, the fertilizer application rates are the average of all field trials.
[2] Means followed by same letters were not significantly different at P<0.05.

3.2. Grain yield and nutrient use efficiency of PFPN

The yields for NE treatment in XJ and ZD were 3.7%-5.2% higher than for FP treatments, but there were no significantly grain yield differences in all sites and years (Table 3). But to PFPN, the NE treatment significantly higher than FP due to the lower fertilizer input, especially the lower N fertilizer input. The partial factor productivity of N for NE treatment was from 37.2 kg kg^{-1} to 47.5 kg kg^{-1} in XJ and ZD in two crop years. The averaged PFPN was 43.2 kg kg^{-1}, nearly double than for the FP treatments. The PFPN for NE in this research was very close to the China national average of 43.0 kg kg^{-1}[3] but higher than the North China Plain average of 33.9 kg.kg^{-1} [19]. These results showed the NE strategies can be used in the wheat nutrient recommendation to keep grain yield in the Hebei plain.

Table 3. Wheat grain yield and PFPN under different nutrient management.

Year	Locations	Sites	Grain Yield (kg ha^{-1})			PFPN (kg kg^{-1})	
			CK	NE	FP	NE	FP
2010-2011	XJ	31	5965b	7638a	7363a	44.9a	22.7b
2011-2012	XJ	12	5641b	7115a	6846a	47.5a	25.0b
2011-2012	ZD	8	5385b	6765a	6430a	37.2a	26.2b
All sites & years			5664b	7173a	6880a	43.2a	24.6b

3.3. Estimated GHG emission

The N_2O emission had already be recognized as the main component of GHG balance and increased exponentially with nitrogen fertilizer input rate [16-17, 20]. In this research, the estimated winter wheat N_2O emission for FP was 1.85 kg N ha^{-1}, similar to reported 1.3-1.7 kg N ha^{-1} [21], which was tested by continuous air flow enclosure method. The NE significantly reduced the N_2O emission to 0.69 kg N ha^{-1} by reducing the N input than FP (Table 4). The excessive N fertilizer application in the FP treatment caused excessive emissions of direct and indirect N_2O of 0.68 and 0.47 kg N ha^{-1}, respectively.

The GHG emission due to N fertilizer input in this research was 95.0% and 98.2% for NE and FP, respectively. The GHG emissions for NE were 1686 kg CO_2 eq ha^{-1} (Table 4), significant lower than the FP treatments of 3329 kg CO_2 eq ha^{-1}. The NE treatment significantly reduced 49.4% total GHG emissions and 50.4% GHG emission intensity than FP. The high N fertilizer use for FP treatment resulted to significantly higher GHG emission and GHG emission intensity than NE. This result suggested reducing N fertilization rate is very important in reducing GHG emission.

Table 4 GHG emissions by fertilizer input for different nutrient management strategies.

	Treatments	N_2O emission (kg N ha^{-1})			GHG emission (kg CO_2 eq ha^{-1})				GHG emission intensity
		Direct	Indirect	Total	N fertilizer use	N fertilizer production	Others	Total	(kg CO_2 eq Mg^{-1})
XJ2011	NE	0.32b	0.25b	0.58b	270b	1238b	82a	1590b	267b
	FP	0.80a	0.65a	1.45a	681a	2500a	53b	3234a	537a
XJ2012	NE	0.64b	0.34b	0.98b	461b	1269b	86a	1816b	257b
	FP	1.98a	1.11a	3.08a	1444a	2332a	73a	3849a	579a
ZD2012	NE	0.34b	0.31b	0.66b	307b	1509b	90a	1906b	288b
	FP	0.63a	0.50a	1.13a	527a	2036a	70a	2633a	418a
All sites &years	NE	0.41b	0.28b	0.69b	325b	1277b	84a	1686b	252b
	FP	1.09a	0.75a	1.85a	865a	2403a	61a	3329a	508a

3.4. Correlation of GHG emission with PFPN

Significantly linear plus plateau correlations were found between GHG emission and nitrogen use efficiency of PFPN across all sites and years (Figure 1). The GHG emission decreased while the PFPN increasing, but when the PFPN higher than 41.3 kg kg^{-1}, the GHG emission had the trend of keeping at about 1538 kg CO_2 eq ha^{-1}. The GHG emission for NE more closely to the plateau area, and had the possibility of keeping lower GHG emission than FP. And on the other hand, the result suggested the NE treatment had the potential of decreasing the GHG emission through increasing the nutrient use efficiency. The further reducing GHG emission could be contributed to improving crop yield through enhancing the nutrient management technology level [22].

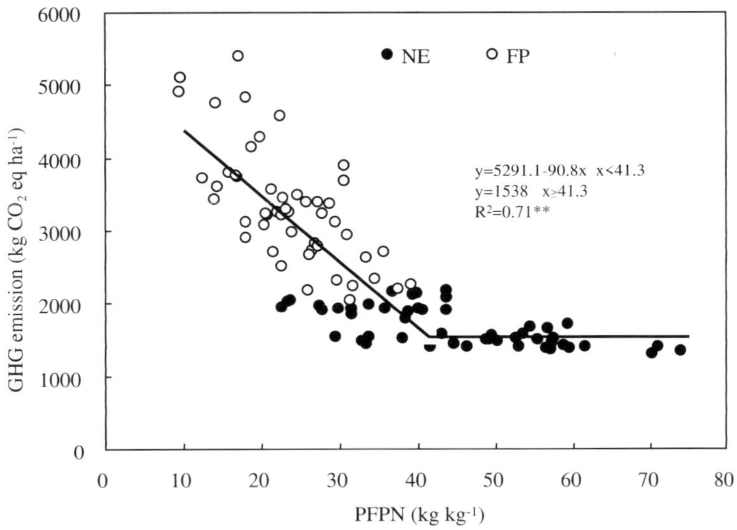

Figure 1. Correlations of GHG emission with PFPN

4. Results

Results from our study have demonstrated that Nutrient Expert system showed greater yields in NE with lower N input through improved fertilizer use efficiency while maintaining low total N_2O emissions and GHG emissions. There were significantly linear plus plateau correlations between GHG emissions with nutrient use efficiency of PFP-N, which suggested the keeping the PFPN higher than 41.3 kg/kg is the pathway to reducing the GHG emissions in the Hebei Plain wheat production. This research in the 51 farmers' field showed the potential of the Nutrient Expert for winter wheat could be easily used in smallholder farmers' fertilizer recommendation, and could significantly decreased the N fertilizer input than farmers' practice.

Acknowledgement

This research was financially supported by National Key Research and Development Programme (2017YFD020170702), Key Technology Support Project of Hebei Province, P.R. China (16226407D), the Youth Innovation Team of Hebei Academy of Agricultural and Forestry Sciences (HAAFS) and the International Plant Nutrition Institute (IPNI).

References

[1] Chinese Agriculture Yearbook (Ed), 2016. Editorial Board of Agriculture Yeabook of China. *China Agriculture Press*, Beijing (in Chinese).
[2] Zhao R F, Chen X P, Zhang F S, Zhang H L, Schroder J, and Roemheld V 2006. Fertilization and nitrogen balance in a wheat-maize rotation system in North China. *Agron. J.*, **98**: 938-945.
[3] Zhang, F S, Wang J Q , Zhang W F, Cui Z L, Ma W Q, Chen X P , and Jiang R F 2008. Nutrient use efficiencies of major cereal crops in China and measures for improvement (in Chinese with English abstract). *Acta Pedo. Sinica*, **45**: 915-924.
[4] Ju X T, Xing G X, Chen X P, Zhang S L, Zhang L J, and Liu X J 2009. Reducing environmental risk by improving N management in intensive Chinese agricultural systems. *Proc. Natl. Acad. Sci. USA*, **106**: 3041-46.
[5] Le C, Zha Y, Li Y, Sun D, Lu H and Yin B 2010. Eutrophication of lake waters in China: cost, causes, and control. *Environ. Manage.*, **45**: 662-668.
[6] Gao Q, Li C, Feng G, Wang J, Cui Z L, and Chen X P 2012. Understanding yield response to

IOP Conf. Series: Earth and Environmental Science **185** (2018) 012011 doi:10.1088/1755-1315/185/1/012011

nitrogen to achieve high yield and high nitrogen use efficiency in rainfed corn. *Agron. J.*, **104**: 165-168.

[7] He, P, Li S T, Jin J Y, Wang H T, Li C J, Wang Y L, and Cui R Z 2009. Performance of an optimized nutrient management system for double-cropped wheat–maize rotations in North-Central China. *Agron. J.*, **101**: 1489-96.

[8] Huang S, Jin J, Zuo Y, Yang L, and Cheng M 2002. Evaluation of agricultural soil nutrient balance for Yutian County and Lingxian experimental area in Huang-Huai-Hai Plain (in Chinese with English abstract). *Plant Nutri. Ferti. Sci.*, **8(2):**137-143.

[9] Chuan L, He P, Pampolino M F, Johnston A M, Jin J Y, and Xu X 2013a. Establishing a scientific basis for fertilizer recommendations for wheat in China: yield response and agronomic efficiency. *Field Crop Res.*, **140**: 1-8.

[10] Chuan L M, He P, Jin J Y, Li S T, Grantc C, Xu X P, Qiu S J, Zhao S C, and Zhou W 2013b. Estimating nutrient uptake requirements for wheat in China. *Field Crop Res.*, **146**: 96–104.

[11] Xu X P, He P, Pampolinoc M F, Johnston A M, Qiu S J, Zhao S C, Chuan L M, and Zhou W 2014. Fertilizer recommendation for maize in China based on yield responseand agronomic efficiency. *Field Crop Res.*, **157**: 27–34.

[12] IPCC. Mitigation. Contribution of Working Group III to the Fouth Assessment report of the intergovernmental panel on climate change, United Kingdom and New York, *Canbridge University Press*, 2007.

[13] Zhang W F, Dou Z X, He P, Ju X T, Powlson D, and Chadwick D 2013. New technologies reduce greenhouse gas emissions from nitrogenous fertilizer in China. *Proc. Natl. Acad. Sci. USA*, **110**: 8375-8380.

[14] Di X, Nie Z, Zuo T. 2005. Life cycle emission inventories for the fuels consumed by thermal power in China (in Chinese with English abstract). *China Environ. Sci.*, **25**: 632-635.

[15] Yuan B, Nie Z, Di X, Zuo T. 2006. Life cycle inventories of fossil fuels in China(Ⅱ):Final life cycle inventories (in Chinese with English abstract). *Modern Chem. Indu.*, **26(3)**: 59-61.

[16] Cui Z L, Yue S C, Wang G L, Meng Q F, Wu L, Yang Z P, Zhang Q, Li S Q, Zhang F S, and Chen X P 2013. Closing the yield gap could reduce projected greenhouse gas emissions: a case study of maize production in China. *Global Change Biol.*, **19**: 2467–2477.

[17] Cui Z L, Yue S C, Wang G L, Zhang F S, and Chen X P 2013. In-season root-zone N management for mitigating greenhouse gas emission and reactive N losses in intensive wheat production, *Environ. Sci.Technol.*, **47**: 6015-22.

[18] Cui, Z L, Chen X P, and Zhang F S 2010. Current Nitrogen management status and measures to improve the intensive wheat-maize system in China. *Ambio*, **39**: 376-384.

[19] Wang X, Li Z Y, Ma W Q, Zhang F S 2010. Effects of fertilization on yield increase of wheat in different agro-ecological regions of China (in Chinese with English abstract). *Scien. Agri. Sinica*, **43**: 2469-2476.

[20] Xu C, Han X, Bol R, Smith P, Wu W L, and Meng F Q 2017. Impacts of natural factors and farming practices on greenhouse gas emissions in the North China Plain: A meta-analysis. *Ecol. Evol.*, **7**: 6702-15.

[21] Tan Y C, Zhang Y P, Liu D X, Wu W L, and Meng F Q 2016. Effect of farmland management on N_2O and CH_4 emission from winter wheat-summer maize rotation system in North China Plain (in Chinese with English abstract). *Acta Scien. Circum.*, **36**: 2638-49.

[22] Tilman D, Balzer C, Hill J, Befort B L 2011. Global food demand and the sustainable intensification of agriculture. *Proc. Natl. Acad. Sci. USA*, **108**: 20260-64.

IOP Conf. Series: Earth and Environmental Science **185** (2018) 012012 doi:10.1088/1755-1315/185/1/012012

Rural development model and influence of endogenous power in Guangdong Province, China

Z H Su[1,2], W F Li[1], G J Xiao[1], Y Y Gan[1], J G Zhang[1], H Y Duan[1] and W H Yang[1]

[1] Institute of Agricultural Economics and Rural Development, GAAS, Guangzhou 510640, China

E-mail: suzhuhua@163.com

Abstract. Innovating the rural reform, development path and stimulating the endogenous of rural development are conducive to solving the problems lying in "agriculture, rural areas and farmers" comprehensively in the aspects of theoretical innovation and practical guidance. In this study, the research paradigm of "literature review + empirical analysis of investigation cases + theoretical framework analysis + countermeasures and suggestions" is used for analysing Guangdong rural development models and the influencing factors of the endogenous power of rural development. Firstly, the domestic and foreign studies on rural development models and endogenous power are summarized. Secondly, the current situation of Guangdong rural development is statistically analysed from the aspects of new rural construction, rural social service system, comprehensive rural reform and income increase of farmers. Thirdly, five Guangdong rural development models are classified and sorted out through the investigation by the research team for more than 1 year, and 9 typical cases in Guangdong selected from the investigated materials are empirically analysed. Fourthly, efforts are made to analyse the internal logical relations among rural property right system reform, modern agricultural management system and mechanism, and rural grass root governance innovation and to analyse their influencing factors by establishing the theoretical model and paradigm of "supporting foundation - key factors - guarantee mechanism". Finally, several countermeasures and suggestions are proposed for promoting rural development and accelerating agricultural modernization by integrating abovementioned factors from aspects of early and pilot implementation and expansion of the promotion of practical achievements.

1. Introduction

The Central Committee of the Communist Party of China and the State have paid high attention to the problems of "agriculture, rural areas and farmers", and introduced a series of important policy documents. In order to highlight rural reform and development vitality and stimulate the endogenous power of rural development, the Central Committee of the Communist Party of China and the State Council have made more efforts to deepen rural reform and development by issuing the *Comprehensive Implementation Plan for Deepening Rural Reform* in October 2015. Then, CPC Guangdong Provincial Committee and People's Government of Guangdong Province quickly made deployment and introduced the *Comprehensive Implementation Plan for Deepening Rural Reform in Guangdong* in March 2016, striving for breakthrough in 5 fields before 2018, namely, rural collective property right system, agricultural management system, agricultural support and protection system, urban-rural integration system and mechanism, as well as rural social governance system. The Report of the 19th National Congress of the Communist Party of China has proposed the rural revitalization strategy and deemed that the rural collective property right system reform is as important as a national strategy.

Many domestic and foreign scholars have currently conducted intensive study on rural development and its endogenous power, and relevant studies are being innovated and developed continuously. Fei [1] first put forward the concept of "model", which means "a characterized development path in certain regions under certain historical conditions". In addition, Du [2] deemed that the basic management system of combination of family centralization and decentralization must be adhered to, perfected and innovated in order to realize sustainable rural development. Wen Tiejun [3], from Renmin University of China, asserts that the rural development can be vitalized by deepening the rural reform in the new era. Chen Xiwen [4] deems that the key is to solve and handle the relation of "people" and "land", and then further give play to the main role of farmers to adapt to the needs of rural transformation and development. Dang Guoying [5] proposes to solve the problem of endogenous power driving rural development by appropriate policies and measures, such as developing family farm management in moderate scale. Xu Xianglin [6] deems that the criterion for determining whether rural development can be driven by its endogenous power is whether farmers obtain reasonable return after inputting various production factors. Zhu Ze [7] points out that it is necessary to innovate agricultural management system, as well as scientific and technological system and mechanism, and solve the problems such as rural hollowing and farmer aging, in order to realize endogenous driving of rural development. Xu Yong [8], from the Institute of China Rural Studies, proposes that the endogenous power of poor regions can be stimulated by improving grass root governance. He Xuefeng [9] proposes that the key measure for settling the rural reform and development in different regions in China is to establish and cultivate a standard rural governance system according to local conditions. Fang Jin [10] proposes that the key issue is how to effectively control the interaction between local and external forces in the development process, and to find appropriate and effective operational methods to achieve rural development goals.

By integrating the study theories and methods of those scholars in this study, the subject team members have conducted tracking investigation (with assistance from competent rural departments of Guangdong) for more than one year in 11 national rural reform experimental areas (distributed in 9 prefecture-level cities) in Guangdong, and carried out special investigation for these regions with remarkable achievements in rural reform and development in the Pearl River Delta region and eastern, western and northern Guangdong on the basis of the extensive investigation materials of rural development models. Consequently, 5 rural development models in Guangdong are classified and sorted out, and 9 typical cases in different social and economic development regions in Guangdong are selected from the investigation materials to try to analyze the influencing factors of endogenous power shared by rural development. Then, several targeted policy suggestions are proposed briefly, hoping to provide some policy application values for competent rural development departments at all levels in the Mainland.

2. Representative model of Guangdong rural reform and development and case demonstration

For this study, the subject team members have conducted investigation for more than 1 year, visited 11 national rural reform test areas (contact points and pilots) distributed in 9 prefecture-level cities of Guangdong and accumulated a large quantity of first-hand materials, and also have been assisted by the competent departments for rural work of CPC Guangdong Provincial Committee and relevant competent agricultural and rural departments in the counties, cities and towns where the rural reform pilot is conducted in respect of case interview, demonstration and data acquisition. Five rural reform and development models are classified based on careful analysis and sorting out of these collected investigation cases and by comprehensive consideration of the social and economic development driving mode, rural development subject driving mode, and emphasis of rural reform and development, which are respectively rural property right reform driving model, primary, secondary and tertiary industries driving models, and collective economy driving model.

2.1. Comprehensive rural property right reform driving model

2.1.1. "1331" rural reform and development model - taking Jiaoling of Meizhou as an example. Jiaoling County of Meizhou was determined as the "National Comprehensive Rural Reform Demonstration Pilot Unit" in 2013, to implement 3 reforms (rural property right reform, system and mechanism innovation, and integrated development of urban and rural areas), create 3 complexes (rural property right trading, agricultural production and management, and new urban-rural development) and realize 1 ultimate goal for rural development of Jiaoling by centering on 1 theme (comprehensive rural reform demonstration pilot), namely, the 1331 rural reform and development model. It is found during investigation that there are 2 noticeable features of Jiaoling's practice for rural development driven by comprehensive rural reform:

On one hand, promoting pilot and reform of rural property right trading and financing guarantee to optimize rural land element resources. Based on completion of land right confirmation and certificate issuing, Jiaoling County established the first county-level rural property right trading center in Guangdong in 2013. Currently, town-level rural land transfer trading service center, county-level rural property right exchange and rural property right financing guarantee platform have been established in 8 towns, and relevant policy measures have been developed, including the *Regulations for Rural Property Right Trading* and the *Trading Rules for Rural Land Contracting Management Right*. As of 2015, 70 businesses have been accumulatively concluded and more than 10,000 mu land (including more than 6,000 mu paddy field) has been transferred in the rural property right trading center of Jiaoling County. The land transfer price traded through the rural property right trading platform is over 20% higher than that before land right confirmation on average.

On the other hand, great efforts on promoting family farm development to increase family business income of farmers [11]. Jiaoling County also supports family farm development by formulating documents including *Regulations for Family Farm in Jiaoling County* and *Interim Procedures for Certifying Demonstrative Family Farm in Jiaoling County*. There are currently 106 family farms approved by the Department of Agriculture, 82 family farms registered with the administration for industry and commerce County, and 12 county-level demonstrative family farms selected in Jiaoling. The area of family farm management is 6,448 mu, with an annual output value of CNY 38.06 million and the average yearly income of CNY 168,000 for family farms. The average yearly income of family farm members is CNY 28,000, which is 186% and 233% of the per capita income of urban residents and farmers respectively in Jiaoling County at the same period.

2.1.2. Innovating rural agricultural development mode by "three integrations" - taking the reform pilot in Qingyuan as an example. As the rural reform pilot unit established by the Central Rural Work Office, Qingyuan has made great achievements in promoting "three integrations" (land integration, fund integration and public service platform integration). In 2013, test projects were fully implemented in administrative villages and natural villages (group of villagers) in first 3 pilot towns, namely Xiniu Town of Yingde, Jiubei Town of Lianzhou and Shijiao Town of Fogang County. Such practice has been currently promoted in Qingyuan and will be promoted in eligible regions of Guangdong after improvement. We deem that the model mainly has 3 features as follows:

Firstly, exploring integration of land resources to accelerate agricultural moderate scale management. The main practice in Qingyuan is to guide farmers to interchange and merge land voluntarily, by which Qingyuan has successively explored the land integration models including "right confirmation after integration and interchange", and "confirmation of land boundary to each cooperative and confirmation of area to each household". As of 2015, the area of integrated land in Qingyuan has been 4.267 million mu, including 1.258 million mu cultivated land, which accounts for about 31% of the total contracted land area in Qingyuan. From the perspective of practice, Yangshan County of Qingyuan has transformed scattered farming on original multiple land parcels into centralized scale production, which can facilitate mechanized production, promote land transfer and avoid land abandonment. In addition, large contiguous land is convenient for mortgage and guarantee to obtain higher transfer value and help solving financing difficulties faced by farmers.

Secondly, exploring integration of agriculture related funds to improve the efficiency of supporting agriculture by financial funds [12]. Such problems as "management by multiple departments without well coordination" and "distribution in a strictly equal way" during utilization of agriculture related funds are obvious. Therefore, Qingyuan has summarized and promoted experience of some villages in the utilization of inclusive funds, including compensation for ecological public-welfare forests and direct subsidy for growing grain, for the construction of infrastructure and the development of public welfare in the village.

Thirdly, exploring integration of agriculture related service platforms to intensify the socialized service system for agricultural production. Qingyuan has established 60 town-level service platforms for rural collective "funds, assets and resources" management, covering more than 70% of towns in Qingyuan. In respect of life services, the improvement of rural e-commerce system has been accelerated through cooperation with e-commerce enterprises such as Alibaba under help of rural network layout of the post, which creates a consumption environment as convenient as that in urban areas for farmers.

2.2. *Primary, secondary and tertiary industries driving models*

Guangdong has been promoting the models of industry driving rural economy, industrial connection with poverty alleviation, One Village One Product, professional town and park, and modern agricultural park supporting rural industrial development since 2000. In addition, combined with the trend of integrative development of primary, secondary and tertiary industries, many new models, including e-commerce village for industrial and agricultural products, agritainment and countryside leisure travel, have emerged in rural areas in Guangdong in recent years. In this Study, empirical analysis is conducted mainly by taking the modern agricultural management subject cultivation and industrial management in Yunfu, the development of Tangerine Peel Village, Xinhui District, Jiangmen Municipality, and the e-commerce village development model in Jiedong District as representative cases.

2.2.1. New agricultural management and development models including "leading agricultural enterprise + family farm" and "enterprise + cooperative + peasant household" explored by Yunfu. After launching the pilot and reform of modern agricultural management system and mechanism in 2011, Yunfu has created great social response in modern agricultural management subject cultivation and connection, and innovating industrial management model as of 2015. As found during investigation, there are mainly 2 rural industrial development mechanisms with remarkable achievements in Yunfu, which have been promoted in the city and learned by some regions in and out of Guangdong.

Firstly, innovating the mechanism of "company + modern family farm". Taking Wens Group in Xinxing County, Yunfu as an example: Wens Group has currently established the social service system perfecting breeding industry, where Guangdong Nanmu Machinery and Equipment Co., Ltd. is responsible for providing machinery and equipment, Wens Research Institute of Internet of Things is responsible for improving the information control system and Wens Group, together with banks, is responsible for providing credits, by which more than 2,900 peasant households in China have been upgraded to family farms (including 505 in Yunfu) . Consequently, the income of peasant households has been increased by times compared with that in 2011, with the increased income about CNY 100,000/household. The raising scale is more than 1.8 times of that of common cooperative peasant households and the income is also 1.8 times of that of common raising households.

Secondly, innovating the mechanism of "company + cooperative (council) + peasant household". As there are many mountains and less land in Yunfu, in order to solve such problems as low degree of agricultural intensiveness, low quality product rate and low organizational degree, it has innovated the management mechanism of "company + cooperative + peasant household" by taking several companies in rice industry in Luoding as test subjects, which has helped solving problems such as land intensiveness, standardized production and agricultural production environment renovation to increase income of farmers by taking the leading agricultural enterprise as the leader, the local natural village council as the bond, and the farmers' professional cooperative and major farmer households as service platforms. As at the end of 2015, 83 companies in Yunfu have applied this mechanism, which connects 122 farmers' professional cooperatives, 172 local councils and 138,400 peasant households. The purchasing price of agricultural products produced by it has increased at a speed more than 5% every year since 2011 and the average annual increase in the income of farmers has been more than 16%.

2.2.2. New model of whole-process industry chain for agricultural industrialization explored in the market of Tangerine Peel Village, Xinhui District, Jiangmen. The whole-process industry chain including planting, production, processing, exhibition, sales, storage, trading and financing has been gradually established in Tangerine Peel Village, Xinhui District by creating the characteristic leading agricultural enterprise integrating the agricultural advantages of Xinhui, modern agricultural technology and Internet finance.

One is the multiple combination models implemented for planting in tangerine peel. Such models include "company + peasant household", "company + professional cooperative" and "direct contracting by company", which have maximized the scale aggregation of planting in tangerine Peel.

The other is the establishment of normalized and standardized production standards for tangerine peel. Firstly, provide guidance on standardization at planting stage to effectively combine traditional experience and modern technology; secondly, carry out procedural operation at storage stage to establish regulations and systems for the temperature, humidity, specification and quality of storage; then, classify and manage the products in tangerine peel according to standards of year, quality and place of origin; meanwhile, completely record the detailed tracing information of products in tangerine peel by the QR code technology of the third party warehousing enterprise.

2.2.3. The model of "Junpu E-commerce Village" in Jiedong District, Jieyang Municipality. The model of "Taobao Village in China - Junpu E-commerce Village" in Jieyang, as the core, has radially driven rapid development of e-commerce industry in surrounding towns and villages to build a large rural e-commerce cluster. By focusing on the construction of "five highlands", namely industry, talent, service, system and culture, Junpu Village of Xichang Town has created an e-commerce road with self-employment of farmers and become the true "Taobao Village in China". There are currently 490 families in Junpu Village, with the total population of 2,695, where more than 350 households and about 2,000 persons have been carrying out online sales activity. There are more than 3,000 Taobao online stores and 300 physical online wholesale stores, with about 1.20 million transactions monthly and the trading volume up to CNY 150 million in peak months.

2.3. Collective economy driving model

2.3.1. Property economy driving - taking the rural collective economic development model in Dongguan as an example. Dongguan governs 556 village-level and 2,288 group-level collective economic organizations. It has been one of the earliest and fastest regions in respect of rural economic development driven by collective economy in Guangdong and many regions in and out of Guangdong have followed and referred to its experiential model. During sorting out the rural collective economic development of Dongguan, we found that it experienced three evolving stages with different features.

The first stage is to emphasize the conservation of collective economy when promoting family co-production contract responsibility system. At the end of the 1970s and beginning of the 1980s, real estates such as ancestral halls were conserved for collective use, based on which the first batches of "the three-plus-one trading-mix" enterprises were established. Professional contracting was implemented for fish ponds.

The second stage is to emphasize and strengthen collective economy during promoting rural industrialization. Dongguan proposed the development strategy of "marching to rural industrialization" in 1984, by which land and other production means would be uniformly managed by the collective and income from land would be retained in the collective for expanding reproduction. In this way, the rural collective economy has been rapidly strengthened in the model of "making profit by land, collecting rent by building factories and rolling development".

The third stage is to promote upgrading of property economy since the end of the 1990s. The district government has introduced reward policies for capital introducing to towns and villages, major tax payers in village manufacturing and village-level key project tax, and has established the special fund of subsidy and reward for industrial upgrading in towns and villages in the amount of CNY 100 million to encourage the collective to introduce quality projects and promote property upgrading. The development model of "development led by cities and towns, and benefits distributed at city, town and village levels" has been implemented to guide villages (communities) to participate in the construction of new urbanization in forms of land and capital. Collective economic organizations have been encouraged to invest idle funds into principal-protected financial products and equity projects with steady earnings and low risks to further develop investment type economy. As at the end of 2015, the collective net assets and net income of Dongguan have been respectively CNY 118.3 billion and CNY 11.4 billion, both of which account for 40% of that in Guangdong. The income of farmers has been increased effectively by enlarging the collective economy and standardizing collective distribution.

2.3.2. The model of shareholding management system for rural collective land - taking Nanhai District of Foshan as an example. Nanhai District of Foshan has proposed the assumptions of accelerating the transfer of collective construction land and building the urban-rural integrated land market in recent years in order to solve problems such as shortage of national construction land, low use efficiency of collective construction land and low-end industrial form. Firstly, establish the transfer mechanism for collective construction land use right. By successively introducing the Measures for Implementation of Collective Construction Land Use Right Transfer in Nanhai District of Foshan and 3 policy documents, Nanhai District has further standardized the transfer of collective land by ways of assignment, sublease and mortgage, initiatively established the standard land price system for collective construction land to provide a price basis for transfer of rural collective construction land use right, formed a relatively complete policy system for transfer from primary market to secondary market, and explored the partitionable sales of commercial service industry carrier projects on collective land. Secondly, explore the construction of village community apartment. The district government introduced the Opinions of Nanhai People's Government of Foshan on Implementation of Further Promoting Village Community Apartment Construction, aimed at meeting the rigid demand of farmers for residential land, guiding farmers living individually and dispersedly to live in multi-storey apartment and improving the new urbanization level of Nanhai District by construction of community apartment. At present, 24,992 parcels of land have been transferred through the asset platform in Nanhai District, involving land area of 183,000 mu and total amount of concluded subjects as CNY 22.23 billion. 74 parcels of collective construction land with the area of 1,731 mu have been sold and 20 parcels of collective construction land with the area of 264 mu have been mortgaged.

3. Analysis of influencing factors for endogenous power of rural development

Agricultural efficiency, rural development and income increase of farmers' are the most important parts of the three rural issues. To solve the key problem of "agriculture", the key point lies in clarifying the ownership of the assets and resources of rural areas and farmers, jointly developing the agricultural industrial management with the new type of management entity, guaranteeing the farmers' rights and the harmonious and stable development of rural area [13]. In other words, only the accurate understanding of the cause and evolution of endogenous power in rural areas, can we realize the rural comprehensive well-off and realize agricultural modernization. As the analysis of the various modes of rural development in Guangdong Province that mentioned above, the factors of endogenous power that affects the formation and strength of rural development model, which must include 3 levels - system foundation, key factors and guarantee mechanism. First, a Rural Collective Property Right System with clear ownership, clear family property and regulated transaction behavior shall be established and improved, which is the supporting foundation for the formation of endogenous power in rural development. Secondly, the innovation of appropriate scale of land transfer, the cultivation of integrated management mechanism of primary, secondary and tertiary industries developed with new management entity and promotion of the flow and optimization of rural factor resource are the key factors to strengthen to strengthen the endogenous impetus of rural development. Finally, this paper will explore the reform paths of villager autonomy, the standardized management of collective economic organizations and the separation of politics and economy, which are the guarantee mechanisms to realize the endogenous power integration of rural development and to serve rural economic and social development.

3.1. Foundation support: Rural property right system reform

The reform of rural property rights system can make the farmers' rights, village cadres' rights and the rural collective property rights (i.e. the rural collective assets capital assets resources) clear, which is the foundation of the rural comprehensive reform and the agricultural economic society development. To the end of 2015,a total of various types of 5,304 rural property transfer management service platforms(formerly known as rural collective "Three-Capital" Management Service Platform") have been built in the whole province (excluding Shenzhen city), the coverage rate in county and town levels is more than 50%, networking rate of 69%, village-level coverage of 20.3%, The initial registration rate of the collective construction land use right and the homestead use right is 96.96% and 95.14% respectively. From the above statistic data, it can be seen that the establishment and consolidation of rural collective property rights system in Guangdong Province, will provide a solid foundation support for the future development of "two rights" mortgage loans (rural land contract management rights and rural homestead use right), the expansion of rural collective economy and the increase of villagers' (residents') income.

3.2. Key factor: agricultural management system and mechanism innovation

The most crucial factors for rural development are land resource, agricultural talent and cooperative organization. Firstly, based on confirmation of rural collective property right, in particular, rural land contracted management right, accelerate transfer of agricultural land in moderate scale, cultivate professional farmers and new management subjects, explore innovative method of modern agricultural management, create comprehensive modern agricultural management system and promote integrated development of primary, secondary and tertiary industries. The People's Government of Guangdong Province introduced the opinions on implementing the promotion of family farm development in 2015, providing a good opportunity for rapid development of new management subjects. As of 2015, various new subjects have emerged, including 2,850 leading agricultural enterprises, 30,000 farmers' cooperatives and 37,500 family farms. By integrating the "Farmers' Income Multiplication Plan" in Xinxing County, Yunfu has cultivated 505 family farms led by Wens Group. Secondly, focus on promoting carrier construction including modern agricultural park. Guangdong has currently established 3 national and 8 provincial modern agricultural demonstration areas, and great achievements have been made in the construction of 8 Guangdong - Taiwan agricultural cooperation parks. Thirdly, innovate the way of providing agricultural socialized services. The People's Government of Guangdong Province introduced the *Implementation Plan for Promoting Rural E-commerce Development in Guangdong* in January 2016. Yangshan County of Qingyuan officially launched the "Village Taobao" Project by cooperating with Alibaba at the beginning of 2015, by which the county-level service center and 19 village-level service stations for "Village Taobao" have been established currently. To sum up, the key factors for intensifying the endogenous power of rural development are the mobility of agricultural land, the cultivation of new management subjects and the management mechanism innovation.

3.3. Guarantee mechanism: rural grass root governance innovation

At present, villagers and village collectives in 11 national rural reform experiment pilots in Guangdong have been first benefited from the development achievements of comprehensive rural reform, including such reform dividends from economic development, income increase and village appearance improvement brought by confirmation of rural land (cultivated land, forest land, residential land, rural collective construction land, etc.) right, rural land transfer and industrial management. However, it is necessary to further guarantee the sustainability of endogenous power of rural development in order to make villagers reasonably and sustainably benefit from and promote reform dividends, for which Guangdong has explored a solid guarantee path for deepening rural grass root governance innovation.

Since 2012, counties, towns and villages of Guangdong, particularly, national rural reform experiment pilots, have been actively exploring rural grass root governance under overall plan and coordination by the CPC Guangdong Provincial Committee and the People's Government of Guangdong Province to stimulate the endogenous power of Guangdong rural development and constantly explore innovation in grass root democratic consultation mechanism. According to incomplete statistics, 52,220 village-level social organizations functioning as a council of villagers have been established in Guangdong, which have participated in or assisted with handling 88,823 village disputes and public welfare construction projects accumulatively. By focusing on cultivation and development of councils of villagers/local councils in natural villages, some cities of Guangdong, including Qingyuan and Yunfu, have established the new pattern of rural grass root collaborative governance cored at the village party organization, based on villagers' self-governing organization and supplemented by rural social organizations. As of 2015, 16,412 councils have been established in Qingyuan alone, which have coordinated with the village committee in constructing 22,887 public welfare projects and mediating 25,192 rural social disputes. At the same time, Guangdong has comprehensively launched standardized management and system reform of rural collective economic organizations. 10,040 villages in 27 counties of Guangdong have carried out reform of rural collective economic organizations, with the total quantitative assets of CNY 292.3 billion. Additionally, Nanhai of Foshan and Xinhui of Jiangmen have carried out the reform of "political and economic separation". As shown by above cases and data, Guangdong has made great achievements in exploring rural grass root governance, which has become the strong guarantee for intensifying the endogenous power of rural development.

4. Brief Policy Suggestions

4.1. Select a reasonable and sustainable rural development model according to local conditions to achieve simultaneous progress in poverty alleviation and establishment of an all-round well-off society

Currently, the regional economic development in Guangdong has not been in balance, where the economy in the Pearl River Delta Region is developed. However, most rural areas have implemented the change of village to residential community, the rural social and economic development system and mechanism have been improved, in particular, some regions have carried out the experiment of "political and economic separation", the collective economic aggregate has been large, village dwelling environment has been improved and the salary income of villagers has been generally more than their management income, which indicate a basic well-off level. The difference of rural economy in eastern, western and northern Guangdong is large, the levels of property right reform, industrial economy and rural social governance are uneven, and some regions are still backward relatively. Some rural areas, especially villages and villagers in mountainous areas, are still below the poverty line of Guangdong, while in some regions, the agricultural industry is developed, the income level of farmers is high and the village appearance is good. Based on comprehensive deep investigation for the rural development in different economic regions of Guangdong, this Study has preliminarily summarized 5 rural development models, which have great practical significance though not perfect. Therefore, the CPC Guangdong Provincial Committee, the People's Government of Guangdong Province and local governments are suggested referring to these models in developing policies for rural development or the planning strategy for rural reform and development. In addition, this Study analyzes the supporting foundation, key factors and guarantee mechanism for the endogenous power of rural development at the level of academic theory.

4.2. Make early and pilot implementation in reform experimental areas more intensive to further deepen comprehensive rural reform

According to the spirits of the Report of the 19th National Congress of the Communist Party of China, the Central Rural Work Conference in 2016 and the Guangdong Rural Work Conference in 2017, as well as the requirements of relevant policy documents including the No. 1 Central Document and the *Comprehensive Implementation Plan for Deepening Rural Reform in Guangdong* issued by the CPC Guangdong Provincial Committee and the People's Government of Guangdong Province, relevant competent agricultural and rural departments of Guangdong are suggested strengthening communication with the superior, and finding out such links where key problems are not tackled thoroughly, the mechanism is not solid and the masses have urgent need in the fields of rural development requiring enhancement in the future, mainly including rural collective property right system, agricultural management system, urban-rural integration system and mechanism, as well as rural social governance system, according to new requirements of the *Comprehensive Implementation Plan for Deepening Rural Reform* issued by the Central Government and by following the principles of "doing no prohibited thing, doing restricted things according to regulations and continuing bold exploration of things without requirement", so as to continue deepening comprehensive rural reform in Guangdong and further stimulate the endogenous power of rural development.

4.3. Expand the promotion of experience and achievements in reform experimental areas to further benefit vast rural areas from reform achievements

Relevant competent departments of the CPC Guangdong Provincial Committee and the People's Government of Guangdong Province are suggested leading and organizing provincial agricultural social science research institutes to summarize the practices of these regions having made achievements in rural development in Guangdong since the issuance of "the 12th Five-Year Plan", especially the great achievements, mature experience and feasible practices of 11 national rural reform experimental areas in aspects of rural collective property right system, agricultural management system and rural social governance system, and to profoundly and systematically summarize the rural reform and development achievements in Guangdong by the method that "an achievement is summarized and promoted once it is mature". In addition, the CPC Guangdong Provincial Committee and the People's Government of Guangdong Province are suggested actively making leading and promoting publicity on national and provincial media based on the summary of excellent achievements in experimental areas, so as to achieve effective connection of project achievements with their promotion and conversion.

Acknowledgments

We thank the supported by grants from the Guangdong Philosophy and Social Science Planning Project (GD16CGL11), Guangdong Soft Science Project (2016A070705037), Guangdong Science and Technology Program (2017A020224020,2014B090908017,2014B070706015), Guangdong Soft Science Project (2015A070704034), President Fund Project of Guangdong Academy of Agricultural Sciences (201532),Guangzhou Soft Science Project (201709020008).

References

[1] Fei X T, 1998, I saw the rural industrialization and urbanization road in China, *Zhejiang Social Sci.*, **7**: 3-6.
[2] Du R S, 1995, Chinese agriculture in the period of economic transformation, *Reform*, **8**: 15-23.
[3] Wen T J and Dong X D. 2010. Rationality of village community: A new perspective of solving the dilemma of "three agriculture" and "three Cures". *J. Party School Central Committee C.P.C:* **14**(4): 20-23.
[4] Chen X W. 2013, Some important issues in the current rural reform and development of China. *Issues Agri. Econ.*, **1**: 4-6.
[5] Dang G Y. The basic tasks of rural reform during the "Twelve-Five Plan" period. 2010. *Chin. Rural Econ.*, **8**: 14-23.
[6] Xu X L, 2008, A new economic mechanism for the development of urban and rural integration, *J. Party School CPC Central Committee*, **6**:28-9.
[7] Zhu Z, 2016, Using the new concept of development to solve the new problem of "agriculture, rural areas and farmers", *Peasant Daily*, 02-15.
[8] Xu Y, 2016, Innovation of rural governance in the process of urban-rural integration, *China Rural Econ.*, **10**:23-26.
[9] He X F, 2014, On the Chinese urbanization and modernization road, *China Rural Survey*, **1**:2-12.
[10] Fang J. 2018. Endogenous rural development: practice,characteristics and reflection. *China Agri. Uni. J. Social Sci. Edition*, **2**: 24-34.
[11] Balezentis T and Krisciukaitiene I. 2012, Application of the bootstrapped dea for the analysis of lithuanian family farm efficiency. *Management theory and studies for rural business and infrastructure development*, **34**(5): 35-45.
[12] Nehring R, Gillespie J and Hallahan C. 2015, The economics and productivity of organic versus non-organic us dairy farms. 2015 Annual Meeting, January 31-february 3, Atlanta, Georgia, [S.1]: Southern Agricultural Economics Association.
[13] Cortez-arriola J, Groot J C J, Massiotti R D A, Scholberg J M S, Aguayo D V M and Tittonell P 2014, Resource use efficiency and farm productivity gaps of smallholder dairy farming in north-west Michoacdn, Mexico. *Agri. Sys.*, **126**: 15-24.

IOP Conf. Series: Earth and Environmental Science **185** (2018) 012013 doi:10.1088/1755-1315/185/1/012013

Price simulation and prediction of leaf vegetable based on gene expression algorithm

Z H Su[1]

Institute of Agricultural Economics and Rural Development, GAAS, Guangzhou 510640, China

E-mail: suzhuhua@163.com

Abstract. Based on the traditional gene expression algorithm, the paper proposed an improved gene expression algorithm, adding "inverted string" and "gene extraction" operator. An improved GEP algorithm is used in the leafy vegetables prices forecast, create the mathematical model through the analysis and evolution of training data, realize the simulation and forecast of the price of leafy vegetables. Through several experiments, the results proved that the new type of gene expression programming algorithm has faster convergence rate and higher precision in predicting the price of leafy vegetables.

1. Introduction

Guangdong is a large vegetable consumption province, and has a strong demand for vegetables. There are a large number of vegetables that enter the market through various channels, and the quality of vegetable circulation is directly related to the quality of life of the residents. Therefore, how to understand and monitor vegetable circulation organizations in terms of industrial circulation quantity, price, quality and safety, and carry out a basic and objective evaluation of operational risk is a key problem to mitigate the risk of vegetable circulation industry.

From the research trend at home and abroad, the price forecast has made great progress in the fields of social economy, energy, transportation and other fields [1-5]. In the market price prediction of agricultural products, Henry made a regression prediction for the US cotton, and the forecast results coincide with the actual results at that time [6]. Cui Guoli found that the chaotic neural network model was more suitable on the price of Chinese cabbage [7]. Zhu Xiaoxia used Markov chain analysis to simulate the predictability of vegetable price fluctuation cycle [8]. To sum up, there are few studies on vegetable price prediction, and the existing methods of prediction are mainly based on econometrics.

Gene expression absorbs the advantages of the two and overcomes the shortcomings of the two ways, and its remarkable feature is that it can be used to solve complex problems by simple coding. It has widely applied to solve spatial location, price prediction, and image recognition and method evaluation and so on, and achieves good results [9-11]. In this study, vegetables leafy vegetables were selected as representatives, and gene expression programming was applied to improve the price prediction of leafy vegetables, so as to provide references for monitoring and early warning decisions of government in the operation of vegetable circulation industry.

2. Application of new gene expression algorithm in price prediction of leaf vegetable

Since the design of the gene expression program was proposed in 2001, many scholars have improved it from different angles. In order to improve the efficiency of the price forecast, a new method of "gene extraction" was proposed in this paper. In the genome, the genome appears when the first gene as a terminator, not only to retain the gene, but also let the back of the hidden gene expression, the solution is to use the gene is extracted, it connected to the rest of the genome, because the head primordium group of a forward finally, the head of a gene complement.. The improved algorithm part process is as follows:

2.1. New new type of gene expression program design

The new genetic operators of gene expression have such parts as single point recombination, two point recombination, and gene recombination.

2.1.1. Single point recombination. At single point recombination, the parent chromosomes match each other and randomly select a position to exchange all the genes behind the two parent chromosome. The following is an example of a single point recombination (a dark blue gene, Figure 1).

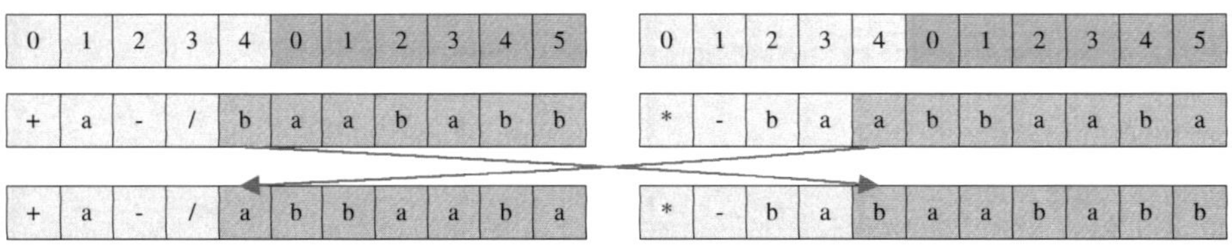

Figure 1. Single point recombination.

Two point recombination. Two point recombination is similar to single point recombination, only two points are randomly selected on the chromosome to exchange the genes between two points. Single point recombination and two point recombination are the most common and most effective genetic operators in traditional algorithms *(*dark blue as the exchange of genes, Figure 2*)*.

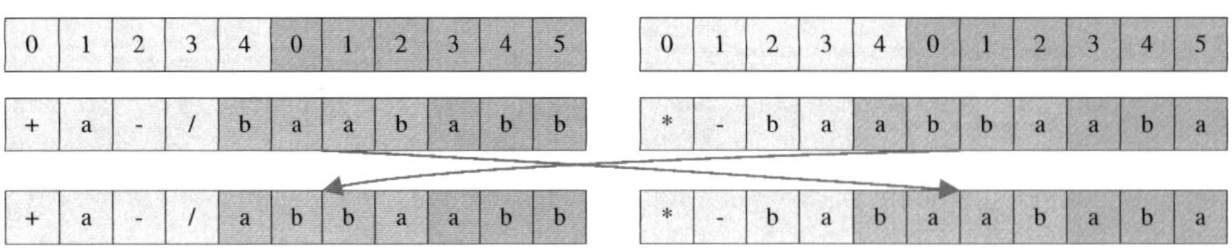

Figure 2. Two point recombination.

2.1.2. Gene recombination. Gene recombination acts on multi genome chromosomes, selects two parent chromosomes, selects a genome randomly, and exchanges two parent chromosomes (the deep blue is the exchanged genome, Figure 3).

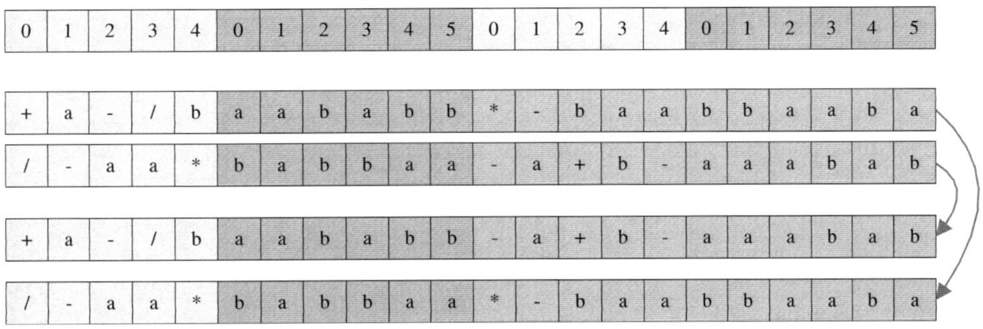

Figure 3. Gene recombination.

2.1.3. Algorithm of gene extraction operator. After input and output, Check the genome and mark the first genomes with the terminator, and turn to 2. If the genome is not found in all genomes, the algorithm ends. Search he first gene is non-terminating, the shortest effective gene and at least two bit shorter than the length of the gene head, if found, turn to 3, otherwise the algorithm is over. Move the genome head back to two bits, then fill in the connector in the first place, and fill in the terminator found in step 1 with the second bit and go to 1.

The important link is the terminator gene extracted in other genes are not at the end of the first and the effective length of the shortest genomic gene, gene behind so as to let the first end of the genome to be expressed, and the effective length of each chromosome genome tend to average.

2.2. Empirical analysis

As leafy vegetables are seasonality, a leaf vegetable listing period is relatively short. Therefore, from the quarterly perspective, it is not suitable for prediction for a long time as accuracy consideration.

2.2.1. Import data. The experimental data selected the average daily average price of leafy vegetables in Guangdong province in July 31st to May 12 in 2015. The data are derived from the Shenzhen Middle Agriculture Data Network (¥/500g).

3.89 3.95 4.05 4.08 4.12 4.19 4.27 4.33 4.45 4.31 4.25 4.20 4.18 4.05 3.98 3.90 3.90 3.85 3.78 3.75

3.70 3.75 3.78 3.80 3.85 3.90 4.05 4.20 4.27 4.35 4.40 4.45 4.56 4.58 4.65 4.65 4.73 4.70 4.70 4.72

4.66 4.68 4.66 4.50 4.55 4.42 4.40 4.33 4.27 4.28 4.19 4.30 4.39 4.56 4.72 4.80 4.69 4.82 4.90 4.95

4.99 4.98 5.12 5.26 5.22 5.34 5.46 5.42 5.31 5.50 5.69 5.67 5.57 5.63 5.69 5.71 5.75 5.65 5.52 5.43

The above total of 80 data, the first 70 data as a training set, the last 10 data as the evaluation of the evaluation of the evaluation set. When the width of the sliding window is 5, 65 sets of training data can be obtained after the data processing.

2.2.2. Parameter setting. In this experiment, the length of the head is 10, the number of terminator is 5, the number of genome is 8, and the number of chromosomes is 10. According to the formula t=h*(n-1) +1, the length of the genomic tail of this chromosome is 11, so the length of the genome is 21. The length of a chromosome is 21 * 8, that is, 168. The operator set {+, -, *, /, Q, S, C, E, L}, terminal set for the {a, B, C, D, e} respectively represent the five consecutive days of price. After the data processing, the training set has 62 sets of data, because the selection range is 10, so the maximum fitness is 620.

2.2.3. Traditional algorithm evolution. When the data is imported and the parameters are set, the traditional algorithms are beginning to evolve. The groups of genes through variation, inverted string, insert string, string root genetic operators such as modification, continue to move toward the direction of maximum fitness evolution.

2.2.4. End of evolution and get the prediction formula. When the evolutionary algebra is set, the traditional algorithm stops evolving. The optimal output is selected as the optimal solution under the setting condition. In this experiment, the optimal formula for the prediction of chromosomes and leaf vegetables is as follows:

```
SL+**a//b*baedbadaaecS~SSb+-dbQeebcacadcbbSc/C~Ee/Seeedbdbbadcd~Q*cQS~
SSLbacccdddbcdS-L+-S/+-*decaccadeabCLca~cS*/Cddbdbddddac+c-EcLc~S*dbcda
aecdbd~-ee/cSeCScbbcedbbcba
```

The final formula for predicting the price of leaf vegetables can be obtained by connecting each expression with the symbol "+". As shown below:

$$f_i = \sin(\log(f_{i-5} * f_{i-1} + f_{i-5} * f_{i-4} / f_{i-1})) + \sin(-\sin(\sin f_{i-3}))$$
$$+ \sin f_{i-3} - \sqrt{f_{i-3} * \sqrt{\sin(-\sin(\sin(\log(f_{i-4}))))}} + \sin(\log(-(f_{i-3} + f_{i-5})) -$$
$$(\sin(f_{i-5} * f_{i-2}) + f_{i-2} / f_{i-1})) + \cos(\log f_{i-3}) + (f_{i-3} + e^{\log f_{i-3}})$$

2.3. Comparison and analysis of prediction results

In order to verify the feasibility of the application of the traditional algorithm in the prediction of the price of leaf vegetable and the superiority of the improved algorithm, a series of experimental analyses are carried out in this paper.

2.3.1. Comparison of gene utilization ratio. The advantage of the improved algorithm is to improve the efficiency of gene utilization. Figure 4 shows the 200 generation in one experiment, and the gene utilization ratio of each generation's traditional algorithm and improved algorithm. The average utilization rate of the improved algorithm is 18%, while the traditional algorithm has a gene utilization rate of 12%. It can be seen that the gene utilization rate of the improved algorithm is obviously higher than that of the traditional one.

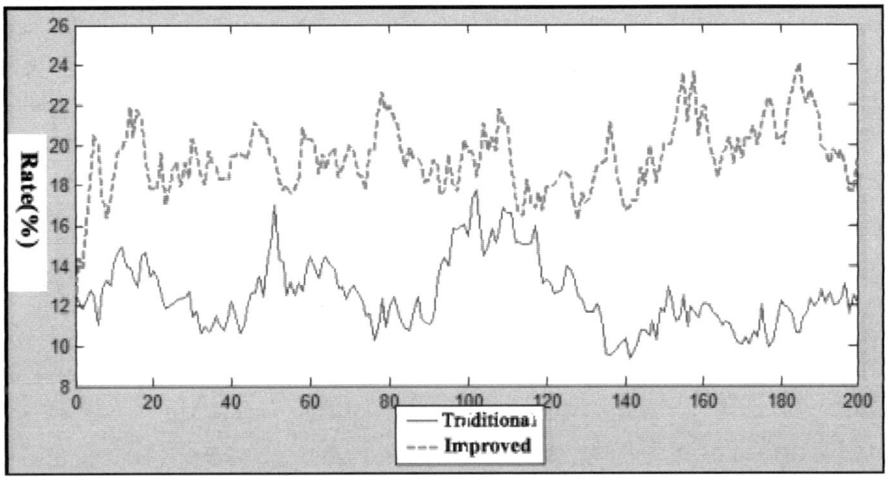

Figure 4. Comparison of gene utilization ratio between GEP and IGEP.

2.3.2. Effect of the increase in gene utilization on evolution. In order to find out the effect of the increase in gene utilization on evolution, a comparative experiment was designed in this paper. First, the conditions of evolution end are set, and the evolution is stopped when the given error range is reached. For example, when the set error is 10%, the maximum fitness 620 of 90% is 558, that is to say, the evolutionary algebra that the traditional algorithm and the improved algorithm achieve 558 respectively. The following are the average evolutionary algebra required for the different error ranges in the 100 experiment, the traditional algorithm and the improved algorithm.

Table 1. Algebraic contrast of evolution needs.

Error (%)	Traditionalalgorithm	Improved algorithm
10%	72	98
8%	96	136
6%	166	192
4%	288	368
2%	512	384
0.8%	702	466
0.6%	783	683
0.4%	843	742
0.2%	1124	926
0.1%	1440	1020

The results in Table 1 showed that the evolutionary speed of the improved algorithm is not faster than that of the traditional algorithm when the error range is large. The main reason is that the fitness evaluation function used in the experiment is absolute error, and the price value of the leaf vegetable is small, which leads to a very high value of adaptability very quickly. The improved algorithm is an increase in genetic operators on the basis of the traditional algorithm, the algorithm is equivalent to the basic role of chromosome, the improved algorithm for further modification, resulting in the error is low, but because of improved rather than basic algorithm takes the evolution algebra. Therefore, in the price prediction of the designated evolutionary algebra as the end condition, the improved algorithm's evolution solution, that is, the prediction function, will be better than the traditional algorithm.

In order to get more accurate conclusions, 10 prediction experiments were carried out in this paper, and the conditions of each experiment were consistent. Then the average value of the prediction is taken and the basic algorithm and the improved algorithm are evaluated according to the prediction error. The traditional algorithm parameters in the experiment are as follows (Table 2):

Table 2. Gene expression algorithm parameter.

Parameter	numerical value
Maximum evolutionary algebra	1080
Population size	30
Function set	+ - * / ~ Q S C L E
Terminator set	a b c d e f g h
Head length	30
Number of genes	20
Connector	+
Mutation probability	0.005
Single point recombination probability	0.25
Two point recombination probability	0.20
Gene recombination probability	0.15
IS insertion probability	0.20
RIS insertion probability	0.20
Gene string probability	0.15
Gene extraction probability	0.55

After 10 experiments, 10 sets of predicted values were obtained and their average values were as follows (Table 3).

Table 3. Comparison of prediction results (¥/500g).

Price	1	2	3	4	5	6	7	8	9	10	Relative error (%)
Real price	4. 43	4.48	4.58	4.52	4.66	4.72	4.64	4.51	4.62	4.44	
Traditional	4.29	4.34	4.42	4.61	4.76	4.89	4.83	4.39	4.78	4.52	14.03
Improved	4.48	4.52	4.54	4.55	4.68	4.82	4.75	4.43	4.56	4.40	7.25

The results showed that in leafy vegetable price prediction, the relative error between predicted price and real price based on gene expression operator and inverted string operator is 7.25%, but the traditional relative error is 14.03%. The improved is higher than that of GEP and closer to the real leafy vegetable price. It also shows that the new genetic expression programming algorithm is feasible in vegetable price prediction and is of practical reference value.

3. Conclusion

In view of the shortcomings of traditional gene expression programming in the process of gene utilization, a new improved method based on "gene extraction" operator and "inverted string" operator is proposed, and the algorithm is applied to leaf vegetable price prediction. The feasibility of gene expression programming in vegetable price prediction and the superiority of the new gene expression program design are proved by a number of comparison experiments.

The research shows that gene expression programming has certain practical value in vegetable price prediction, but this paper only uses time series prediction method, without considering other factors. There are many factors that influence the price of vegetables, such as temperature and weather. In the next step, we can introduce these factors into the prediction of gene expression program of vegetable price.

Acknowledgments

We thank the supported by grants from the Guangdong Philosophy and Social Science Planning Project (GD16CGL11), Guangdong Soft Science Project (2016A070705037), Guangdong Science and Technology Program (2014B090908017, 2014B070706015), Guangdong Soft Science Project (2015A070704034), President Fund Project of Guangdong Academy of Agricultural Sciences (201532).

References

[1] Nguyen D H and Widrow B, 1990, Neural networks for self-learning control system, *IEEE Con. Sys. Mag.*, **12**: 18-23.
[2] Dieng A, 2008, Alternative forecasting techniques for vegetable prices in senegal, *Intl. Conf.Power Sys. Tech. IEEE*, **1**: 5-10.
[3] Li G Q, Xu S W and Li Z M, 2010, Short-term price forecasting for agro-products using artificial neural network. *AASP*, **1**: 278-87.
[4] Ju J, Zhao L and Wang J, 2013, Forecasting the total power of China's agricultural machinery based on bp neural network combined forecast method, *CCTA*,**4**: 85-93.
[5] Li Z M, Xu S W and Cui L G, 2012, The short-term forecast model of pork price based on CNN-GA, *Adv. Mater. Res.*,**11**: 396-401.
[6] Moore H L, 1997, Forecasting the yield and the price of cotton, *Mac Comp.*, **15**:100-13.
[7] Li C G,2013, Short-term prediction of vegetable prices in China based on chaos neural network mode.Chin. *Academy Agri. Sci.*, **22**:101-15.
[8] Zhu X X, 2012, Markov chain analysis and prediction of vegetable price fluctuation cycle, *Productivity*, **8**: 143-6.

[9] Zhou A M, Cao H Q and Kang L S, 2003, Automatic modeling of complex functions based on genetic programming, *J. Sys. simu*, **6**: 44-6.
[10] Xie J Y, Wang M Z, Zhou Y, Gao H C and Xu S Q, 2018, Research on the differential expression gene selection algorithm of non-equilibrium gene data, *J. Comp. Sci.*, **5**:1-18.
[11] Wang Z L, Wang H K, Li X S, Gu F, Zhou Y and Ma W, 2016, Research on bi-directional clustering algorithm based on gene expression data, *J Natural Sci. Heilongjiang Uni.*, **33**: 444-9.

IOP Conf. Series: Earth and Environmental Science **185** (2018) 012014 doi:10.1088/1755-1315/185/1/012014

Study on precise positioning and construction of improved varieties of forest seed base

J Zhao[1, 5], Q F Lei[1], H W Du[2], Y Zhao[3], F Y Wang[1] and D D Liu[4]

[1] Liaoning Forestry Vocational-technical College, Shenyang, 110101, China

[2] Jinzhou Medical University, 121001, Jinzhou, China

[3] Medical College of Jinzhou Medical University, 121001, Jinzhou, China

[4] Liaoning Laotudingzi Forest Natural Reserve Administration Bureau, Benxi, 117200, China

Email: gisgis@126.com

Abstract: In this paper, the cultivars of Pinus koraiensis and Japanese larch varieties in Liaoning Province were studied. Aiming to solve the problem of data update difficulty and lack of information management, the Kriging interpolation method and UAV (Unmanned Aerial Vehicles) and GPS technology positioning method were used for fine positioning of the improved varieties of forest seed base. Using linear comparison and distributed comparison method for accuracy assessment, 100 Pinus koraiensises were sampled by liner comparison method among which 10 were selected to be tested by distribution comparison method. With the help of the high density total station, the samples were measured and analyzed by the coordination. It showed the positioning accuracy was less than 1m, which met the need of the production of the forest seed base. On this basis, the tree seed production management information base database was established and the database structure design and structure design was made. Therefore, the managers in the forest natural reserve could use the database to plan and do the daily management, measure the area, the circumference, and the coordination of the forest natural reserve. In this way, the field work was lightened and it also provided support for the seed base management and production.

1. Introduction

The effective management of the good planting base of forest trees is the guarantee of the development of forestry construction. The organization and management of the good seed base is a huge and complex project[1]. The managers of good seed base should find out the problems in forest management and production in time [2]. We should fully recognize the necessity and importance of planning guidance for forest seed production [3]. The high tech means of information management is used to manage the good species of forest trees, enhance the scientific and technological input of forest tree seed [4].

Literatures on domestic forest seed base reveals the drawbacks like data update and management difficulty, digitization and informatization deficiency [5]. Furthermore, some of the existing drawings of the seed base do not match the authentic seed base. Some drawings are lost, resulting in the impossibility to trace back the male parent and female parent improved varieties. Other researchers only focus on the construction and countermeasures, the spatial arrangement, and benefit evaluation of the forest tree seed base, lacking the precise positioning and informatization. In this case, we take the cultivars of Pinus koraiensis and Japanese larch varieties in Liaoning Province as subject, and focus on the precise positioning and database constructing.

2. Research content and method

The indoor interpretation and field investigation, combined with the UAV technology and forestry management techniques are used to collect the basic data in the study area (seed orchard thinning design, seed orchard layout). The technology of geographic information system (GIS), UAV Remote Sensing Technology (RS) and the global satellite positioning technology (GPS) were used as well. Then, the data were processed by Smart 3D capture and pix 4D mapper software to generate results and data report of Laotudingzi forest natural reserve seed base for precise positioning and precision evaluation. Therefore, the production management information database is established.

2.1. Research area survey

The research area—Liaoning Laotudingzi Forest Natural Resere is located in Fushun, Liaoning province. Up to 2013, the scale of the improved forest seed base was 3480 hectares, which was divided into the following functional group areas: the determination group, the breeding colony area, the seedgrowing area and the production group. The breeding area has 720 mu of Japanese larch junior clone seed orchard, 195 mu red Pine seed orchard, 150 mu Japanese larch 1.5 generation clonal seed orchard, 105 mu of mother forest of cleft leaf elm, 810 mu of the mangrove forest, 60 mu of yew trees, 90 mu of rowan trees and 45 mu of the mother woods of tianmu [6].

Liaoning Laotudingzi Forest Natural Reserve has rich tree seed base archives, thematic map complete, and detailed survey data, which provides favorable and rich data support for the research.

2.2. Data collection

The collection of materials: Planting position of the seed orchard, tree planting location maps, the 1.5 generation seed orchard of Japanese larch, Xiaonancha red pine, Miaojia Gou configuration distribution map. UAV image collecting scope: The shooting has two main areas, respectively: 124 degrees to 124 degrees 54 '15.2 "56' 12.48" E, 41 degrees to 41 degrees 24 '22.514 "25' 18.47" N; 124 degrees to 124 degrees 54 '40.33 "54' 48.24" E, 41 ~ 25 '50.63 "to 41 degrees 25' 57" N, an area of about 40.7hm^2.

2.3. Kriging interpolation tree vertex extraction

The digital terrain model and the canopy height model were generated under the ArcGIS software, the forest canopy with moderate to low canopy density was selected as the research object, combined with the high-precision DSM and DOM models. Open the Arc Toolbox, select the "3D analysis tool", and use the subtraction in grid computing. The data settings are as follows: the DEM input is the subtraction, the DSM input is reduced, and the grid generation after the subtraction operation becomes CHM data.

Good seed vertex extraction: (1) Using the spatial analysis module, the neighborhood analysis below the block statistical function, the CHM data as the input data of the operation. (2) Using the 3D analysis module, the subtraction function is calculated under the grid, and the output image after the operation is the vertex image of the good base tree. (3) Using the space analysis module to reassign 0 and 1 assignments, and turn raster image of tree species into base vector data of tree species, and save them.

As the crown surface is curved and irregular, sometimes larger seed crown on multiple tree vertices, which means multiple maximum. In this paper, we use the method of vertex extraction from Wang Wei's scholars' tree base at Beijing Forestry University, and delete some pseudo tree vertices using the algorithm of the scholar, so that the new vertex map obtained is true vertex [7-9].

2.4. Unmanned Aerial Vehicle (UAV) and GPS technology positioning method
In addition to the middle and low canopy density, the improved seeds are positioned by the combination of UAV technology and GPS technology. In order to ensure the accuracy requirements, UAV carries out high-precision control at the early stage of data acquisition. The quality of unmanned aerial vehicle (UAV) aerial film is improved by the way of post processing [10-11]. The basic process is as follows:

(1) To guarantee the high precision, one reference station in the known location of Laotudingzi forest natural reserve is established. Data is distributed through the network and recorded to the local area at the same time

(2) The high precision module record data is added to the UAV and the high precise position and IMU information of each photo are obtained by using the way point software.

(3) Finding some high precision phase space points on the ground, combined with the location information of aerial photographs, we use smart 3D software to process high resolution plane data and three-dimensional data.

(4) Adding field work in done: the high-precision hand-held machine is used to collect high precision trees with high precision and high precision. The results of the UAV positioning are shown in the following chart.

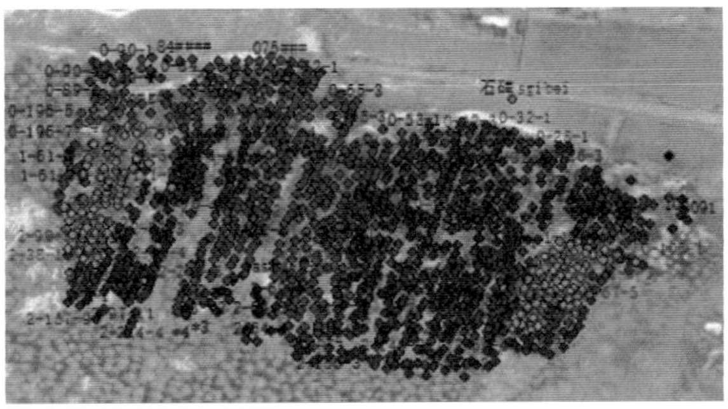

Figure 1. UAV positioning results.

3. Accuracy evaluation
The precision positioning data of the good species base are verified by using the linear comparison method and the distributed comparison method.

3.1. Linear analysis method
A sample of 100 red pine trees in a small area of the good species base was selected to be tested (open, severely sheltered and adjacent to a number of trees). A high precision prism free total station was used to measure the sampled trees. The coordinates of 100 improved species were corrected by aerial photograph, and compared with the coordinates of the total station.

As shown in Figure 2, when the information collected by GNSS high-precision equipment is not fully, the accuracy can reach 1m, and the accuracy will exceed 3m when occlusion is fully. After using the aerial film to correct the error, the precision can be less than 1m. The location accuracy of individual trees is due to the tree skew growth, which leads to the difference between the center point and the tree vertex coordinates.

Figure 2. Airplane correction X error, (linear analysis).

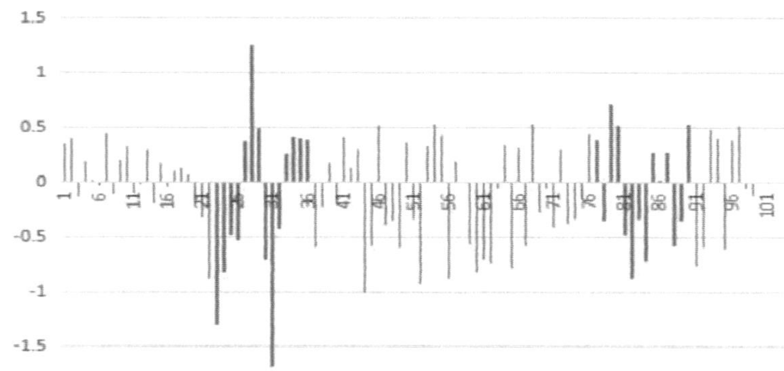

Figure 3. Airplane correction Y error (linear analysis).

3.2. Discrete distribution test

10 representative points were selected randomly in the large area, such as severe occlusion, remote mountains, large chest diameter, small chest diameter, and a number of trees growing next to them. Only a few trees shade and aerial edge in the forest, the error is about 4m. GNSS equipment in the areas of severe occlusion can achieve sub-meter accuracy, as shown in Figure 4, after Aerial corrected, in the forest and open areas error within 1m, reached the seed base management precision [12].

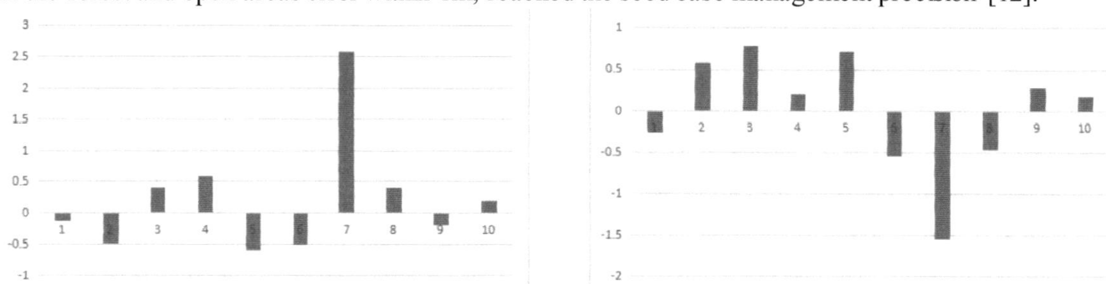

Figure 4. The X-deviation, Y-deviation (discrete distribution).

4. Establishment of information base for good species base

4.1. Database structure design

Use arc catalog to create a spatial database, the desired layer is created according to the business needs: Plant germplasm (point layer map), cutting orchard, nursery, parent trees, experimental forest, seed orchard, progeny forest, and each layer involved in the field.

The field of state-level demonstration base of forest tree improvement includes: basic situation, auxiliary production engineering (including irrigation water source engineering, drainage engineering, road engineering, power supply communication engineering, protective engineering and so on), as shows in the following table:

Table 1. Laotudingzi Forest Natural Reserve layer attribute structure

Serial Number	Field Name	Field Type	Length	Decimal	Unit
1.	County	text	6		
2.	establish time	date			
3.	introduction	text	500		
4.	water source project	text	20		
5.	drainage works	text	20		
6.	road project	text	20		
7.	power supply communication engineering	text	20		
8.	protection engineering	text	20		
9.	area	double	16	2	hectare

The tree seed base (cutting orchard, seed orchard, progeny forest, parent trees), the fields include: the basic situation, clonal disposition, project, scale, clonal origin, pest control, quantity, tree management, fire protection management.

The nursery fields include the basic situation, the nursery area, the road engineering, and the production supporting facilities and so on.

The production engineering fields include: basic conditions, water supply engineering, power supply communication engineering, protection engineering and production facilities and other fields.

4.2. Data loading of good species base

Design survey of forest resource planning (second class investigation) sub-compartment data layer, as a reference, to determine the scion, nursery, parent trees, experimental forest, seed orchard, forest layer range according to the image and data of sub-compartment.

Spatial data and attribute data entry: the collected data is processed into SHP format by software. After data preprocessing and excel connection, the data is loaded and integrated into the individual germplasm layer.

4.3. Information base function

The characteristic functions of the information base of the good species base are as follows:

(1) Statistical analysis:

Through spatial overlay analysis, how many seeing germplasm can be analyzed in a seed orchard (or mother forest); how many classes within the area, what are the dominant tree species among the classes, the accumulation of information, the DBH(diameter at breast height) also can be analyzed. It can also can realize the commonly used information statistical function: maximum analysis value, minimum value, average value and so on. As shown in the following figure:

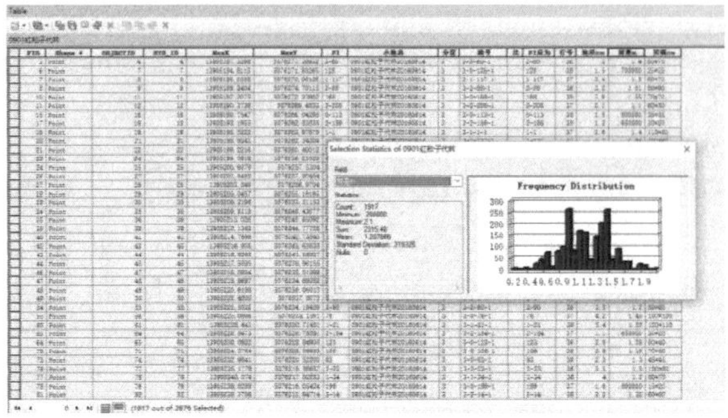

Figure 5. Analysis of statistical results.

(2) Distance measurement:

We can query the distance between tree and tree, the distance from tree to road, and the distance between nursery (or seed orchard, mother forest) and road (river).

(3) Query browsing:

Spatial query, the selection of germplasm in the area of interest can be selected according to the location.

Attribute query, attribute query basic information through query interested accessions, including small names, tree species, tree height, DBH, crown width, and so on.

(4) Three dimensional display function:

By using the three-dimensional display function, we can simulate the growth of each improved variety, and assist the managers of the reserved area to plan and manage the improved seed base. It is also possible to measure the area, circumference and coordinate of the good species base in the reserved area. And reduce the workload of field work measurement. The measurement of area and perimeter is as shown in the following chart.

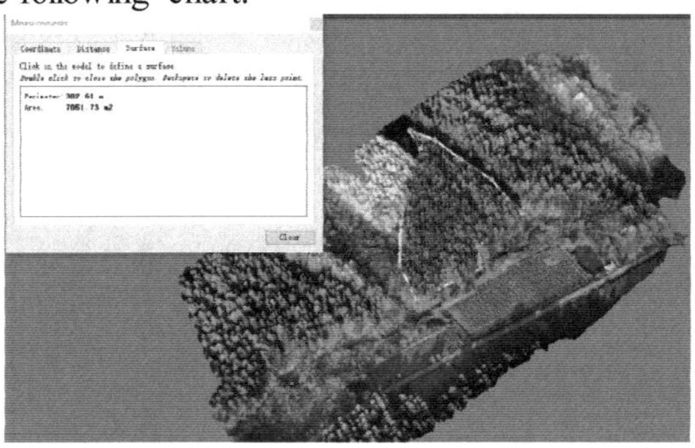

Figure 6. Circumference of the measurement.

5. Conclusion

(1) Tree vertex extraction can be done by using the Kriging interpolation method, combining with UAV and GPS positioning method, the completion of the seed base of precise positioning and precision evaluation is realized.

(2) After precise positioning of the improved seed base, using linear comparison and distributed comparison method for accuracy assessment, 100 Pinus koraiensises were sampled by liner comparison method. 10 out of 100 Pinus Koraiensises were selected to be tested by distribution comparison method. With the help of the high density total station, the samples were measured and analyzed by the coordination. It showed the positioning accuracy was less than 1m, which met the need of the production of the forest seed base. The high positioning precision also meets the needs of field work investigation and management of the improved tree seed base.

IOP Conf. Series: Earth and Environmental Science **185** (2018) 012014 doi:10.1088/1755-1315/185/1/012014

(3) The establishment of Laotudingzi forest natural nature reserve seed base information database, Therefore, the managers in the forest natural reserve could use the database to plan and do the daily management, measure the area, the circumference, and the coordination of the forest natural reserve. In this way, the field work was lightened, and it also provided support for the seed base management and production. It has realized the management mode from the original paper data management to the digitalization.

Reference:

[1] Shen X H 2004 Opinion on the current work fine breed base in Liaoning *Liaoning Fores. Sci. Tech.* **01**:29-31

[2] Om P Rajora 2001 Challenges and opportunities for conservation of forest genetic resources. *Euphytica Springer J.* **4**: 136-7

[3] Tiit Maaten 2010 About management and research of forest genetic resources *Fores. Studies Metsanduslikud Uurimused* **02**: 56-7

[4] Hu C 2014 Study on GIS-based Planning and Layout of Metasequoia glyptostroboides in Hubei Province *J. Northwest Forest. Uni.* **02**: 70-2

[5] Wang X J 2005 Study on tree species recognition and plant location method based on large scale aerials *J. Northeast Fores. Uni.* **05**: 60-5

[6] Li J J, Liu D D and Lv R 2016 Old bald sub-national nature reserve scientific research report *Shenyang Press,* Shenyang, China :11-65

[7] Wang W 2015 *J. Beijing Forest. Uni.* **04**:23-5

[8] Vera R 2011 a gis-based mapping and estimation the current forest landscape state and dynamics *J. Landscape Ecol.* **4**: 34

[9] Julia 2008 Factors affecting the accuracy of forest clear-cut area estimation on medium spatial resolution satellite winter images *Forestry Studies Metsanduslikud Uurimused* **05**: 122-4

[10] Zhao J, Liu D L, Zheng X X, Zhao X M and Zhao H S 2010 Application of GIS to Study on Multi-function Operation in Jingouling Forest Farm *J. Northwest Forest.Uni.* **25**: 207-9

[11] Stere K. 2016 UAV and GIS based tool for collection and propagation of seeds material-first results *Remote Sensing Spatial Inform. Sci.* **07**: 213

[12] Liu D D and Zhao J 2004 The precise location research of the good breeding base -- Taking the old bald nature reserve as an example *Liaoning Forest. Sci. Tech.* **01**: 29-31

Correlation research between gas production characteristics and CNCPS components for roughages

M X Sui[1]

Key laboratory of biochemistry & molecular biology in universities of Shandong, Weifang University, Weifang, Shandong, China

E-mail: suimx63@163.com

Abstract. The experiment was to study the correlations between gas production characteristics and Cornell Net Carbohydrate and Protein System (CNCPS) components of seven roughages, which shortened the experimental cycles. The ruminant fluid was collected and cultivated in vivo. The results showed negative correlations between cumulative gas productions at 12, 24, 36, 72 h ($P < 0.05$), 48 h ($P < 0.01$) and Crude protein (CP), but positive correlations between gas production at 12, 24, 36, 48, 72 h ($P< 0.01$) and Neutral Detergent Soluble (NDS)/CP, between gas production at 12 h ($P< 0.05$), 24, 36, 48, 72 h ($P< 0.01$) and (NDS-ASH(crude ash))/CP. Significant positive correlations were found between parameters a, b, a+b, c and NDS/CP ($P< 0.01$), as well as between parameters a, b, a+b ($P< 0.01$), c ($P< 0.05$) and (NDS-ASH)/CP. However, significant negative correlation was found between b, a+b, c and CP ($P< 0.05$). The experimental results inferred a positive correlation between degradation rate of Neutral Detergent Fiber (NDF) ($P < 0.01$), Acid Detergent Fiber (ADF) ($P< 0.05$) and CP, but negative correlation between degradation rate of NDF ($P < 0.05$), ADF ($P< 0.01$) and NDS/CP, (NDS-ASH)/CP.

1. Introduction

The cardinal appraisal procedure for feed nutritional values includes in vitro, in vivo and in-situ methodologies. In Vivo and in situ methodologies are time-consuming and hard sledding, and apparently limited to experimental time and animal quantities when used to evaluate massive samples. Although the CNCPS, which used to reflect comprehensive feed nutritional value, contains plenty of indices and can, its assessment procedure is complex. In Vitro gas production technique, which was first applied by Menke *et al* [1] successfully, is simple, economical and rapid, and widely used to evaluate feed nutritional values. Seven roughages, which are commonly used in north-eastern China, combined with modified gas production apparatus of In Vitro fermentation, were selected and used to predict gas product volumes, as well as to verify the correlations between gas production characteristics and CNCPS components.

2. Materials and methods

2.1. Experimental design

Four Cannula sheep were used as the source of rumen fluid, which was cultured In Vitro simulating rumen fermentation. Seven roughages used are corn straw silage, Alfalfa, Corn straw, Bean straw, Chinensis Rice, straw, as well as specific silage corn straw.

All experiments were quadruple replicated. The blank control was set at different times for each roughage.

2.2. *Estimation of gas production profiles*
The rumen fluid was prepared by following procedures: two hours after morning diet, each rumen content was gathered through fistula, and placed promptly into a vacuum flask filled with carbon dioxide (CO_2), finally taken back to the laboratory.

Each rumen content was filtered by four layers gauze. Before fermentation, Menke culture solution was fully filled with CO_2, which indicated by the vanished pink of resazurin. Four hundred mg substrate and sixty mL mixed media was put into each culture flask respectively, then placed into a shaking bath with constant temperature of 39 degree for further culture. Before fermentation, residual gas in each flask due to thermal expansions was tired out.

2.3. *Kinetic of gas production*
The gas production values of different roughages at different time were plugged into the following formula of gas production model (φrskov *et al.*, [2]):

$$p=a+b(1-e^{-ct}) \quad (1)$$

where t is fermentation time, the duration from the beginning of culture to the recording time of gas production, and fixed to 72h in this research; p is gas production at time.

2.4. *Data processing*
The experimental data were processed by Excel. The gas production was fitted to the General Linear Model (software SAS V8.1) and compared by analysis of variance. The parameter estimation was carried out by nonlinear regression (NLIN) process; finally the linear correlation and regression analysis was conducted with CORR and REG procedures.

3. Results analysis

3.1. *The component of crude protein and carbohydrates in roughages CNCPS*
The analysis of dry material (DM), crude fat (EE), and ash (ASH) were carried out according to the method descried in <Feed analysis and quality testing technology>, neutral detergent fiber (NDF), lignin (ADL), neutral detergent insoluble nitrogen(NDLP) and acid washing insoluble nitrogen (ADIP) was conducted with methods developed by Van Soest *et al.* [3] , while the soluble protein (SP) was analyzed by the procedure developed by Krishnamoorthy [4]; analysis of starch was carried out by AACC(American Association of Cereal Chemists) method. The nutritional ingredients of different roughages were shown in table 1.

Table 1. Nutritional ingredient of roughages.

	Corn Straw Silage	Alfalfa	Corn Straw	Bean Straw	Chinensis	Rice Straw	Specific Silage Corn Straw
DM	91.99±0.27^c	90.10±0.05^e	90.65±0.03^d	88.53±0.05^f	91.84±0.02^c	92.78±0.05^b	96.46±0.06^a
ASH (%DM)	6.71±0.05^d	10.22±0.13^b	8.09±0.09^c	5.32±0.93^d	5.72±0.08^d	14.60±0.08^a	6.12±0.22^d
CP (%DM)	10.16±0.12^b	20.49±0.15^a	7.35±0.19^d	7.21±0.04^d	8.18±0.26^c	4.73±0.10^e	4.87±0.07^e
SP (%DM)	5.41±0.03^b	9.00±0.31^a	3.14±0.08^c	2.92±0.12^d	2.57±0.07^d	1.64±0.05^e	1.82±0.02^e
ADL (%DM)	2.39±0.18^c	1.51±0.38^d	5.13±0.29^b	6.86±0.36^a	4.67±0.29^b	1.20±0.14^d	2.80±0.09^c

ADIP (%DM)	0.51±0.12[c]	2.99±0.22[a]	0.58±0.16[c]	0.72±0.14[bc]	0.79±0.16[b]	0.53±0.12[c]	0.59±0.01[c]
ADF (%DM)	36.19±0.24[c]	29.28±0.83[d]	42.71±0.56[b]	41.71±0.54[b]	42.03±1.07[b]	44.24±0.14[a]	37.70±0.76[b]
NDIP (%DM)	1.40±0.19[b]	3.94±0.84[a]	1.79±0.20[b]	1.36±0.52[b]	1.49±0.21[b]	1.37±0.19[b]	1.29±0.09[b]
NDF (%DM)he	65.65±0.75[b]	51.95±0.87[c]	66.85±0.74[b]	70.52±1.12[b]	63.78±2.24[b]	70.69±2.83[b]	78.91±0.70[a]
NPN (%DM)	4.16±0.05[a]	4.63±0.20[a]	1.85±0.11[b]	2.16±0.26[b]	1.62±0.09[b]	1.02±0.12[b]	1.43±0.08[b]
EE (%DM)	4.46±0.52[a]	1.84±0.84[c]	0.28±0.01[c]	0.97±0.85[c]	2.56±0.24a[b]	1.03±0.86[c]	0.59±0.57[c]
Starch (%DM)	13.99±0.86[a]	2.51±0.47[d]	9.71±1.04[b]	5.72±0.54[c]	9.56±0.57[b]	4.44±1.07[c]	9.18±1.05[b]

a, b, c, d, e: different letters of data superscripts in the same row mean different significantly ($P < 0.05$)

The results calculation was conducted according to the method used by Sniffen *et al.* [5]

$$PA(\%CP)=NPN(\%SP)\times0.01\times SP(\%CP) \quad (2)$$

$$PB1(\%CP)=SP(\%CP)-PA(\%CP) \quad (3)$$

$$PC(\%CP)=ADIP(\%CP) \quad (4)$$

$$PB3(\%CP)=NDIP(\%CP)-ADIP(\%CP) \quad (5)$$

$$PB2(\%CP)=100-PA(\%CP)-PB1(\%CP)-PB3(\%CP)-PC(\%CP) \quad (6)$$

where PA is non-protein nitrogen, PB1 is rapid degrading protein, PB2 is moderate degrading protein, PB3 is slow degrading protein, PC is conjugated protein, CP is crude protein, NPN is non-protein nitrogen. Therefore

$$CHO(\%DM)=100-CP(\%DM)-EE(\%DM)-ASH(\%DM) \quad (7)$$

$$CC(\%CHO)=100\times[NDF(\%DM)\times0.01\times ADL(\%NDF)\times2.4]/CHO(\%DM) \quad (8)$$

$$CB2(\%CHO)=100\times\{[NDF(\%DM)-NDIP(\%CP)\times0.01\times CP(\%DM)-NDF(\%DM)\times0.01\times ADL(\%NDF)\times2.4]/CHO(\%DM)\} \quad (9)$$

$$CNSC(\%CHO)=100-CB2(\%CHO)-CC(\%CHO) \quad (10)$$

$$CB1(\%CHO)=Starch(\%NSC)\times[100-CB2(\%CHO)-CC(\%CHO)]/100 \quad (11)$$

$$CA(\%CHO)=[100-Starch(\%NSC)]\times[100-CB2(\%CHO)-CC(\%CHO)]/100 \quad (12)$$

where CHO is carbohydrate, NSC is non-structural carbohydrate, CA is saccharide, CB1 is starch and pectin, CB2 is available fiber, CC is non-available fiber. The amount of non-digestible fiber in different roughages was about 2.4 times than that of lignin. CNCPS components of seven roughages were shown in Table 2.

Table 2. CNCPS component of seven roughages.

	Corn Straw Silage	Alfalfa	Corn straw	Bean Straw	Chinensis	Rice Straw	Specific Silage Corn Straw

PA(%CP)	40.96	22.58	25.24	29.95	19.86	21.70	29.26
PB1(%CP)	12.32	21.37	17.53	10.61	11.53	13.00	8.18
PB2(%CP)	32.93	36.82	32.90	40.66	50.32	36.29	36.16
PB3(%CP)	8.79	4.69	16.41	8.73	8.61	17.74	14.29
PC(%CP)	5.01	14.55	7.92	10.06	9.69	11.28	12.10
CA(%CHO)	32.41	39.22	23.38	19.38	28.76	8.93	15.97
CB1(%CHO)	17.78	37.19	11.52	6.62	11.45	5.58	10.38
CB2(%CHO)	42.59	17.21	50.50	54.94	46.35	82.88	66.05
CC(%CHO)	7.23	5.38	14.61	19.05	13.43	3.62	7.61

CNCPS divides feed carbohydrate into CA (rapid degrading component), CB2+ CB1 (moderate degrading component) and CC (non-degrading component). The CA value from alfalfa was scientifically higher than the other groups, which demonstrated a higher CA content of available carbohydrate in alfalfa and a better nutritional quality. The CA value of corn straw silage was slightly lower than that of alfalfa, but CB2+CB1 value was high, which also indicated a good nutritional quality.

3.2. Gas production profiles with in Vitro formation and NDF, ADF degradation rate

The gas production characteristics of 7 roughages after 72h in Vitro fermentation and NDF, ADF degradation rate at 48h were shown in table 3.

Table 3. Gas production characteristics and NDF, ADF degradation rate at 48h of seven roughages.

	Corn Straw Silage	Alfalfa	Corn Straw	Bean Straw	Chinensis	Rice Straw	Specific Silage Corn Straw
12h(ml)	17.98 ± 1.02^{d}	12.03 ± 1.22^{e}	23.28 ± 0.56^{c}	27.47 ± 3.80^{bc}	21.19 ± 0.98^{c}	38.17 ± 0.34^{a}	31.81 ± 0.91^{b}
24h(ml)	22.81 ± 0.47^{c}	21.63 ± 1.52^{c}	35.27 ± 1.32^{b}	36.27 ± 4.50^{b}	33.14 ± 1.00^{b}	50.51 ± 1.51^{a}	49.79 ± 1.23^{a}
36h(ml)	39.10 ± 1.14^{c}	36.25 ± 0.63^{d}	45.24 ± 0.77^{b}	47.22 ± 4.16^{ab}	45.18 ± 2.83^{b}	55.64 ± 0.46^{a}	52.66 ± 2.04^{a}
48h(ml)	43.66 ± 1.90^{c}	37.27 ± 1.31^{d}	48.32 ± 0.98^{bc}	50.83 ± 5.27^{b}	45.29 ± 2.62^{c}	55.89 ± 0.65^{a}	53.97 ± 3.05^{ab}
72h(ml)	43.73 ± 1.92^{c}	40.95 ± 2.16^{c}	49.98 ± 1.04^{b}	51.70 ± 5.81^{b}	48.51 ± 3.07^{bc}	56.07 ± 1.25^{a}	55.52 ± 3.57^{a}
a	1.96 ± 0.15^{d}	1.51 ± 0.29^{d}	2.35 ± 0.04^{d}	4.18 ± 0.27^{c}	2.33 ± 0.36^{d}	6.56 ± 0.13^{a}	5.12 ± 0.28^{b}
b	53.13 ± 0.55^{c}	50.62 ± 0.29^{d}	58.72 ± 0.96^{bc}	62.50 ± 0.24^{b}	55.18 ± 0.35^{c}	66.40 ± 0.38^{a}	63.39 ± 0.41^{b}
a+b	55.46 ± 0.15^{c}	52.13 ± 0.33^{d}	60.42 ± 0.56^{bc}	66.68 ± 0.24^{b}	57.04 ± 0.69^{c}	72.96 ± 0.38^{a}	68.51 ± 0.28^{b}
c	0.0405 ± 0.23^{d}	0.0299 ± 0.28^{e}	0.0678 ± 0.34^{bc}	0.0691 ± 0.36^{b}	0.0652 ± 0.67^{c}	0.1034 ± 0.98^{a}	0.0701 ± 0.45^{b}
NDF degradation rate (%)	39.75 ± 1.23^{b}	44.64 ± 1.79^{a}	37.77 ± 0.86^{b}	36.82 ± 0.58^{b}	38.07 ± 5.38^{b}	28.64 ± 0.46^{c}	35.34 ± 0.48^{b}
ADF Degradation rate (%)	20.69 ± 0.92^{ab}	22.55 ± 0.94^{a}	19.86 ± 0.43^{bc}	19.29 ± 0.70^{c}	20.38 ± 1.90^{b}	18.29 ± 0.05^{c}	18.94 ± 0.09^{c}

$^{a, b, c, d, e}$: different letters of data superscripts in same row mean different significantly (P <0.05)

As shown in Table 3, rice straw had a highest a (gas production parameter) value, which mean the rapid degradation component in rice straw could generate most gas production quantity, and more than that of the other roughages (P< 0.05). Gas production of slow degradation component from rice straw was significantly higher than that of bean straw and specific silage corn straw, with lowest value from alfalfa. Meanwhile, rice straw had the highest potential gas production (a+b) and gas production constant (c) value, which were significantly higher than the other 6 roughages ($P<0.05$).

3.3. Correlations between different nutritional ingredients and gas productions with time and 48h NDF, ADF degradation rate

The correlations between different nutritional ingredients and gas productions with time and 48h NDF, ADF degradation rate were shown in Table 4.

Table 4. Correlations between nutrition, gas productions with time and NDF, ADF degradation rate at 48h.

	12h	24h	36h	48h	72h	NDF Degradation Rate	ADF Degradation Rate	a	b	a+b	c
CP	-0.87*	-0.88*	-0.89*	-0.93**	-0.87*	0.96**	0.86*	-0.77	-0.83*	-0.82*	-0.83*
ADF	0.81	0.85*	0.90*	0.88*	0.85*	-0.89*	-0.89*	0.73	0.78	0.76	0.90*
NDF	0.67	0.71	0.70	0.75	0.66	-0.79	-0.66	0.52	0.59	0.58	0.59
NDS/CP	0.95**	0.94**	0.96**	0.93**	0.94**	-0.91*	-0.95**	0.98**	0.97**	0.97**	0.96**
NDS	-0.67	-0.71	-0.70	-0.75	-0.66	0.79	0.66	-0.52	-0.59	-0.58	-0.59
HC	0.52	0.54	0.51	0.58	0.47	-0.63	-0.46	0.35	0.42	0.41	0.36
(NDS-ASH)/CP	0.92**	0.88*	0.92**	0.95**	0.96**	-0.93**	-0.87*	0.92**	0.98**	0.96**	0.91*

*: Correlation, **: Significant correlation.

3.4. Regression model between gas production at 12, 24, 36, 48, 72h of different roughages and components of 400 mg substrate

Absolute amounts of nutrition in carbohydrate and protein according to CNCPS (400 mg) were shown in table 5.

Table 5. Absolute amounts of nutrition in carbohydrate and protein according to CNCPS (400 mg)

	Corn Straw Silage	Alfalfa	Corn Straw	Bean Straw	Chinensis	Rice Straw	Specific Silage Corn Straw
PA(mg)	1.56	1.67	0.98	0.79	1.06	0.38	0.53
PB1(mg)	0.81	1.58	0.52	0.47	0.55	0.23	0.24
PB2(mg)	2.12	2.72	1.39	1.22	1.51	0.64	0.68
CA(mg)	10.41	11.58	6.81	5.77	7.06	2.64	3.36
CB1(mg)	5.20	11.68	3.27	2.97	3.49	1.42	1.50
CB2(mg)	12.52	6.50	15.06	16.34	14.33	24.29	22.47

Multiple regression analysis were conducted between the gas producing ingredients after fermentation, PA, PB1, PB2, CA, CB1, CB2, and gas production after different in Vitro culture durations, and we could inferred the regression model between gas production (Yt, mL) under time t and nutritional ingredients of different roughages, as shown below:

$$Y_{12} = 144.44\ PA + 1844.76PB1 - 114.15\ PB2 - 41.58CA\ - 204.07CB1 + 31.6\quad (R^2 = 0.973\ ,\ n = 5) \quad (13)$$

$$Y_{24} = 165.35PA\ + 1211.32PB1 - 127.76PB2 - 31CA\ - 130.08CB1 + 57.39\quad (R^2 = 0.974\ ,\ n = 5) \quad (14)$$

$$Y_{36} = 79.62PA + 753.99PB1 - 44.58PB2 - 21.33CA - 83.49CB1 + 55.36\quad (R^2 = 0.988\ ,\ n = 5) \quad (15)$$

$$Y_{48} = 40.77PA + 726.42PB1 - 64.63PB2 - 11.26CA - 79.61CB1 + 57.45\quad (R^2 = 0.989\ ,\ n = 5) \quad (16)$$

$$Y_{72} = -15.68\ PA\ - 77.75\ PB1\ - 4.01\ PB2\ + 2.89\ CA\ + 9.02\ CB1 + 62.02\quad (R^2 = 0.991\ ,\ n = 5) \quad (17)$$

The feeds selected for modeling were alfalfa, corn straw, bean straw, chinensis and rice straw. The remained feeds (corn straw silage and specific silage corn straw) were used to validate the model. The results were shown in Table 6.

Table 6. Validation of regression model.

	Corn Straw Silage			Specific Silage Corn Straw		
	Measured Data	Predicted Data	Deviation(%)	Measured data	Predicted Data	Deviation(%)
PA	1.56	1.48	5.13	0.53	0.50	5.66
PB1	0.81	0.82	1.23	0.24	0.24	0.00
PB2	2.12	2.12	0.00	0.68	0.69	1.47
CA	10.41	10.41	0.00	3.36	3.32	1.19
CB1	5.20	5.22	0.39	1.50	1.46	2.67

Deviation =| (data of prediction-data of measure)/data of measuring|×100%

4. Discussions

4.1. Analysis of parameter patterns
In this research, all the parameters, a, b, a+b and c in the model p=a+b (1-e^{-ct}) had real biological implication and showed certain regularity. Parameter a was the intercept on axis Y of the cumulative gas production curve when t= 0, which also represented the instant gas production of rapid degradation components (the soluble component) of feeds; parameter b represented gas production of insoluble component but fermentable components (slow degradation component) in feed; parameter c was the gas production constant during whole fermentation, which also represented the consumption rates of potential digestive components per hour; parameter a+b was the potential gas production volume. The rice straw in table 2 had the highest a+b and c values among seven roughages.

4.2. Correlation between gas production characteristic and ingredients
It was reported the gas production of roughages originated from carbohydrate fermentation [6]. The accumulative gas production of roughages in Vitro fermentation was dominated by the ratio between soluble non-structural carbohydrate and crude proteins [7-8]. The higher ratio infers a higher accumulative gas production and increasing in Vitro fermentation. Tan *et al* [9] inferred an appropriate ratio between non-structural carbohydrate and structural carbohydrate in sheep diet. If so, the ratio could benefit degradation and digestion of fibers in gastrointestinal tract. These results also validated our research.

4.3. Verification of regression model by prediction results between gas production and roughages nutrition

It could be speculated from the regression model that, gas production at different time were mainly depending on contents of PA, PB1, PB2, CA and CB1, less impacted by CB2 in roughages.

The deviation of predicted results from the regression model between gas production and roughages nutrition in table 5 were less than 1.23%, 1.47%, 1.19%, 2.67% for PB1, PB2, CA, CB1, respectively; however, the deviation for PA was 5.66%. The determination of PA was the difference between crude and true protein. The latter was greatly impacted by the poor repetitiveness of quantitative filter paper used.

The CNCPS methodology is manpower and resource consuming [10-11], however the determination time is long. Using in Vitro batch fermentation for modelling rumen to predict gas production for each roughage is simple and practical, and has significant positive correlations with CNCPS methodology for prediction of diet nutrition [12]. Therefore, this research was used in Vitro batch fermentation methodology to determine the gas production of seven roughages at different times. The values from five roughages were adopted to establish the regression model between gas production and diet nutrition, the remained two roughages values were used to validate and calibrate the model, which was beneficial for practical operation.

Whether the regression model established in this research could be used to predict rumen gas production for single or mixed diet still needs further verification by in Vivo results. However, the regression model for sorting nutrition content is practical.

5. Conclusions

This research verified significant correlation between gas production characteristics, NDF, ADF degradation rate for each roughage and CP, NDS/CP, (NDS-ASH)/CP. This verified in Vitro gas production methodology could be used to evaluate the digestibility of nutrient. Combined with fermentation parameters of artificial rumen, this methodology could be used to evaluate diet digestibly accurately, even guide the use of these roughages scientifically.

The regression model developed in this research was time-saving and could be used to predict CNCPS nutrient of each roughage. However, the roughages were often fed in combination with no less than two different roughages, therefore further verification for mixed roughages should be conducted.

Acknowledgements

The research was supported by the Natural Science Foundation of Shandong Province (ZR2017QC009).

References

[1] Menke K, Raab L and Salewski A 1979 *J. Agri. Sci. (Cambridge)* **93**: 217-222
[2] φrskov E and McDonald I 1979 *J. Agri. Sci. (Cambridge)* **92**: 499-503
[3] Van Soest P, Sniffen C and Mertens D 1981 *Stillwater: Div of Agric, Oklahoma State Univ.* MP-109, 265
[4] Krishnamoorthy U, Sniffen C and Stern M 1983 *British J. Nutri.* **50**: 555-560
[5] Sniffen C, O'Connor J and Van Soest P 1992 *Ani. Sci.* **70**: 3562-3577
[6] Ru C, Zan L and Ma S 2006 *Chi. J. Ani. Sci.* **42(11)**: 42-44
[7] Tang S, Jiang H and Zhou C 2005 *Acta Prataculturae Sinica* **14(3)**: 72-77
[8] Zhao G, Yao X and Li Y 2008 *Small Ruminant Res.* **13(2)**: 133-139
[9] Tan Z, Lu D and Hu M 2002 *Asian-Austra. J. Ani. Sci.* **15(11)**: 1591-1598
[10] Hao X, Gao H and Zhang X 2016 *Chi. J. Ani. Nutri.* **28(3)**: 842-850
[11] Kelzer J, Kononoff P and Tedeschi L 2010 *J. Dairy Sci.* **93(6)**: 2803- 2815
[12] Du X,Wang X and Chai S 2016 *Jiangsu Agri. Sci.* **44(1)**: 260-263

Spent mushroom substrates as component of growing media for lettuce seedlings

C J Liu#, Y L Duan#, R Z Jin, Y Y Han, J H Hao and S X Fan[1]

Beijing Key Laboratory of New Technology in Agricultural Application, National Demonstration Center for Experimental Plant Production Education, Beijing University of Agriculture, Beijing 102206, China

#These authors contributed equally to this work

E-mail: fsx20@163.com

Abstract. Spent mushroom substrates (SMS) instead of peat for lettuce seedlings could reduce the costs of natural capital. SMS, peat and vermiculite were combined in various ratios. The effects of different mixed seedling substrates on lettuce seedling growth and nutrient content were investigated. The results show that, with increasing SMS content in mixed substrate, height, root length index of lettuce seedling increased, number of leaves was decreased, phosphorus and potassium content of lettuce seedlings was increased gradually. The root length, leaf number, shoots fresh weight, biomass of 50% SMS in mixed substrate was better than other treatments (T3). The above results indicate that SMS mixed substrate can improve the growth of lettuce seedlings, and 50% SMS in mixed substrate was suitable for lettuce seedlings.

1. Introduction

Growing media were applied apt environment for plant growth, including organic materials such as peat, and inorganic materials such as perlite [1]. Usually, the physical and chemical properties of single substrate were not satisfactory for plant growth, so mixing substrates was popular in soilless culture, such as peat and perlite; peat and compost [2]. Growing media should have large particles with adequate pore space between the particles [3].

Throughout the world, the raw materials used vary based on their limited availability [4]. With the rapid development of soilless cultivation, growing media has become the biggest limiting factor. Peat becomes the most commonly used horticultural substrate, because it has stable physical and chemical properties [5]. However, peat prices are high; the global resource of peat is limited and cannot be updated in a short time. Therefore, many countries began to restrict the exploitation of peat. Researchers use compost, agricultural organic waste, sewer sludge and the like as substitutes for peat.

The annual output of edible mushrooms in China was about 1×10^7 tons (2006), accounting for more than 70% of the world's total output [6]. Each year, a large amount of edible mushroom waste is produced, which accumulates or burns away, which leads to waste of resources and serious environmental pollution. Using the mixed substrate of spent mushroom substrates, the results show that the bulk density and porosity of SMS are within the appropriate range of nursery substrate, and the total nitrogen (N), phosphorus (P), potassium (K) content was significantly higher than that of peat mixed substrate [7]. The use of SMS not only addresses the issue of waste of resources in the production of mushrooms, but also effectively reduces costs. Mushroom residue matrix as the mushroom cultivation process has been fully decomposed, the composition of the structure tends to be stable, and the structure of the mushroom residue was granular analogous to soil aggregate structure. Mushroom residue is a good substrate to replace peat, but due to the conductivity (EC) of mushroom residue substrate, the pH value is too large, which is not conducive to the nursery. In this study, the use of mushroom residue and peat physical and chemical properties of the experimental analysis of the mushroom residue matrix was improved. Peat pH, EC value is too small and mushroom matrix mixed, you can learn lessons from each other. Using mushroom residue, peat and vermiculite, five treatments were established up to determine the effects of physical and chemical properties, seed germination index and morphological growth and nutrient content of lettuce seedlings, slag organic substrate. Despite this, there is little research on their use as a growing media. Therefore, the main purpose of this experimental study was to mix commercial peat and various concentrations of spent mushroom substrates bases into a series of mixtures and to study the properties and effects of the mixtures, thereby linking the growth and nutrient content of the mixtures and lettuces of different concentrations.

Lettuce (*Lactuca sativa L.*), belonging to the family Compositae, is a one or two-year-old herb; its common name is lettuce [8], is a widely grown and popularly consumed vegetable worldwide. Lettuce is a low-sugar, low-fat vegetable that is not only rich in vitamin C, vitamin E, calcium, phosphorus, iron and other minerals and vitamins but also contains malic acid, tartaric acid, lactic acid, mannitol and the active thyroid hormone; thus, it has a high nutritional value [9]. The stem and leaf extract of lettuce can enhance gastric juice, stimulate digestion, increase appetite, and has a hypnotic analgesic effect. In addition, lettuce extract can lower cholesterol, assist in the treatment of neurasthenia, and have other effects, such as the risk of chronic diseases such as cancer, cardiovascular disease, and aging-related functional decline. As a result, lettuce has become a popular vegetable, and its productivity has increased every year [10]. Epidemiological studies have shown that there is a correlation between increasing the consumption of vegetables and reducing the risks of chronic diseases such as cancer, cardiovascular disease and age-related functional decline. Lettuce has a shallow root, a larger leaf area, and does not tolerate drought. It has very high moisture content, drought conditions will significantly decrease its quality and yield, causing major economic losses. These health benefits are considered as related to macronutrients, micronutrients and bioactive compounds in vegetables. The use of spent mushroom substrates as a component of the hybrid substrate for lettuce seedlings growth instead of peat was studied.

2. Materials and methods

2.1. Experimental material

Spent mushroom substrates were mixed with commercial peat in unlike proportions. Among them, vermiculite was bought on the market and SMS was fermented from the plant medium after mushroom cultivation. T5 with peat (66.7%) and vermiculite (33.3%) was verified as a control (CK). To prepare the substrates, the residues were dried at 45 °C in an air forced oven for 48 h and then they were ground to 5 mm to obtain homogeneity. The variety of lettuce was 'Gemini', lettuce seedlings was planted in plug seedlings in greenhouse, Beijing University of Agriculture. The volume ratio of spent mushroom substrates was screened based on the lettuce seedlings morphological growth and nutrient content.

The seeds of lettuce were sown on different growth media in a dish and one seed was sown per cell. The treatment of this experiment (growth medium) was formed in a completely randomized design with three replicates per treatment. Germination is held in a light incubator at 27±1°C and 90-95% relative humidity. When the seedlings reached the size of the viable transplant, respectively, the plants were harvested and the aerial parts of all the plants were weighed to determine the fresh weight. The plants were then washed, dried (in a forced air oven at 60°C for 72 hours) and ground to 0.5 mm for analysis.

Six treatments were established for the experiment as Table 1. T1 was full of SMS; T2 was consisted of 75% SMS and 25% peat; T3 included SMS and peat each accounted for half; T4 was consisted of 25% SMS and 75% peat; T4 included One hundred percent of peat; 66.7% peat and 33.3% vermiculate was set up as control (CK).

Table 1. Proportion of for experimental seedling substrates (percentage by volume).

Treatment	SMS (%)	Peat (%)	Vermiculite (%)
T1	100	0	0
T2	75	25	0
T3	50	50	0
T4	25	75	0
T5	0	100	0
CK	0	66.7	33.3

2.2. Morphological growth of lettuce seedlings
When the seedlings reached a commercial transplant size, 15 seedlings were randomly harvested from each treatment at 45 days of lettuce. In the stem, the seedling height (H) was evaluated from the root collar to the shoot tip and measured using a ruler; (LFW) and fresh weight (RFW) of the fresh weight (FW). Finally, the seedlings were placed in an air forced oven at 105 celsius degrees. Dry for 24 hours to determine the dry weight (DW), including dry aboveground weight (SDW) and root dry weight (RDW) [11].

2.3. N, P and K content of lettuce seedlings
Some chemical properties were assessed: N using the Dumas method [12], phosphorus as per Murphy and Riley [13]. Assimilable elements were assessed in 1:5 (v/v) suspension, using water to extract N and P, ammonium acetate to extract K. In the corresponding extracts, N was measured by refl optometry in RQflEx 10 colorimeter by Merck, phosphorus as per Murphy and Riley (1962). All elemental analyses were repeated three times.

2.4. Data processing and analysis
For all data, homogeneity of variance assumptions were tested utilizing a Levene's test. If assumptions were met, an ANOVA and Tukey post-hoc test were utilized. The statistical analysis was based on the one-way analysis of variance of the mean values of each parameter for each treatment (substrate), in order to test the statistically significant differences, to compare the differences between specific treatments, the Tukey-b test at $P < 0.05$ were used. Analysis was performed using SPSS.

3. Results and analysis

3.1. Effects of different SMS mixed substrate on morphological indicators of lettuce seedlings.
As can be seen from Table 2, the overall trend showed that with peat content reducing, the plant height, root length index was gradually increased. With the increasing of peat content, the content of SMS gradually decrease, and the number of leaves was gradually increased. The difference of stems in each treatment was not significant.

The root length, leaf number, above ground fresh weight, whole plant dry weight and other indicators of T3are better in all treatments. The fresh weight, root fresh weight and root fresh weight of different treatment seedlings was different, but fresh weight and root dry weight decreased with the increase of peat content, root weight, roots fresh weight increasingly. With the decreasing of the content of SMS, the content of peat increasing, the fresh weight of shoot and the dry weight of shoot slightly increased, the same trend of change was shown.

Table 2. Effects of different SMS mixed substrate on the morphological indicators of lettuce seedlings.

Treatment	H(cm)	LN	RL(cm)	D(mm)	SFW(mg)	RFW(mg)	SDW(mg)	RDW(mg)	DW(mg)
T1	12.3ab	2.7c	1.8b	0.81a	173.65b	6.80a	5.42b	0.27a	5.68b
T2	13.6a	3.2ab	1.5bcd	0.85a	306.02ab	11.85a	9.77ab	0.43a	10.20ab
T3	13.2a	3.3ab	3.3a	0.81a	376.87ab	11.15a	12.20a	0.58a	12.78a
T4	13.2a	3.0bc	1.3cd	0.85a	368.57ab	10.27a	10.43ab	0.40a	10.83ab
T5	11.5b	3.5a	1.1d	0.86a	294.72ab	8.90a	10.50ab	0.37a	10.87ab
CK	13.2a	3.3ab	1.6bc	0.80a	390.23a	9.45a	13.05a	0.43a	13.48a

Note: Different letters showed significant differences ($P<0.05$)

3.2. Effects of different SMS mixed substrate on nutrient contents and accumulation of lettuce seedlings

Macronutrient content in the graft is statistically affected by the type of waste and the proportion of the mixture. Nitrogen, phosphorus and potassium are critical nutrients for plant growth. The supply status of N, P and K directly affects the yield and quality of crops and is the major limiting factor in crop production. As can show in Table 3, phosphorus content and potassium content in lettuce seedlings increased with the increase of SMS. Because the accumulation of lettuce seedlings was equal to the elemental content of lettuce seedlings multiplied by the whole plant dry weight, change range of the dry weight of each treatment was less than that of the nutrient elements in lettuce seedlings. However, dry weight changes in the same tendency.

Table 3. Effects of different SMS mixed substrate on the nutrient contents and accumulation of lettuce seedlings.

Treatment	Nitrogen		Phosphorus		Potassium	
	Concentration($g.kg^{-1}$)	Accumulation($mg.Plant^{-1}$)	Concentration($g.kg^{-1}$)	Accumulation($mg.Plant^{-1}$)	Concentration($g.kg^{-1}$)	Accumulation($mg.Plant^{-1}$)
T1	2.69b	15.30b	8.93ab	50.81b	5.78a	32.65b
T2	3.08ab	31.44ab	9.68a	98.71ab	5.91a	60.26ab
T3	2.77b	35.43ab	8.86ab	113.44a	5.11b	65.14a
T4	3.31a	36.21ab	8.67b	93.96ab	4.31c	46.70ab
T5	2.97ab	32.33ab	8.61b	92.91ab	4.02cd	43.89ab
CK	3.12ab	42.29a	9.30ab	125.14a	3.72d	50.19ab

The nutrient contents of lettuce were influenced by the proportion of SMS mixing substrate. K contents of SMS mixing substrate were greater than CK. In addition, N and P content of SMS mixing substrate was similar to CK, and no significant differences were found among the different treatments.

4. Discussion and conclusion

Peat is the most commonly used horticultural cultivation substrate, but the high price, limited resources, a lot of mining will do harm to the environment. Mushroom residue as edible fungus waste, its processing, can be a superior cultivation substrate.

Liu et al. [14] showed that the use of vinegar residue as organic substrate should be combined with not less than 25% peat and should be safe for vegetables, nurseries and cultivation. The results showed that seedling root length, shoot fresh weight, dry biomass and other indicators of 50% SMS mixed substrate were preferable to other treatments. It was not only can reduce costs and improve the quality of lettuce seedlings for 50% SMS mixed substrate instead of peat substrate for lettuce nursery. It is a good alternative to peat organic substrate mixed with SMS. From the results obtained, it can be concluded that the medium containing SMS had sufficient physical properties and significant phytonutrient content. Regarding the most suitable substrate for plant growth, any substrate refined with waste was found to be ample for lettuce seedling production. Therefore, all of these conclusions indicate that the use of discarded mushroom matrices in professional horticulture serves to treat them in an environmentally friendly manner while reducing the need for peat.

Acknowledgment
This research was financially supported by the National Key Research and Development Program of China (2016YFD0201010) and the earmarked fund for Beijing Innovation Consortium of Agriculture Research System (BAIC07-2018).

References

[1] Grunert O, Perneel M and Vandaele S 2008 Peat-based organic growbags as a solution to the mineral wool waste problem. *Mires & Peat*, **3**: 1-5
[2] Nair A, Ngouajio M and Biernbaum J 2011. Alfalfa-based organic amendment in peat-compost growing medium for organic tomato transplant production. *Hortscience*, **46**: 253-9
[3] Albano J P, Warren S L, Bilderback T E and Jr J S O 2005. Healthy substrates need physicals too!. *Horttechnology*, **15**: 747-51
[4] Schmilewski G. 2009. Growing medium constituents used in the eu. *Acta Horticul.e*, **819**:33-46
[5] Pane C, Spaccini R, Piccolo A, Scala F & Bonanomi G 2011. Compost amendments enhance peat suppressiveness to pythium ultimum, rhizoctonia solani and sclerotinia minor. Biolog. Control, **56**: 115-124
[6] Li X Q, Bu C X and Guo S R 2006 Effects of compound substrate of mushroom residue on growth of some vegetable seedlings. *J. Shenyang Agri. Uni.*, **3**: 517-20. DOI: 10. 3969/ j. issn. 1000- 1700. 2006. 03. 063
[7] Gruda N 2012. Do soilless culture systems have an influence on product quality of vegetables?. *J. Applied Bot. Food Qual.*, **82**: 141-7
[8] Martindiana A B, Rico D, Frias J, Gtm H, Mulcahy J and Barat J M, *et al.* 2006. Effect of calcium lactate and heat-shock on texture in fresh-cut lettuce during storage. *J. Food Engineer.*, **77**: 1069-77
[9] Pepe G, Sommella E, Manfra M, De N M, Tenore G C and Scopa A, *et al.* 2015. Evaluation of anti-inflammatory activity and fast uhplc-dad-it-tof profiling of polyphenolic compounds extracted from green lettuce (lactuca sativa l.; var. maravilla de verano). *Food Chem.*, **167:** 153-161
[10] Vanisree M, Mathieu N, Ajay N, Zhang Y, Adityal G and Muraleedharang N 2010. In vitro evaluation of red and green lettuce (lactuca sativa) for functional food properties. *Food Chem.*, **118**: 300-6
[11] Ceglie F G, Bustamante M A, Ben A M and Tittarelli F 2015. The challenge of peat substitution in organic seedling production: optimization of growing media formulation through mixture design and response surface analysis. *Plos One*, **10**: e0128600
[12] Walls G L 1943. The vertebrate eye and its adaptive radiation. *Ameri. J. Ophthalmo.*, **26**: 204-5
[13] Bremner J M and Breitenbeck G A 1983. A simple method for determination of ammonium in semimicroâ kjeldahl analysis of soils and plant materials using a block digester1. *Communi. Soil Sci. Plant Analysis*, **14**: 905-913
[14] Liu C J, Guo S R, Wang C Y, Shu S, Liu S R and Cheng Y J 2010. Vinegar residue substrate as component of mixed substrate for pepper seedling growth. *Acta Horticul. Sinica*, **37**: 559-66

IOP Conf. Series: Earth and Environmental Science **185** (2018) 012017 doi:10.1088/1755-1315/185/1/012017

Nicotine content of tobacco leaf estimated by UV spectrum

X N Wei, Y Liu and Y L Tang[1]

College of Physics, Guizhou University, Guiyang 550025, China

E-mail: tylgzu@163.com

Abstract: The spectral data of different grades flue-cured tobacco sample have been determined by UV spectrometer. Their differences were compared. The spectral prediction models of nicotine content were established through correlation analysis between the content of nicotine and spectral data of flue-cured tobacco with the stepwise multivariate regression method. The statistical results showed that the nicotine content of flue-cured tobacco were very significantly correlated to their UV spectrum. There were also very significantly correlations between the predicted and experimental values. This indicates that it would be feasible to estimate the nicotine content in flue-cured tobacco by UV spectroscopy and UV spectroscopy can be used to determine conveniently and quickly nicotine content of flue-cured tobacco leave samples without pollution.

1. Introduction

The nicotine, one of the main chemical compositions in tobacco leaf, has the function of excited stimulation to the human body. Too much nicotine can harm the health of smokers. The near infrared (NIR) technology has been widely applied to rapidly analysis of tobacco chemical composition, tobacco leaf classification, production directions analysis, cigarette real false identification, etc [1-6]. But, the UV spectrum method is relatively less applied to study of tobacco leaf. Applying NIR to forecast cellulose content in the tobacco leaf [7] and flame atomic absorption spectrometer to continuous determination method of Cu, Fe contents in tobacco have been used [8]. UV spectrometer has been applied to measure nicotine content of tobacco leaf [9]. The near infrared reflectance spectroscopy (NIRS) has been used to analyze the protein content in flue-cured tobacco leaves. The mathematical model of NIRS for the protein content in flue-cured tobacco leaves has been established with stoichiometry method [10]. The near infrared spectroscopy has forecasted the tobacco leaf structure and oil [1]. The spectral method can forecast the tobacco leaf nicotine, K and total N, so as to establish spectra prediction models [11]. A mathematical prediction model of Ca in tobacco with FT-NIR spectrometry has been established [12-13]. 4,8,13-duvatriene-1,3-diol diastereoisomers have been identified in the cuticular wax of fresh burley tobacco leaves [14]. A reliable and rapid method based on high-performance liquid chromatography (HPLC-UV) and positive ion electrospray-time of flight mass spectrometry (ESI-TOF/MS) have been developed for the characterization and quantification of solanesol in extracts of tobacco leaves from different sources [15]. A mathematical model of nicotine in total particulate matter has been established by scanning tobacco power with FT-NIR spectrometry [13]. A new dual-excitation fluorescence light detection and ranging (DE-FLIDAR) has been designed for remote quantitative assessment of epidermal UV absorption of tobacco leaves and canopies from chlorophyll (Chl) fluorescence (ChlF) measurements [16]. Nicotine has been extracted from tobacco by ultrasonic extraction method [17]. Determination of nicotine content in tobacco specifically by a reverse phase HPLC has been established [18]. Based on the linear relationship of nicotine content on the short wave absorption, the difference of different grades of tobacco leaf spectrum was compared, and the nicotine content of tobacco leaf was measured. A method of UV spectroscopy is discussed the feasibility of the determination of the content of nicotine in tobacco in this paper.

2. Materials and methods

The sample of flue-cured tobacco is K326 from the Tobacco Company of Guiyang in Guizhou Province (respectively, top, middle, next three part). The sun-cured tobacco came from market. The spectra and chemical component of samples were measured at Guizhou University.

The UV spectroscopy is American Agilent 8453E UV Spectroscopy system with wavelength from 190 nm to 1100 nm and slit width 1 nm. Before measurement, the tobacco leaf sample was dried to constant weight at 80 ℃ constant temperature in a constant temperature box. It was ground and sieved through 200 μm mesh. The 4 mg sample was placed in a 10 ml cuvette; which was joined 80% ethanol dissolution and vibrated approximately 1 min to dissolve completely. The solution was washed and filtered to a 25 ml capacitance flask with 80% ethanol. Then, the sample solution was put into quartz vessel to measure the UV spectrum from 190 nm to 1100 nm with Agilent 8543E UV spectroscopy. Before each spectral measurement, the instrument was adjusted with 80% ethanol. The characteristic absorption peak was in the spectrogram wavelength 210~360 nm. The nicotine contents were measured by chemical method.

3. Experimental results and analysis

3.1. The different grade flue-cured tobacco and sun-cured tobacco spectral analysis

In wavelength 210~360 nm, the information of the flue-cured tobacco was very abundant, but signal intensity was quite high. The spectral difference of different grade flue-cured tobacco sample was shown in Figure 1. The spectra of the same leaf position for flue-cured tobacco were also very different. The spectra represent the information of pyridinium, pyrrole, adenine, etc. Different grade sun-cured tobacco sample has different spectrum, showed in Figure 2. This explains what was flue-cured tobacco or sun-cured tobacco, because different grade sample had different nicotine content. The highest absorbance, the most nicotine content the tobacco contains. From Figure 2 and Table 1, we may see differences of nicotine content and spectral curve in different sun-cured tobacco samples. The SY-1 indicated the high nicotine content, while the SY-2 indicates the low nicotine content.

Table 1. Nicotine content of flue-cured tobacco and sun-cured tobacco.

Sample	Flue-cured tobacco										Sun-cured tobacco	
Tobacco leaf grade	B3F	B3L	C2F	C3F	C3L	C4F	X2F	X2L	X3F	X3L	SY-1	SY-2
Nicotine content (10%)	3.01	2.00	2.68	1.81	1.83	1.77	1.75	0.84	1.33	1.23	4.47	2.55

3.2. Correlation analysis of ultraviolet spectrum and nicotine content of flue-cured tobacco

It has been separately calculated for correlation coefficient between the spectrum of flue-cured tobacco of different grade and its nicotine content (%). The conclusion indicated that there is a significant correlation between the spectra of flue-cured tobacco of different grades and their nicotine content (%). By using stepwise regression equation in multivariate analysis, the best wavelength band and variable dimension of regression equation were chosen to fit the best mathematical model. The best wavelength band was 210~360 nm, and the best regression equation variable was two variables. From Table 2, the correlation coefficient was 0.8978.

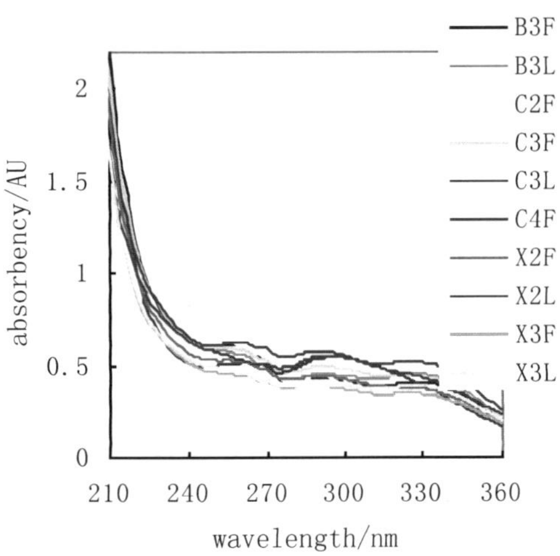

Figure 1. Ultraviolet absorption of 10 kinds of different samples flue-cured tobacco.

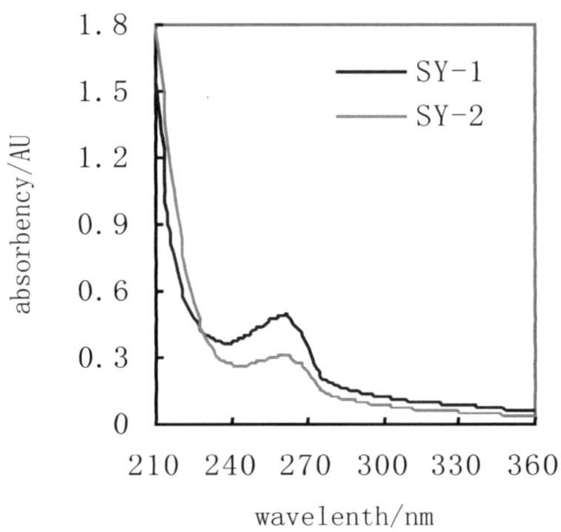

Figure 2. Ultraviolet absorption of 2 kinds of different samples sun-cured tobacco.

Table 2. Stepwise multiple regression equation between nicotine content (%) and flue-cured tobacco leaf spectra.

Sample	Regression model	R^2	F(=MSR/MSE)
Flue-cured tobacco	$Y = -2.5365+2.9584x_1-3.3598x_2$	0.8978^{**}	30.7359

Note: ** indicates that $p < 0.01$ significance test has been passed.

Thirteen samples were extracted in flue-cured tobacco leaf samples. This UV spectral examination model was carried out to examine their nicotine content. These values were compared to UV spectral calculated value. The conclusion demonstrated that two sets of data were very identical, showing that UV accuracy approaches the chemical method very much. From Table 3, it can be seen that the nicotine content of flue-cured tobacco can be predicted by multivariate stepwise regression method according to their ultraviolet absorption spectra. The prediction accuracy is more than 90%.

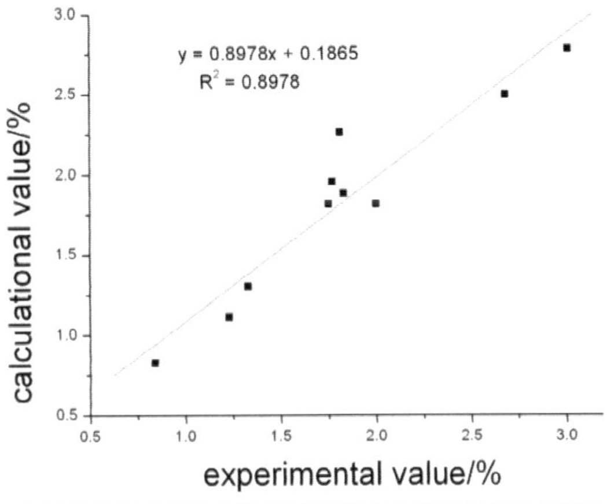

Figure 3. Comparison of predicted value and measured value of the nicotine content of flue-cured tobacco leaf.

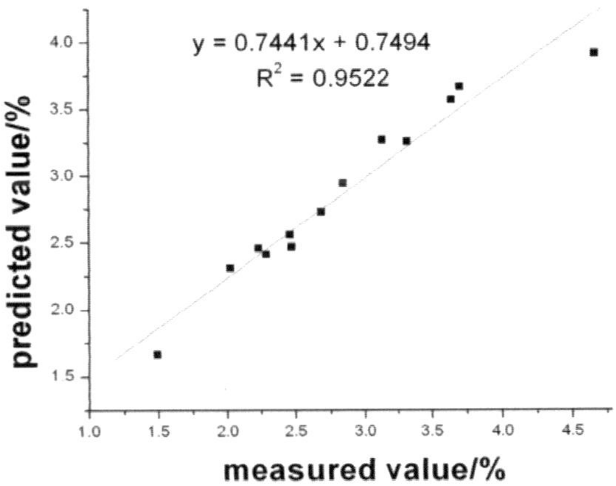

Figure 4. Comparison of predicted value and measured value of the nicotine content of flue-cured tobacco leaf.

Table 3. Comparison of the predicted value of the ultraviolet absorption spectrum and measured value of nicotine content (%) in flue-cured tobacco.

Sample number	Measured value	Predicted value	Absolute error	Relative error (%)
1#	2.29	2.41	0.12	5.24
2#	2.02	2.31	0.29	14.36
3#	3.64	3.57	0.07	1.92
4#	3.32	3.26	0.06	1.81
5#	3.70	3.66	0.04	1.08
6#	4.66	3.91	0.75	16.09
7#	3.13	3.27	0.14	4.47
8#	2.46	2.56	0.10	4.07
9#	2.84	2.94	0.10	3.52
10#	2.23	2.46	0.23	10.31
11#	1.49	1.66	0.17	11.41
12#	2.47	2.47	0.00	0.00
13#	2.68	2.73	0.05	1.87

$$\text{Absolute error} = \left| predicted\ \ value - measured\ \ value \right|,$$

$$\text{Relative error} = \left| \frac{predicted\ \ value - measured\ \ value}{measured\ \ value} \right| \times 100\%$$

4. Conclusions

The UV spectroscopic analysis of was developed to set up the spectral examination model of nicotine content of flue-cured tobacco. The results of the model were consistent with the experimental data. UV spectroscopic analysis for nicotine content of flue-cured tobacco sample has advantages of simple operating and quick survey speed for massive samples.

The spectral detection model proposed in this paper can accurately predict the nicotine content in flue-cured tobacco, but it requires more samples to improve its accuracy.

Acknowledgments

This work was supported by the National Natural Science Foundation of China (Grant numbers 11164004), Guizhou industrial breakthrough project (Grant numbers GY [2012]3060) and Nomarch Foundation of Guizhou Province.

References

[1] Zhou H P, Wang X M, Song J Z, Cai X J, Xi J Q and Yin Q S 2006 Prediction of structure and oil of flue-cured tobacco with near-infrared reflectance *Tobacco Sci. Tech.* **2006**(1): 10-4

[2] Qin Z Q, Cai S S, Xie H, Si H, Zhan J L and Yan T J 2007 Prediction of total nicotine, total sugar, reducing sugar and total nitrogen in tobacco with AOTF-NIR *Tobacco Sci., Tech.* **2007**(2)30-4

[3] Fu Q J, Wang S S, Dou Y Q and Liu G L 2006 Near-infrared analysis of nitrogen, potassium, calcium, and magnesium in tobacco root *Tobacco Sci. Tech.* **2006**(10) 35-37

[4] Jiang J F, Li L and Zhao M Y 2006 Rapid determination of main chemical components in tobacco leaf by NIR diffuse reflectance spectroscopy *Acta Tabacaria Sinica* **12** (2) 8-12

[5] Shao Y N, He Y and Wang Y Y 2007 A new approach to discriminate varieties of tobacco using vis/near infrared spectra *Eur Food Res. Tech.* **224** 591-6

[6] Zhang Q J, Tian K D, LI Z H, Lyu Y Q, Xiong Y M and Min S G 2014 A study of recognition of tobacco aroma style using near infrared spectroscopy *Modern Instru. Medi. Treat.* **20**(4) 66-8

[7] Duan Y Q, Kong X Y, Li Q Q, Tang D Y and Liu W 2006 Prediction of cellulose content in tobacco with near infrared spectroscopy *Tobacco Sci. Tech.* **2006**(8) 16

[8] Lin X F, Yang S Y, Lin J and Zhang Q J 2003 Continuous determination of Cu , Fe in tobacco by flame atomic absorption spectrometry *J. Qujing Normal College* **22**(3) 14-6

[9] Chen S Z 2003 The nicotine measurement in tobacco leaf by UV-spectrophotometer *EIC* **10**(6) 16-7

[10] Zhou S P, Li W H, Cheng G M and Tian X X 2004 The mathematical model of near infrared reflectance spectroscopy analysis for protein content in flue-cured tobacco leaves *Guizhou*

Agri. Sci. **32**(6) 25-7

[11] Liu L Y, Jin Z W, Wang J H, Yang T Z, Huang W J and Zhao C J 2005 Prediction of nicotine, potassium and nitrogen in tobacco with spectrometry *Tobacco Sci. Tech.* **2005**(6) 26-9

[12] Qiu J, Wang Y B, Fu Q J, Zhang H B and Wang X L 2006 Rapid determination of calcium in tobacco with Near Infrared spectroscopy *Tobacco Sci. Tech.***2006**(2) 31-2

[13] Qiu J, Wang Y B, Zhang H B, Fu Q J, Guo X F and Ai H L 2006 Prediction of nicotine in total particulate matter by near infrared spectroscopy *China Tobacco Sci. Tech.* **2006**(2) 12-3

[14] Chang S Y 1976 Duvatriendiols in cuticular wax of Burley tobacco leaves *J. Lipid Res.* **1976**(17) 7-11

[15] Chen J H, Liu X P, Xu X Q, Lee F S C and Wang X R 2006 Rapid determination of total solanesol in tobacco leaf by ultrasound-assisted extraction with RP-HPLC and ESI-TOF/MS *J. Pharmaceutical Biomed. Analy.* 2006(9) 1-7

[16] Ounis A, Cerovic Z G, Briantais J M and Moya I 2001 Dual-excitation FLIDAR for the estimation of epidermal UV absorption in leaves and canopies *Remote Sens. Envir.* **76**(1) 33-48

[17] Chu Z B, Zhou X G, Shui H F and Chen J M 2007 Determination of nicotine content from tobacco by HPLC *Tech. Develop. Chem. Indus.***36**(2) 46-8

[18] Chu Z B, Zhou X G, Shui H F and Chen J M 2006 Determination of nicotine content in tobacco by a reversed phase-high performance liquid chromatography(RP-HPLC) *J. Agro-Envir. Sci.* **2006**(25) 751-4

IOP Conf. Series: Earth and Environmental Science **185** (2018) 012018 doi:10.1088/1755-1315/185/1/012018

Preparation and application of chelating multi-metal chitosan selenite

X H Zeng[1,2,3], Z Wu[2], Y C Liu[2], S M Wen[2] and H M Wang[1,2,4]

[1] Affiliated Dongfeng Hospital, Hubei University of Medicine, Hubei 442008, P.R. China;

[2] Hubei Key Laboratory of Wudang Local Chinese Medicine Research, School of pharmacy, Hubei University of Medicine, Hubei, 442000, P.R. China;

[3] Wudang Animal Pharmacy Co., Ltd. Hubei, 442100, P.R. China;

E-mail: meirwang@126.com

Abstract. [Objective] The aim is to introduce a production processes for the preparation of chelating multi-metal chitosan selenite and its different dosage forms, which is suitable for the fields of food and feed chemistry, fertilizer chemistry or fine chemical technology, and also belongs to the field of biomedical engineering technology. [Method] Disperse chitosan in water, and stir it in water bath at 50-70°C under heating. Add trace elements and rare earth metal salts with matching ratio to make them dissolve, where ratio of chitosan and metal salts is 1:0.1-0.3 (W/W). After 2-5 hours of stirring, chitosan chelated by multi-metal can be obtained. Chelating multi-metal chitosan selenite can be made after adding selenium oxide aqueous solution. [Result] The same molecules of target products produced according to the process contain a variety of metal elements and selenium elements and have biological synergistic effect with its biological solubility and biological activity significantly improved. [Conclusion] Chelating multi-metal chitosan selenite presented in this paper is a kind of trace element and/or rare earth preparation, which is environmentally friendly, easy to absorb and rich in selenium without toxic side effects. Therefore, it could be used as a new preparation to supplement selenium and trace elements in medicine, health care products, cosmetics and agricultural fields.

1. Introduction

With the improvement of life quality, more attention was attracted by the relationship between trace elements and health. Selenium is more prominent and enjoys reputation of “Fire of Life”, “Guard Angle of Heart” and “King of Anti-cancer”. Medical studies have shown that low selenium intake is associated with heart and cellular diseases and with high incidence of breast, thyroid, prostate, lung, and colon cancers [1]. WHO stipulated that 50-250 micrograms of selenium should be contained in daily diet of human beings. Unfortunately, China is a big country lacking selenium, and daily selenium intake of most people per day is only about 50 micrograms [2]. Inorganic selenium has the advantages of high selenium content and low price, but absorption and utilization rate of inorganic selenium (such as Na_2SeO_3) is not ideal with low bioavailability, heavy toxicity. Besides, there is a very small gap between the poisoning amount and requisite amount; therefore, its use should be strictly limited.

Molecular structure of chitosan [3] [shortened as CTS, molecular formula as $(C_6H_{11}O_4N)_n$, also named as chitosan, scientific name as β(1,4)-2-amino-2-deoxy-D-glucose] is as follows:

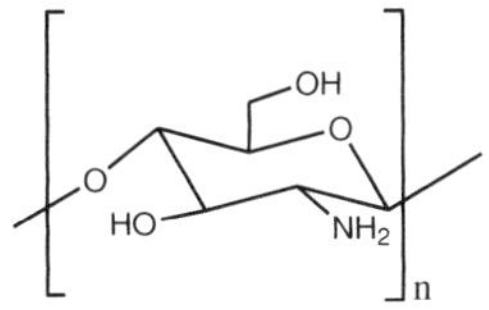

Chitosan, which has no damage to the environment, people and livestock, is able to regulate plant growth, induce plant disease resistance and improve yield and quality [4-5]. Therefore, it is generally accepted that chitosan will become the leading product for the development of green agriculture in 21st century.

Chitosan molecules contain a large amount of hydroxyl and amino groups, which are good ligands to metal ions. Many researches showed that chitosan chelating metal could significantly enhance antioxidant and antibacterial activities of chitosan, and improve biocompatibility of metal ions and the absorption and utilization rate *in vivo* [4]. It has potential application prospects in industry, agriculture, medicine, food and so on. The two parallel molecules interact with each other through chelating between chitosan and metal at 2-NH_2 and 3-OH of glucose monomer. Chitosan, as a macromolecular compound, can have as high as tens of thousands or even hundreds of thousands of molecules and many glucose monomers that can connect many metal ions [6].

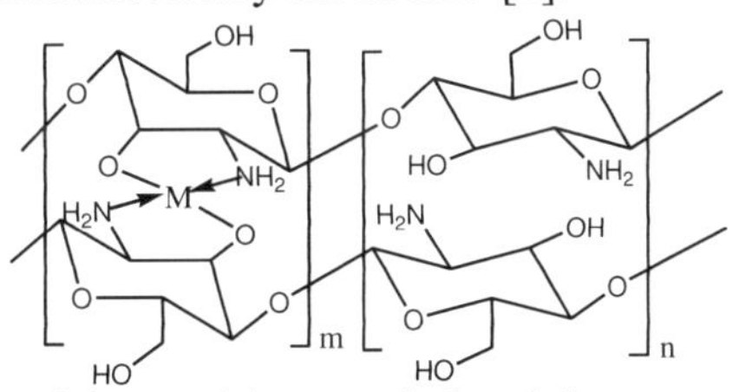

If we add non-monolithic metal ions, chitosan chelated by non-monolithic metal ion could be obtained[5-12]. Chitosan chelated by non-monolithic metal ions is different from compound chelated by the monolithic metal ions. However, a variety of metals exist in the same molecule; with their synergistic effects, its antioxidant & antibacterial activity, the biological compatibility of metal ions and *in vivo* rate of utilization and absorption will be greatly improved. 2-NH_2 on the glucose monomers not coordinated in the chitosan molecules and seleninic acid are combined to produce salt, which can improve the effective absorption of selenium, increase its antioxidant and anti-tumor activity, and will not produce toxicity of inorganic selenium [13-16]. Therefore, it can be used as a new kind of selenium supplement in the fields of medicine, health care, cosmetics and agriculture, etc.

This paper aims to provide a method for preparing a chelating multi-metal chitosan selenite and different dosage forms. The chelating multi-metal chitosan selenite is an environmentally friendly preparation, free of toxic and side effects, easy to absorb, rich in selenium, trace elements and/or rare earth [13]: chitosan selenite chelated by trace element can be used as a new class of preparation supplementing selenium and trace elements in the field of medicine, health products, cosmetics and agriculture[8]. For chelated chitosan selenite containing rare earth and trace elements, enzyme activity of rare earth can be used as feed additives to significant gain weight of fish, shrimp and poultry and can improve the animal's anti-bacterial and anti-virus ability [7]. High concentration of multi-element liquid fertilizer prepared by chelated chitosan selenite containing rare earth and trace elements can enhance the vitality of animals and plants with significant effect of fertilizer.

2. Materials and methods

The production processes for the preparation of chelating multi-metal chitosan selenite and its different dosage forms is as follows:

Disperse chitosan in water, stir it in water bath at 50-70 °C under heating. Add trace elements and rare earth metal salts with matching ratio to make them dissolve, where ratio of chitosan and metal

IOP Conf. Series: Earth and Environmental Science **185** (2018) 012018 doi:10.1088/1755-1315/185/1/012018

salts is 1:0.1-0.3 (W/W). After 2-5 hours of stirring, chitosan chelated by multi-metal can be obtained. chelating multi-metal chitosan selenite can be made after adding selenium oxide aqueous solution.

Chitosan (Degree of deacetylation 99%) was provided by Shanghai Zhanyun Chemical Co., Ltd. (Shanghai, China). Another chemical reagent (AR) was provided by Sinopharm chemical reagent Co., Ltd. (Shanghai, China).

*2.1. Preparation of chelating multi-metal chitosan **1a***

Firstly, the chitosan (10g) is dispersed in water (1000mL) and stirred at 50-70℃. 3g salts, including $CaCl_2$, $FeCl_2$ and $ZnCl_2$ with appropriate ratio that is 10: 1: 1 (W/W), are added to the solution of the chitosan. Centrifugation is conducted after stirring for 2-5 hours. The supernatant is discarded to obtain chitosan crude chelated by multi-metal. The crude product is washed with distilled water until the supernatant is colorless. The pure product is obtained after being dried.

The product of chelating multi-metal chitosan **1a**, which is easily absorbed and metabolized, can be used in medicine, health products, cosmetics and agriculture and other fields. Synergistic absorption of three ions of calcium, iron and zinc in the same molecule is better than the absorption effect of single ion preparation, and a variety of ions do not affect absorption due to mutual restraint because it is a simple compound preparation.

*2.2. Preparation of chelating multi-metal chitosan selenite **2a***

Chelating multi-metal chitosan **1a** is added to an aqueous solution of seleninic acid, wherein the ratio of chitosan chelated by calcium iron zinc to seleninic acid is 1:0.01-0.1 (W/W). The aqueous solution of chelating multi-metal chitosan selenite is obtained after stirred for 1.5-3, and the mass concentration is 10-45%. Ethanol is added to precipitate, filter and wash with ethanol, then dry to obtain the pure product of chelating multi-metal chitosan selenite **2a**.

*2.3. Preparation of chelating multi-metal chitosan selenite **2b***

The steps are the same as point 2.1, while salts including calcium iron zinc are replaced with salts of calcium magnesium iron zinc manganese rare earth. It consists of $CaCl_2$, $MgCl_2$, $ZnCl_2$, $FeCl_2$, $MnCl_2$ and rare earth chlorides including $LaCl_3$ and $CeCl_3$, whose mass ratio is 10: 2: 1: 1: 0.1: 0.1 (W/W). The other steps are the same, so chelating multi-metal chitosan **1b** is obtained. The steps are the same as in point 2.2, the pure product of chelating multi-metal chitosan selenite **2b** is obtained.

*2.4. Preparation of water quality suspension agent of chelating multi-metal chitosan selenite **2b***

The pure product of chelating multi-metal chitosan selenite **2b** is dispersed in water to obtain an aqueous solution of 10-45%. Xanthan gum and diatomaceous are added in proportion, of which the ratio of chitosan selenite chelated by calcium, magnesium, iron, zinc, manganese and rare earth, xanthan gum and diatomaceous earth is 10: 1: 1 (W/W). The water quality suspension agent of chitosan selenite chelated by calcium, magnesium, iron, zinc, manganese and rare earth is obtained after being stirred for 2 hours.

3. Results and discussion

Chelating multi-metal chitosan selenite **2b** can be used as raw material of feed additive. The effects of chelating multi-metal chitosan selenite **2b** as feed additive are revealed in Table 1. The synergistic effect is significant. It supplements trace elements, rare earth and selenium elements, improves animal antibacterial and anti-viral capacity and selenium content of animal products; the enzymatic activity of rare earth can significantly gain weight of fish, shrimp and poultry.

Table 1. The effects of chelating multi-metal chitosan selenite **2b** as feed additive

No.	area of shrimp pond	Mass concentration in fodder	Effects

1	10 acres of land	0	Gaining weight commonly and containing selenium generally.
2	10 acres of land	1.0 %[a]	Gaining weight quickly, and containing slightly more selenium[b].

[a]. Mixing the feedstuff and chelating multi-metal chitosan selenite **2b** by mass ratio; [b]. Just compare with example 1.

The function mechanism is as follows [5, 7]:

1. Promote endogenous growth hormone (GH), thyroxine (T4), insulin (ISN) synthesis and release in animal body.
2. Activate growth factors in animal body; promote calcium influx; activate calcium channels.
3. Improve activity of intestinal protease, lipase and amylase; promote nutrient digestion and absorption in animal body.
4. Enhance the activity of superoxide dismutase (SOD); inhibit and remove free radicals in animals; prevent and inhibit cancers.
5. Significantly increase the natural killer vitality of immune cell and cell vitality; prevent coagulation and promote tissue repair; promote harmful toxins excreted from the liver and intestinal canal.
6. Chitosan chelated by calcium, magnesium, iron, zinc, manganese and rare earth in the chitosan selenite chelated by calcium, magnesium, iron, zinc, manganese and rare earth is positively charged. It can combine with negative bacteria, inhibit pathogenic bacteria and promote the growth of beneficial bacterium such as bifidobacterium and lactobacillus. It has the effect of class antibiotics without pollution and monomer.
7. Regulate cholesterol metabolism; reduce lipid deposition; improve animal product quality; improve product quality.
8. Chelate heavy metal ions; purify water quality.

In conclusion, chelating multi-metal chitosan selenite presented in this paper is a kind of trace element and/or rare earth preparation. It is environmentally friendly, easy to absorb and rich in selenium without toxic side effects. Therefore; it can be used as a new preparation to supplement selenium and trace elements to be used in medicine, health care products, cosmetics and agricultural fields.

Acknowledgments

We gratefully acknowledge financial support of this work by the Open Project of Hubei Key Laboratory of Wudang Local Chinese Medicine Research (Hubei University of Medicine, Grant No. WDCM009), the Hubei Province health and family planning scientific research project (Grant No. WJ2015Z113), the scientific research project of Educational Commission of Hubei Province of China (Grant No. B2018111), the Shiyan Municipal Science and Technology Bureau Science and technology project (Grant Nos. 18K79, 18Y01), and the scientific research projects of Hubei University of Medicine (Grant Nos. 2017QDJZR09, 2014QDJZR10, 2014CXZ01).

References

[1] Tao X. 2006 Three auspicious jewels in the human body: zinc, copper, selenium. *Heal. Med.* (10), 41-43.

[2] Chen D, Du Y and Liang F. 2014 Research and development status of domestic selenium-enriched foods and analysis of development trend, *J. Trace Elem. Heal.* **31** (1), 76-78.

[3] Zarzycki R and Modrzjewska Z. 2003 Use of chitosan in medicine and biomedical engineering. *Polim. Med.* **33** (1-2), 47-58.

[4] Domard A. 1987 pH and c.d. measurements on a fully deacetylated chitosan: application to Cu^{II}-polymer interactions. *Int. J. Biol. Macromol.* **9**(2), 98-104.

[5] Anker P, Mulcahy H and Strou M. 2003 Circulating nucleic acids in plasma and serum as a noninvasive investigation for cancer: time for large scale clinical studies? *Int. J. Cancer.* **103**

(5), 149-52.
[6] Senlick S. Binding sites of Cu^{2+} in chitin and chitosan. 1986 An electron spin resonance study. *Macromol.* **19**(1), 192-5.
[7] Hu R, Cheng J, Pu D and Li F. 2013 Rare earth-chitosan chelate: functions and applications in animal production. *Chin. J. Anim. Nutr.* **25**(8): 1703-7.
[8] Zeng X, Wang H, Deng S and Wang X. 2017 A synthetic antitumor drug 6-hydroxy-selenite chitosan copper. *CN Patent*, CN 106519071 A.
[9] Liu Y, Zhang A, Wang X, Xu J, Zeng X and Wang H. 2017 A new type of polymeric heavy metal complexing precipitant used as fishery disinfectant and antiparasitic drug. *IOP Conf. Ser.: Earth Environ. Sci.* **82**, 012081.
[10] Zeng X, Gao H, Wang H and Wang X 2016 Iron -dithiocarbamate used as fishery drug. *Lect. Notes Earth Sci.* **4**, 70-1.
[11] Wang H, Li H, Zeng X and Wang X 2016 A new type of copper salt used as disinfectant and fishery antiparasitic medicine. *Lect. Notes Earth Sci.* **4**, 72-3.
[12] Wang A, Shao S, Zhou J and Yu X. 2000 Synthesis and characterization of chitosan and Cu^{II} complex. *Acta Polym. Sin. Chin.* (3), 297-300.
[13] Li Z and Lin M. 2005 Preparation and characterization of chitosan selenide. *Strait Pharm. J. Chin.* **17**(3), 10-3.
[14] Schrauzer G and White D. 1983 Elemental selenium in organic selenium compounds their chemistry and biology. *Bioinorg. Chem.* **8** (3), 303-5.
[15] Zhao Y, Zhang S, Wang P, Fu S, Di W and Liu A. 2017 Seleno-short-chain chitosan induces apoptosis in human non-small-cell lung cancer A549 cells through ROS-mediated mitochondrial pathway. *Cytotech.* 1-13.
[16] Zhang S J, Liu A J, Wang P F and Zheng G Q. 2017 Preliminary study of seleno-chitosan-induced apoptosis in A549 cells through a mitochondrial signaling pathway. *Modern Food Sci. Tech.* **33**(4), 46-51.

Study on three major effects and guarantee system for quality management of edible mushroom products

L Zhou[1], S Guo[1,4], M Shu[2], L Liang[1] and Y Li[3]

[1] Institute of Edible Fungi, Shanxi Academy of Agricultural Sciences, Taiyuan, Shanxi 030031, China

[2] Shea Labagh Dobberstain, San Francisco CA, California 94111, USA

[3] Shanxi Xiu Tong Biotechnology Co., Ltd., Taiyuan, Shanxi 030031, China

E-mail: nkysyj301@126.com; gs0351@sohu.com

Abstract. In order to provide consumers with more safe and edible fungus products and to create a faithful social environment of food quality, the edible fungus products quality safety events, as well as their evoked factors and effects on edible fungus industry were discussed in depth in this paper. The performance of edible fungus products quality management is a comprehensive reflection of technical effect, control effect and implementation effect. Among them, the implementation effect is the most important, which refers to the actual implementation effect of quality and safety rules by practitioners of edible fungus industry. Thus, the initiative implementation effect in quality management will be improved, and a quality assurance system in which practitioners are volunteered to participate and implement initiatively the quality standards and regulations so as to provide safe and high-quality products will be established, thereby achieving the best actual implementation effect in quality management.

1. Introduction

Edible fungi (mushroom) are an important class of food consumed by Chinese people, and their safety quality is related to the health of the people and the social stability of China. China is the largest producer, consumer and exporter of edible fungi in the world, and also an emerging power source for the development of the world's edible fungus industry. According to relevant statistics, the annual output of edible fungi in China increased from 58,000 t in 1978 to 31,679,000 t in 2013, accounting for 80% of the world's total production and ranking first in the world [1-2]. China has the largest number of researchers on basic science and applied technology of edible fungi [3]. However, the quality/safety issues of edible fungus products, such as fluorescent brightener event, formaldehyde event and industrial citric acid event have seriously affected the daily life of consumers [4]. What the status of quality/safety of edible fungus products is, where the quality/safety problems are, and how to deal with the problems are professional and social issues that deserve great attention.

Quality/safety management of edible fungus products is huge and complex system engineering, involved in agriculture, commerce and logistics industry, and related to the community, business and individuals. It has many influencing factors. The effectiveness of safety/quality management of edible fungus products is a combination of technical effect, control effect and implementation effect. Among them, the initiative of implementation effect of quality/safety rules by the practitioners is the most important. In order to study the initiative implementation effect of quality/safety management of edible fungus products and to mobilize the enthusiasm of edible fungus industry practitioners for concerning and implementing the product quality rules, the quality management aspects of edible fungus industry were discussed, and an implementation system was put forward in this paper, thus improving the implementation effect in quality safety management of edible fungus products.

2. Quality/safety issues of edible fungus products and influencing factors

In China, edible fungus industry is initially dominated by self-employment and has now developed into industrialization system with a certain scale, including individual producers, production cooperatives, manufacturing enterprises, product distribution companies, farmers markets, supermarkets and other entities in the product supply chain, as well as administrative departments at all levels, research institutions, tertiary institutions, food hygiene testing departments, consumer associations and other edible fungus products-related practitioners. Due to the high degree of industrialization and the complexity of all aspects of product supply, some problems emerge in the entire edible fungus industry, such as unsatisfactory cooperation and disjointed links. Edible fungus industry in China still has small production scale, low production technology content, low product quality and poor market competitiveness [5]. In addition, there are great hidden dangers in the quality of edible fungus products. For example, abuse and improper application of pesticides lay a hidden danger for the quality safety of products; and sodium sulfite and citric acid added during the processing links of edible fungus products, such as pickling, drying and freezing drying, are prone to exceeding relevant standards [6].

Many quality/safety issues occurred in the edible fungus industry in China, so that consumers have a great distrust for edible fungus products. On December 2, 2009, the Trade and Industry Bureau of Yangzhou City found poisonous fungi that had been coated with phosphorus in the sudden inspection in farmers market; in July 2010, Zhang Hao, a primary school student from Beijing, found fluorescent brightener spots on the surface of 12 edible fungus samples under the guidance of Gao Ruifang, a doctoral student at China Agricultural University; in September 2010, the business sector of Zhengzhou City investigated and dealt with an edible fungus wholesale point, and about 20,000 kg of poisonous mushrooms in a total of 1,500 barrels were confiscated; in June 2011, 22 companies in Jixian, Tianjin were ordered to suspend production for rectification; and in April 2012, formaldehyde-soaked mushrooms were found in Qingdao[4]. Ping *et al.* [7] found the over-limit rates of pesticides, sulfur dioxide and heavy metals in shiitake mushroom and black fungus sampled from the circulation link were higher than those sampled from the production process. According to the information of Zhejiang Entry-Exit Inspection and Quarantine Bureau [8], more than 80 cases of quality/safety violations were notified in the edible fungus products exported to Japan, the United States and the EU from April 2008 to November 2011, greatly influencing the export trade of China. The main causes of these edible fungus product quality safety incidents lie in outdated technology, imperfect management system and unsatisfactory implementation of quality and safety standards.

From a technical point of view, exceeding relevant standards of heavy metals and pesticide residues is mainly caused by improper operation of mushroom production. According to the report of China Network (news.china.com.cn), the Ministry of Agriculture conducted a two-consecutive-year monitoring on the environment of the major wheat-, corn-, rice- and soybean-producing areas (non-mining areas) in 13 provinces during 2003-2004. It was found that 6.4% of the investigated arable land (2,530,000 hm^2) was contaminated by heavy metals, and this phenomenon was more serious in paddy fields, of which the over-standard rate was up to 14% [9]. The main cultivation substrate of edible fungi is agricultural and forestry product scraps. Heavy metals in the soil will be enriched in crops during their growth process, resulting in higher heavy metal residues in the cultivation substrate for edible fungi. Even worse, edible fungi are not sensitive to heavy metals. Therefore, the contents of heavy metals in mushrooms usually exceed relevant standards [10-11]. Pesticide residues exceeding standards have a great impact on the export trade of edible fungus products in China [12-13]. In general, pesticide residue exceeding standard is caused by improper chemical control of diseases and pests in edible fungi, which often occurs in the production management process. In China, the pesticides allowed to be applied to edible fungi are strictly controlled. But in actual production, the phenomenon of indiscriminate use or abuse of pesticides in edible fungus production is more common [13], which is one of the important factors affecting the quality and safety of mushrooms. Chemical pollution is mainly from the practical technology and operation in the processing, storage and transportation links of edible fungi. Microbial contamination refers to the contamination of edible fungus products caused by harmful microorganisms which can occur during all processes of production and marketing of edible mushrooms, leading to the deterioration of the sensory traits and decline of nutritional value and even spoilage. Disqualification of edible fungus products exported from China, such as brine mushrooms, black fungus stripes, freeze-dried mushrooms, mushroom, boiled shiitake mushroom, dried mushroom and ganoderma powder is mainly caused by that all or part of the edible fungus products contain grimed, decaying and decomposed substances or have other problems that are not suitable for consumption [14], which are mainly caused by disqualified packaging labels and improper management procedures.

From the viewpoint of quality control, there are 33 national technical standards and 72 industrial technical standards in China on edible fungi, but they are still imperfect and are lagged behind compared with those in developing countries [2, 4, 14]. For example, some of the pesticides of which the residues in edible fungi are considered excessive by importing counties have no residual hygiene standards, such as chlorpyrifos, fenpropathrin and bifenthrin [15].

In terms of practical implementation, although the administrative departments at various levels in China have formulated various quality management standards for edible fungi, there are still many problems in the overall implementation, such as inadequate awareness (producers, sellers and consumers of edible fungi do not understand relevant standards [16-17]), inactive implementation (as producers and sellers of edible fungi do not understand relevant standards, they have no intention for implementation at all), inadequate implementation strength (none of the raw materials quality inspection, production process management and products quality testing is implemented effectively) and weak quality assurance awareness (although the quality of edible fungi can be tracked, the guarantee and compensation measures are absent, so some of the producers and sellers evade regulation, resulting in quality management chaos).

3. Technical effect, control effect and implementation effect of product quality safety management

Based on the analysis of quality and safety incidents of edible fungus products that occurred in China and factors that affect the quality and safety of edible fungi, it is considered that the quality/safety management of edible fungus products is huge and complex system engineering, of which the management effectiveness is a combination of technical effect, control effect and implementation effect. Under the current technical and policy conditions, the most crucial factor in quality/safety management is the actual implementation effect (Figure 1).

The technical effect of quality management mainly refers to the impact of quality safety-related professional technology and quality testing techniques on the quality of products in the entire production process. For example, crop straw is widely used as the main raw material of production substrate for edible fungi which contains a certain amount of heavy metal elements. If these contaminants are not removed from the production process, they may become one of the reasons for exceeding limiting levels of heavy metals in edible fungus products. Therefore, the technology for removing heavy metals in crop straw affects the quality safety of edible fungus products.

Application of pesticides for controlling pests and diseases in edible fungi is a general farming operation, but improper types and doses of pesticides will cause pesticide residue. Therefore, before the product is sold, rapid and simple determination of pesticide residues in edible fungus products is one of the important techniques for edible fungus producers to control pesticide residues and ensure quality safety. Amano *et al.* [18] quickly determined pesticide residues at the production site using a simple method which makes a rapid determination of pesticide residues possible before shipping and ensures that the pesticide residues in the products are controlled within the allowable ranges.

Control effect of quality management mainly refers to the impact of laws, regulations and standard rules developed by relevant government departments to supervise and guide the commercial activities of practitioners on the quality of products. There are currently 33 national technical standards, 72 industrial technical standards, as well as local standards and procedures and corporate standards on edible fungi in China. In accordance with relevant laws, regulations and technical standards and relevant departments conduct management, supervision and guidance for quality of products. He *et al.* [19] investigated the residual amounts of thiabendazole, carbendazim and prochlorazin in edible fungi of woodland in Beijing. According to GB 2763-2012 (the maximum residue limit of thiabendazole in fresh shiitake mushrooms is 5.0 mg/kg) [20] NY/T 749-2012 (the maximum residue limits of carbendazim and prochloraz in fresh edible fungi are 1.0 and 2.0 mg/kg, respectively) [21], the residual levels of these three kinds of fungicides were in line with the standards for green edible fungi. These standards are developed for mushrooms practitioners and related service personnel, and their implementation will be supervised and regulated. People who violate laws and regulations will be punished.

The implementation effect of quality management mainly refers to the impact of business activities of the members that engaged in product production, distribution, sales, consumption, business management and technical services on the quality of products. For example, in edible fungus production, excessive use of pesticides, or the use of banned pesticides often occurs to prevent pests and diseases. Producers do not understand the quality management standards on the use of pesticides and types and regulations of banned pesticides, or they do not follow the relevant provisions, resulting in product quality problems. Sellers of edible fungus products usually use some preservatives to maintain the freshness of the products. The improper use of preservatives will also bring problems to product quality.

In this complex quality/safety management system engineering of edible fungi, technology, controlling and implementation are closely linked to each other. Sometimes, they are relatively independent, with their own characteristics. These closely-linked but relatively-independent factors need to be considered comprehensively. Different stakeholders are combined with each other, and the resources are integrated, forming synergy.

In today's edible fungi industry, product quality and safety management often focuses on the management and supervision of production technology and administrative regulations, which is insufficient. We must importance to the initiative effects and the actual effects of practitioners.

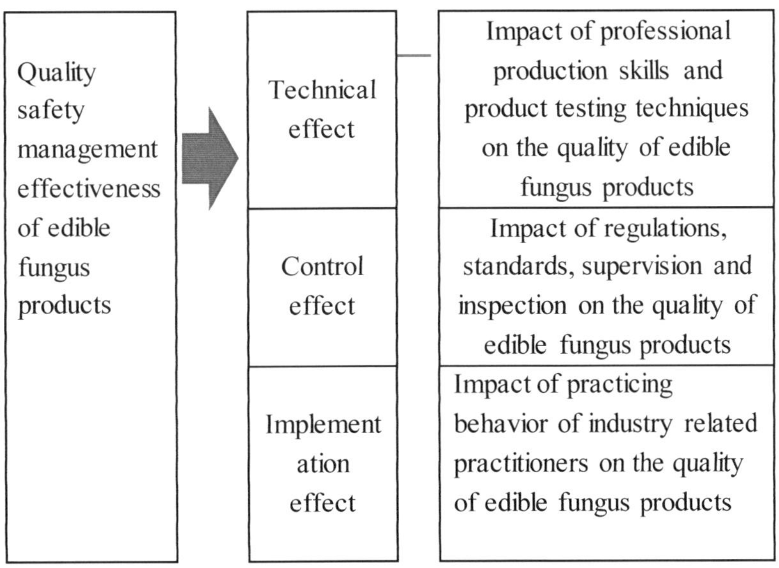

Figure 1. Influencing factors of quality safety management effectiveness of edible fungus products.

4. Passive effects and initiative implementation effect in quality safety management

Edible fungus industry gave great importance to quality safety of products. However, the edible fungus industry and product supply chain are quite different from other agricultural products industries, and its quality safety management has a certain degree of difficulty. As long as it does not violate the laws and regulations, few edible fungus industry practitioners take the initiative to implement the relevant hygiene standards, naturally forming a passive form. For example, product certification is more difficult from the level of technical expertise, making actual practitioners discouraged; market regulation and consumer complaints cause individual distributors to escape and confront; and production enterprises and self-employed managers are lacking in professional skills and guidance on product quality management. The passive form of product quality management is also manifested by many consumers' lack of understanding on the standards and the problems of the quality of edible fungi. Pan *et al.* [17] investigated the awareness of some residents in Haidian District, Beijing on edible fungus products. They found that a considerable number of the residents had no knowledge on the standards for safety quality of mushrooms.

Playing the initiative of edible fungus industry practitioners, mobilizing the enthusiasm of the administrative, commercial and technical departments related to quality management and integrating consumers to create a proactive safety quality management, is a powerful measure to solve the problems with edible fungus products. Some scholars, both at home and abroad, have proposed to pay attention to the influence of the actual operation behavior of practitioners on the safety quality of products. Nishisato [22] proposed quality assurance in the study of quality management. For producers, distributors and sellers, quality evaluation, quality grading system, quality policies and quality assurance must be considered, that is, in the quality management, practitioners should take the initiative to implement the quality safety rules and quality assurance. Some researchers and enterprises in China have carried out this effort. Many agricultural and mushroom enterprises have taken the initiative to put forward the quality safety assurance and product quality certification and improve product quality and integrity. In order to improve the quality of agricultural products, the Helen Farm in Heilongjiang Province established a traceability system of agricultural product quality and built an information transfer system and established a social integrity of product brand [23]. Yuan *et al.* [14] proposed that enterprises should establish and improve the industry self-discipline mechanism for food safety, such as responsibility statement of business sector and food business and establish enterprise self-discipline, industrial and commercial supervision and social supervision trinity regulatory model. Jia*et al.* [15] proposed to raise the level of standardized implementation of mushroom enterprises, and both producers and processing enterprises are the subject of the implementation of the standards for edible fungi and should seriously implement the national and industrial standards for edible fungi. In the construction of the regulatory system for agricultural product quality in Leshan, Sichuan, it was put forward that the awareness of a safety quality of agricultural products has to be improved [24].

Producer-processor, producer-wholesaler and producer-supermarket sellers are the common patterns of combination to determine sale agreement. . However, this pure and simple commodity contract does not have stability. When the commodity market changes, it often leads to a large number of default phenomena, exacerbating the risk of product sales of producers and increasing the operational risk and transaction costs of sales enterprises [25]. The cooperation mode of farmer-supermarket, as well as farmer-cooperative-supermarket is a way of selling the agricultural products. With benefit as a fundamental point, farmers (producers) and supermarkets are matched directly. Based on this, the intervention of cooperatives makes sure that the interests of both parties can be guaranteed. On the basis of government subsidies and supermarket grants for agricultural cooperatives, docking cooperation is easy to set up and the stability of cooperation between cooperatives and farmers is also improved [26].

Yasaka Yasuyoshi *et al.* [27] investigated the concern of consumers in Tokyo, Japan for food labeling that indicates food safety. They found that the Japanese consumers are very concerned about food safety. In addition, they proposed to strengthen the management of food labeling, supervise the amount and accuracy of the content of food labeling, and strengthen the supervision of the government departments and the functions of the inspection departments so as to enhance the trust of consumers. In order to enhance consumers' concern and awareness of food safety and increase consumers' sense of security and trust, quality tracking and tracing system is developed in agricultural product circulation. To a certain extent, more accurate feedback of product information can be obtained. However, at present, China is lacking inaccurate quality indicators in production, processing, distribution and consumption of agricultural products. In China, sub-regulation is employed, which cannot be collaborated and cannot directly trace back to the issue. Even worse, the design of traceability information has limitations. Thus, all aspects cannot be well butt, making full traceability impossible [28].

In the current agricultural quality traceability system, all practitioners have different interests, including producers, processors, sellers and consumers. They lack of understanding the quality traceability of agricultural products and their understanding are inconsistent. As a result, no obvious price difference has been formed between the agricultural products, with and without traceability system. This situation of high quality but low price requires certain investment. The practitioners that had took part in the traceability system did not receive the corresponding benefits, thus the stakeholders who originally had the willingness to participate also lost their enthusiasm, eventually leading to all stakeholders' losing the willingness [29]. This situation also exists in the circulation of agricultural products abroad. The practitioners, especially producers who spend high costs into the traceability system, could not get the corresponding benefits, so their willingness to implement gradually disappears. As the current agricultural product quality traceability system has not been standardized and it is first promoted among producers, there are still many problems that should be pondered [30].

In order to solve these problems better, a quality safety assurance system, based on the actual practitioners of the edible fungus industry is advocated to mobilize the enthusiasm of practitioners in production, circulation, sales and service linked tio the edible fungus industry. Relying on their own integrity, the national quality standards and regulations should be strictly implemented, and high-quality products with clear responsibility distinction will be provided, thus the practitioners will get higher economic benefits and consumers will get safe and healthy mushrooms.

5. Implementation system for improving implementation effect in quality safety management Ensuring quality safety of products and building brand integrity with integrity of practitioners who are volunteered to participate

It is urgent to establish a quality safety assurance system based on the actual practitioners of the edible fungus industry to make the practitioners in production, circulation, sales and service links of the edible fungus industry participate voluntarily. Relying on their own integrity, the national quality standards and regulations will be implemented strictly, and high-quality safe products with clear responsibility distinction will be provided, thus the producers and sellers can get higher economic benefits, consumers can get safe and healthy edible fungus products and brand integrity is built. Figure 2 showed the basic model to improve the implementation effect.

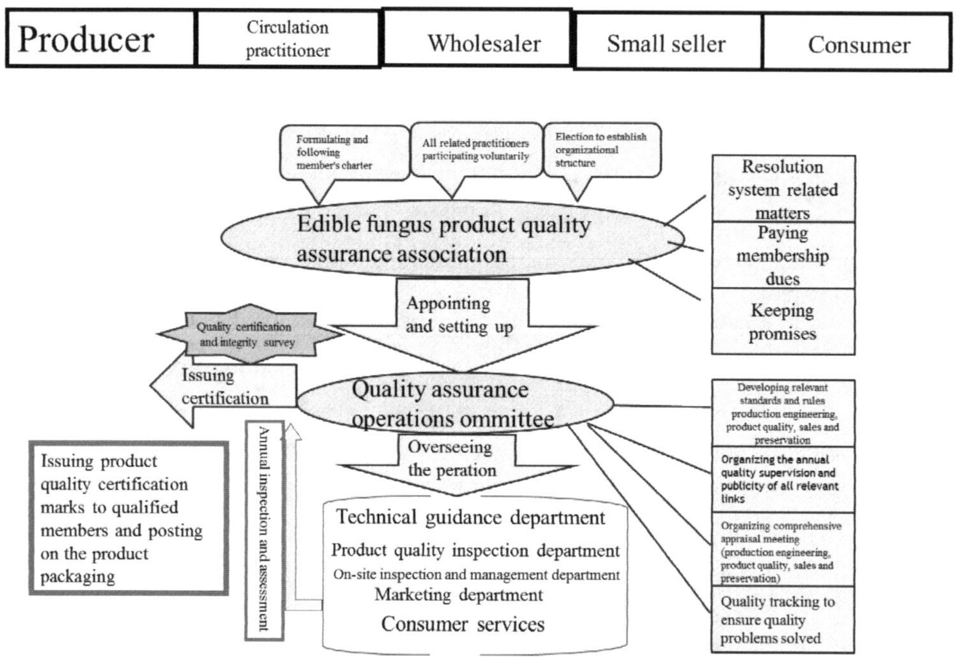

Figure 2. Implementation effect system of edible fungus product quality.

5.1. Replacing complaint and punishment system by responsibility system that ensures quality safety

In the product sales, consumers often do not understand the product. Usually, the quality problems cannot be detected with the ordinary naked eyes. Thus, a blind sense of crisis emerges, causing great social adverse effects. Coupled with the media exposure of some product quality incidents, the majority of consumers have a sense of insecurity. Different from industrial products, unqualified edible fungus products cannot be returned and replaced, which is the special feature of edible mushrooms. Therefore, it is very necessary to establish a system of compensation that protects the interests of consumers. In the advocated safety quality assurance system, consumers must participate. They will monitor the production and marketing aspects and improve the level of product quality. If bad products are found in the market or defective products are purchased, losers can be compensated from this quality safety assurance system. In addition, producers and sellers can be spurred and supervised, the mutual cooperation will be strengthened, and the integrity will be enhanced.

5.2. Holding edible fungus product quality safety appraisal meeting regularly to make the quality of the products public and to monitor the quality safety of the products relying on the wisdom of the majority of stakeholders

The laws and regulations and technical knowledge of product quality safety need to be popularized. Product quality situations also need to understand by vast number of consumers. The producer's production status and product quality needs testing. These can be achieved through regular or irregular control. Thus, the understanding and supervision of consumers on edible fungus product quality is enhanced, the high-quality products, as well as quality safety knowledge and technology are publicized. Eventually, high quality, high benefit, eased consumption and healthy living are achieved. In this way, the edible fungus industry in China can be raised to a whole new level, making the quality of edible fungus products reliable and assured.

References

[1] Zhang J X, Chen Q, Huang C Y, Gao W and Qu J B 2015. History, current situation and trend of edible mushroom industry development. *Mycosystema*, **34**, 524-40.
[2] Li H, Xu X H and Wang X G 2015. Present situation, problems and countermeasures of technical standards of edible fungi in China. *Edible Fungi Chi.*, **34**, 1-6.
[3] Zhang, J X 2014. Powerful supports of scientific research are indispensable to the development of mushroom industry. *Mycosystema*, **33**, 175-82.
[4] Meng X H and Zhang J B 2013. Measures for quality control of edible fungus products. *J. Changjiang Vege.*, **14**, 1-5.
[5] Lin G Z 2015. Analysis on the development of Chinese mushroom industry. *Northern Horticulture*, **3**, 163-64.
[6] Li G B 2013. Potential safety problems of edible fungus product quality and preventive measures. *Edible Medici. Mushrooms*, **21**, 337-38.
[7] Ping H, Wang J H, Ma Z H, Li Y, Li B R, Wang B H, and Fu W L. 2016 Investigation and analysis of quality and safety of edible fungi in Beijing and Hebei. *J. Food Safety Quality*, **7** 478-83.
[8] Xu, L. H., Zhang, Y. Z., Wang, G. J., Wang F, Zhang Y, Wu Y M and Chai Z L 2007. The quality and safety of edible fungi from Zhejiang Province. *J. Agro-Envir. Sci.* **26**, 679-85.
[9] News Center - China Network. Heavy metal pollution enters the high incidence period with annual loss of food for 40 million people. http://www. china. com. cn/news/env/2012-06/13/content_25631940. htm, (2012-06-13) [2016-09-11].
[10] Huang Q, Li W, Guo X, Wang T T and Gui M Y 2014. Research progress on accumulation of heavy metal in edible fungi. *Edible Fungi Chi.*, **33**, 4-6.
[11] Yoshiyuki B, Masayuki K and Koichi F 2013. Effect of the concentration of heavy metals in shiitake cultivation medium on the accumulation of heavy metals in fruiting bodies. *Forest Appl. Res.*, **22**(2), 25-30
[12] Guan D P and Hu Q X 2008. Maximum pesticide residues limits for edible fungi product quality and safety. *Edible Fungi Chi.*, **27**, 3-6.
[13] Xing Z T and Yu Q H 2014. Analysis on the obstruction of export of canned white mushroom in 2012. *Edible Fungi*, (1), 1-3.
[14] Yuan J P and Xiao Z 2006. Study on institutional factors and countermeasures for food safety. *Food Sci.*, **27**, 563-67.
[15] Jia S M and Liu G J 2010. Edible fungus product quality and safety standards and implementation status in China. *Zhejiang Shiyongjun*, **18**, 17-20.
[16] Tao S X and Li L 2016. Regulation effect of agricultural product quality and safety standards on

farmers' production behavior. *Rural Econ.*, **2**, 563-67.

[17] Pan Z Q, Xu T, Zhang D J Zheng B, Lan F Y, Li L and Xu Y J 2015. Detection of heavy metals in two kinds of edible fungi available in market and a related KAP survey in Haidian district in Beijing. *J. Food Safety Qual.*, **6**, 2361-67.

[18] Akiko A and Narita I 2010. Application and dissemination of simple pesticide residue analysis system at agricultural production site. Journal of the Japanese Pesticide Soc., **35**, 516-20.

[19] He M, Luo M M, Wang J H, Ma Z H, Liu Y and Wang S X 2015. Residue and safety evaluation of three fungicides in edible fungi in plantation woodlands of Beijing. *J. Food Sci.*, **36**, 213-16.

[20] National Pesticide Residue Criteria Committee. 2012. GB 2763-2012 *Maximum residue limits for pesticides in food.* Beijing: China Standard Press.

[21] Li, Q W, Wang Q P, Liu H C, He L Z, Wang L X and Mei W Q 2012. NY/T 749-2012 *Green food: Edible fungi.* Beijing: China Standard Press.

[22] Nishi T 2005. Quality assurance management system. *J. Okayama Commer. Soci. Res. Instit.*, **26**: 45-48.

[23] Li D D 2014. Problems in quality and safety of agricultural products and research on quality tracking and tracing construction. *Farm Econ. Manag.*, **1**, 73-74.

[24] Mei C F 2015. Promoting the construction of agricultural products quality and safety system in Leshan. *J. Party School Leshan Munic. Commi. O C. P. C.*, **17**, 72-75.

[25] Wang X P and Zhang X F 2013. Agricultural products traceability system behavior of enterprises and farmers under the game theory. *China Busi. Market*, **9**, 94-99.

[26] Jin J X 2014. Agricultural products "agricultural super-docking" mode development mechanism and policy research (doctor's thesis). Chongqing: Southwest University.

[27] Kageyama Y, Ishida A, and Yokoyama S, 2007. Consideration on consumer consciousness and its attributes in 10 items of food labeling. *Agri. Market Res.*, **16**, 120-27

[28] Huang H B 2016. Current status, problems and countermeasures of agricultural products traceability system construction. *Agri. Tech.*, **36**, 179.

[29] Zhang F, Niu J and Gao F 2012. The construction of agricultural product quality tracing system: the situation, the problems and the way-out. *Chin. Agri. Sci.Bull.*, **28**, 186-89.

[30] Miyamori M 2006. The marketing effect of traceability. *J. Industry Inform. Sci.*, **2**, 11-20.

IOP Conf. Series: Earth and Environmental Science **185** (2018) 012020 doi:10.1088/1755-1315/185/1/012020

Effects of different improvement measures on physicochemical properties of saline-alkaline soil and ryegrass growth and yield in Dingbian

Q L Wang[1, 2, 3, 4, 5], Z Pang[1, 2, 3, 4] and D W Zhu[1, 2, 3, 4]

[1] Shaanxi Provincial Land Engineering Construction Group Co., Ltd., Xi'an 710075, China

[2] Key Laboratory of Degraded and Unused Land Consolidation Engineering, the Ministry of Land and Resources, Xi'an 710075, China

[3] Shaanxi Provincial Land Consolidation Engineering Technology Research Center, Xi'an 710075, China

[4] Institute of Land Engineering and Technology, Shaanxi Provincial Land Engineering Construction Group Co., Ltd., Xi'an 710075, China

E-mail: wql199008@126.com

Abstract. In order to explore the effect of different improvement measures on saline-alkali soil, related experiments were carried out in the field with four treatments sand covering(SC), straw mulching(SM), saline-alkali land amendments(SG) and control group(CK). The results, compared with the control, showed as follows: Soil physical and chemical properties were improved, soil salt content reduced by 1.16-2.22 g kg^{-1}, soil pH reduced by 0.05-0.94, soil density declined by 0.03-0.10g cm^{-3} after improvement measures. Furthermore, the growth of ryegrass had been greatly improved, the ryegrass emergence rate increased by 18.4% to 31.7%, and the total yield of fresh grass increased by 35.24%-113.52%. The soil Hg, Pb, As, Cr contents and the plant Pb, As, Cr contents were significantly influenced by desulphurization gypsum. Whereas the heavy metal content in soil did not exceed Grade-II Soil National Standard for Agricultural Environment Quality, and the ryegrass Hg, Cd, Pb and Cr contents were within National Standard for Feed Qygienic Quality. This research suggested that the application of desulphurization gypsum, among the three improvement measures, produced the best effects. It was therefore recommended for wide application in similar saline-alkali soil conditions.

1. Introduction

With the rapid development of social and economic and the accelerating progress of urbanization in recent years, the number and quality of cultivated land was declining. Saline-alkali soil is the largest reserve land resource in China at present. According to the second national soil survey data, there were 16 million hectares of saline land, 866 thousand and 600 hectares of alkali land and 16 million hectares of all kinds of salinized alkaline land in China. So far, about 80% of the saline soil had not been effectively exploited and utilized [1]. At present, there were still about 140 thousand hectares of saline alkali land in Shaanxi Province, 79.7% of which was cultivated. At present, there were still about 140 thousand hectares of saline-alkali land in Shaanxi Province, 79.7% of which was cultivated land. Salinization degree of saline-alkali land in Shaanxi tended to be serious. Moderate-severe saline-alkali land accounted for about 75 %. There were more than 300 saline-alkali land with an area of more than 1,000 mu [2]. Stepping up efforts to control saline-alkali land could have great significance to increase the income of farmers, guarantee the effective supply of important agricultural products and protect the ecological security.

In recent years, many studies had been made on saline-alkali land in Northern Shaanxi, and certain achievements had been obtained. Shaanxi Academy of Agricultural Sciences found that sand-covering measure had been widely used in the saline-alkali soil improvement in Yulin. Especially in areas with low temperature, high groundwater level and heavy soil salinization, the improvement effect was particularly significant. The thickness of sand covering was determined by the degree of soil salinization. It was better to take 5 to 10cm in moderate-severe saline-alkali land. [3] Li Furong showed that the straw surface coverage could effectively reduce the phreatic evaporation, increase soil water storage space and infiltration capacity, improve water utilizing efficiency, have obvious, soil moisture conservation and salt restraint effects, conducive to crop growth and accelerates growth and development of crop through field test [4]. Xiao Guoju showed that the application of desulphurization gypsum could significantly improve the salt content, alkalinity, total alkalinity and pH of moderate and severe saline alkali soil, and the best application amount of desulphurization gypsum was 20000 kg hm^{-2}[5]. The formation of salinized soil was a long and complicated process. Due to great differences among soil matrix, natural environment and other factors of saline alkali soil in different regions, the effect of different improvement measures was quite different [6,7]. In this study, improvement tests of severe saline-alkaline soil were carried out in Dingbian. The dynamic change of soil physical and chemical properties and the growth status of crops under different improvement measures were comparatively analyzed. The research played the certain roles of providing theoretical reference and technical basis for the improvement and utilization of salinized soil in Northern Shaanxi.

2. Materials and methods

2.1. General situations of study region

The research is located at 36°58′ N and 107°42′ E in Dingbian county, Shaanxi province, and belongs to the semi-arid temperate continental monsoon climate, with average elevation of 1605m, annual average temperature of 7.9℃, average annual rainfall of 316.9mm, annual average frost free period of about 141d, average annual sunshine 2600h. It can fully meet the need for the light and heat of the crops, and is also one of the best ryegrass growing areas. Dingbian country had 192 thousand hectares of farmland, 73 thousand and 300 hectares of saline, in which there were 42 thousand and 700 hectares of severe saline alkali land, 19 thousand and 300 hectares of moderate saline alkali land and 11 thousand and 300 hectares of mild saline alkali land [2]. The background of soil was measured before the test in 2016, and the results were shown in Table 1.

Table 1. The background of the soil.

Depth (cm)	pH	Salinity content (g kg^{-1})	organic matter (%)	Total N content %	Available P content (mg kg^{-1})	Available K content (mg kg^{-1})
0-20	9.21±0.14	6.84±0.12	2.47±0.05	0.25±0.04	0.25±0.04	32.77±3.02
20-30	9.19±0.09	7.06±0.10	2.28±0.03	0.46±0.04	0.34±0.02	33.02±2.14
30-50	9.01±0.21	7.11±0.11	4.47±0.03	0.32±0.03	0.39±0.03	38.75±2.03
50-70	9.14±0.12	6.98±0.07	2.53±0.05	0.31±0.05	0.43±0.07	19.56±1.14

2.2. Test materials and design

Four treatments, including blank control (CK), sand covering (SC), straw mulching (SM) and saline-alkali land amendments desulphurization gypsum (SG) were established in 2016 in Dingbian. The thickness of sand covering, the amount of straw mulching and the amount of desulphurization gypsum referred to the best application rates of medium-severe saline-alkali land, 10cm, 7500kg hm-2 and 20000kg hm-2 [8-9], respectively. The experiment was conducted as completely random block design. Each processing was set with 3 repetitions; the area of experimental plot was 3 m×5 m. The cultivar of ryegrass was Yilite developed by the Prairie Research Institute of Chinese Academy of Agricultural Sciences.

Planted ryegrass in September 10th, and the panting density was 6,000-12,000 plants per hectare. Phosphorus diamine fertilizer 200 kg hm^{-2} was applied when snowing. Urea 50 kg $hm^{-2}$and potassium chloride 35kg hm^{-2} were applied at the tillering stage. On the second day after mowing of ryegrass, using the mixture of urea 150 kg hm^{-2}, potassium chloride 10kg hm^{-2} and water to irrigate the test field, and controlled for insects and weeding.

2.3. Sample collection andprocessing

2.3.1. Soil samples. From September 10, 2016 to October 23, 2017, soil sample were collected from each experimental plot, and the soil total salt, pH and bulk density were measured in the different periods of Ryegrass. After the harvest in October 2017, soil samples were collected in CK and SG to determine its contents of heavy metals such as Cd, Hg, Pb, Se, As and Cr.

2.3.2. Plant samples. 16 parts of the ryegrass leaves were collected in CK and SG, and the leaves were used to determine the content changes of heavy metals such as Cd, Hg, Pb, As, Cr and so on. The collected samples were cleaned with tap water and deionized water in the room, killed out at 90℃ for 30 minutes, and then dried to constant weight at 75 ℃.

2.3.3. Sample detection. Soil salt content was tested by mass method, soil pH by potentiometric method(Soil and water ratio was 5 to 1), soil bulk density by cutting ring method, and heavy metals in soil and ryegrass leaves by inductively coupled plasma mass spectrometry (ICPMS).

3. Interpretation ofresult

3.1. Effects ofdifferent improvement measures on the physical and chemical properties of saline-alkali soil

3.1.1. Soil salt content. Salt in saline-alkali soil has the movement rule of "salt along with water", and the salt content of soil is greatly influenced by precipitation and irrigation [10]. Figure 1 illustrated the change of soil salt content of soil plough layer under different improvement measures in 2016 and 2017. The salt content of SC, SM and SG showed a slow decreasing trend. The average salt content of CK, SC, SM and SG 2016 was 4.82, 4.08, 3.55 g and 2.89 g kg^{-1}, respectively. Compared with the natural control, the salt content of 0-20 cm soil layer of SC, SM and SG decreased by 1.16 g kg^{-1}, 1.47 g kg^{-1}, 2.22 g kg^{-1}.

It could be seen from the Figure 1 that the decrease of salt content in the soil of the plough layer was fairly small in 2016, and the soil salinity in the late growth period (after reaping of ryegrass) was even slightly increased. On the one hand, the soil water permeability was poor, the background value of salt was high, the leaching effect was inconspicuous in the first year of reclaiming wasteland, and there was a lot of water on the surface. On the other hand, due to the high water level, the high degree of salinity and the large amount of evaporation, the deep soil salinity had been strongly evaporated by the surface and moved to the ploughing layer.

The soil salt content in 0-20 cm soil layer in 2017 was obviously lower than that in 2016. On the one hand, tillage measures and improvement measures increased the 0-20cm soil nutrient content, improved soil structure and physicochemical properties, and the irrigation could effectively extract soil salt to deeper soil layer. On the other hand, the groundwater level in the study area from May 2017 to September is low, and the high salinity groundwater had little influence on the salt content in 0-20cm soil layer.

From 2 years of monitoring results of soil salinity, three measures of sand covering, straw mulching and desulphurization gypsum could conduce to reduce soil salinity. The order of decreasing soil salinity is: desulphurization gypsum, sand covering and straw mulching.

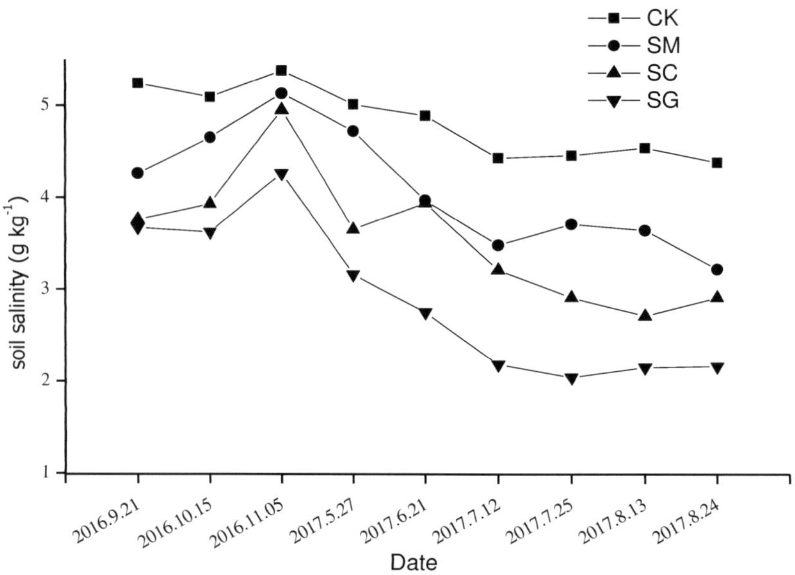

Figure 1. Change of soil salt content in 2016 and 2017 with different improvement measures.

3.1.2. Soil pH. Table 2 was the monitoring result of the soil pH in the plough layer. The soil pH under different improvement measures showed a decreasing trend in the ryegrass growth period in 2016 and 2017. The soil pH of SC, SM and SG was 0.27, 0.43 and 1.13 lower than that of CK. The results of variance analysis showed that the soil pH had no significant difference among SC, SM and CK in the same monitoring period. The soil pH of the plough soil treated with desulphurization gypsum was significantly lower than that of the CK. In 2016, the average decreased by 0.63, and the average decreased by 0.96 in 2017 compared with the control, which indicated that the application of desulphurization gypsum to the salt-alkali land still had an effect on the soil for second years.

Table 2 The soil pH in the plough layer with different improvement measures.

treatment	2016			2017					
	9-10	10-15	11-05	5-27	6-21	7-12	7-25	8-13	8-24
CK	9.13b	9.12b	9.11b	9.22c	9.18b	9.11b	9.09b	9.14c	9.01b
SC	9.11b	9.07b	9.15b	9.09b	9.11b	9.03b	8.98b	8.87b	8.83b
SM	9.13b	9.05b	9.07b	9.12b	9.11b	9.03b	9.01b	8.98b	8.96b
SG	8.62a	8.51a	8.35a	8.32a	8.21a	8.11a	8.20a	8.05a	8.07a

3.1.3. Soil bulk density. The saline-alkali soil had the character of high bulk density, compact structure, more aggregate, poor air and water permeability. Therefore, it was difficult to improve and utilize [11]. The results of soil bulk density in each layer after ryegrass harvest in 2016 and 2017 in Table 3 showed that: Soil bulk density and physical properties could be effectively ameliorated by improved measures. According to the data in 2016, straw mulching had no significant effect on soil bulk density of 0-60cm soil layer. Sand cover and application of desulphurization gypsum significantly reduced soil bulk density of 0-20cm soil layer, but these had no significant effect on bulk density of deeper soil. In addition, from the data in 2017, it could be seen that 3 improvement measures had certain effect on soil bulk density, and as time goes on, soil structure and soil depth would be improved.

Table 3 The soil bulk density (g cm^{-3}).

Years	Treatment	Depth (cm)			
		0-10 cm	0-20 cm	20-40 cm	40-60 cm
2016	CK	1.46	1.54	1.43	1.45
	SC	1.39	1.46	1.46	1.43
	SM	1.45	1.52	1.43	1.46
	SG	1.37	1.47	1.44	1.44
2017	CK	1.45	1.52	1.45	1.45
	SC	1.39	1.42	1.41	1.42
	SM	1.42	1.43	1.41	1.43
	SG	1.35	1.38	1.39	1.41

3.2. Effects of different improving measures on growth and yield of ryegrass

It could be seen, from Table 4, that three different improvement measures could significantly improve the emergence rate of ryegrass. The emergence rate of sand covering, straw mulching and application of desulfuration gypsum increased by 54.18%, 43.35% and 69.95%, and the fresh yield increased by 60.84%, 39.24%, 112.48% in 2016, compared with the blank control, respectively. The results showed that the three different improvement measures could improve the emergence rate and yield of ryegrass. The treatment of desulfuration gypsum had the highest fresh grass yield in 2016, and the treatment of desulfuration gypsum also had the highest fresh grass yield after the harvest of 1 stubble ryegrass, The results showed that the improvement measures of first years still had significant effect on production increase of fresh grass for second years. The yield of 3 stubble ryegrass with desulfuration gypsum increased by 5759, 4463 and 4530 kg hm^{-2}, respectively, and increased by 89.66%, 112.48% and 86.19%, compared with CK.

Table 4. The emergence rate and yield of ryegrass with different improvement measures.

Treaments	Tmergence rate		First mowing in2016		Second mowing in 2017		Third mowing in 2017	
	Average (%)	Increase percentage (%)	Average (kg hm^{-2})	Increase percentage (%)	Average (kg hm^{-2})	Increase percentage (%)	Average (kg hm^{-2})	Increase percentage (%)
CK	50.75c	-	5120c	-	5652c	-	5256c	-
SC	72.75b	43.35	7129bc	39.24	7253b	28.33	7135bc	35.75
SM	78.25b	54.18	8235b	60.84	9023ab	59.64	8387b	59.57
SG	86.25a	69.95	10879a	112.48	10115a	78.96	9786a	86.19

3.3. Effect of application of desulfuration gypsum on heavy metal content in plough soil and Ryegrass

As a waste from coal-fired power plants, the desulfuration gypsum contains some heavy metal elements that would enter the soil with the application of desulfurization gypsum to improve saline alkali land [12]. In order to know the effects of desulphurization gypsum on soil environment, ryegrass and forage safety, the contents of Cd, Hg, Pb, As and Cr in the 0-20cm soil layer and the ryegrass leaves harvested in 2017 were detected. The results showed (Table 5) that the content of Hg, Pb, As and Cr in soil and the contents of Pb, As and Cr in the leaves of ryegrass were significantly increased by the application of desulfuration gypsum. The Hg, As, Pb and Cr in the soil increased by 0.02 mg kg^{-1}, 0.92 mg kg^{-1}, 0.48 mg kg^{-1} and 0.77 mg kg^{-1}, respectively. In the leaves of ryegrass, Cd was not detected, the content of Hg was not increased, and the content of Pb, As and Cr increased significantly as the application of desulfuration gypsum, compared to the blank control. The content of Pb in 0-20cm soil and ryegrass leaves was the highest, followed by As, Hg and Cr.

Table 5. Content of heavy metals in the leaves of 0-20cm soil and ryegrass with the application of desulfuration gypsum.

Sample	Treatment	Cd	Hg	Pb	As	Cr
Soil	CK	0.03	0.11	1.19	1.4	1.58
	SG	0.03	0.13	2.11	1.9	2.35
Ryegrass	CK	ND	0.03	0.89	0.01	1.21
	SG	ND	0.03	1.21	0.21	1.51

Note: ND indicated that the content was lower than the detection limit, and the detection limit of Cd was 0.02mg kg^{-1}.

4. Discussion

The fundamental problem of soil salinization was the movement of high salinity water in the soil. Therefore, controlling the rising of groundwater level or lowering the groundwater level and reducing the evaporation of soil water were the key to solve the soil salinization [13]. The surface coverage measures could separate the soil from the atmosphere to reduce the loss of soil moisture, reduce surface evaporation, prevent the rising of soil salt content, regulating soil water heat process, promote nutrient cycling and crop growth, and further make the saline soil ecological process to the benign transformation [14]. Northern Shaanxi was one of the suitable regions of corn and wheat in Shaanxi and even the whole country. The planting area of corn and wheat was large, and the amount of plant straw was abundant [15]. The literature showed that straw mulching could effectively reduce the evaporation of soil surface, improve soil structure, reduce soil bulk density and improve soil permeability. Straw mulching also had important influence on soil moisture conservation, inhibition of salt surface accumulation, promotion of rain drenching salt, and increase of crop yield. In addition, the crop straw contains a large number of nutrients such as carbon, nitrogen, phosphorus, potassium and trace elements which are important sources of recharge of soil nutrients [16].

Sand covering is one of the important measures to improve the secondary salinization of soil, and is also one of the best methods for the development and utilization of saline alkali land in Northern Shaanxi. It could change the physical and chemical properties of soil, improve the ground temperature and prevent the rise of salt,so there was a proverb 'Salinized land covered with sand like the gold brick' [17]. Covering sand on medium and severe saline soil can help restrain salt and alkali content. The pore of sandy soil is large, and the suction cannot be formed, thus limiting the salinity of the saline alkali soil to move upward and distributing it to a certain depth of the soil, which help to achieve the purpose of developing plough horizon. Or due to a certain degree of mixing between the covered sand soil and the saline alkaline soil, the soil viscosity can be adjusted to reduce the bulk density, improve the soil structure, and reduce the salt effect.

As a traditional chemical modifier, gypsum can reduce soil pH value, alkalinity and bulk density, and improve the infiltration rate, soil porosity and desalination rate, thereby increasing the difficulty of the soil surface crust [18]. The successful experience of improving alkaline soil with gypsum had been recognized by many scholars at home and abroad. In general, the soil with high pH is improved by gypsum. Calcium ions in gypsum replace sodium ions adsorbed on Soil Colloids, and sodium ions are moved to the root layer of the crop under the action of leaching, thereby reducing the degree of dispersion of sodium ions on the soil particles, and decreasing soil pH. A good soil water and salt environment are created for the growth of crops in the end [19]. The desulfuration gypsum is mainly composed of $CaSO_3$ and $CaSO_4$, similar to the natural gypsum, contains rich mineral nutrients such as Ca, S, Si and other mineral nutrition, and can replace natural gypsum to improve saline-alkali soil [20]. The research showed that when desulphurization gypsum was used to improve saline alkali soil and grow oil sunflower. The improved depth of desulfuration gypsum was generally 20-30 cm. So, it is suggested that plough after rotary tillage will be beneficial to improve the improvement effect and promote the growth and development of crops [21].

5. Conclusion

The results showed as follows compared with the control: soil salt content reduced by 0.11-1.46, soil pH reduced by 0.11-1.46, soil density declined by after improvement measures. Furthermore, the physical and chemical properties of soil have been improved effectively, and the growth of ryegrass had been greatly improved. The ryegrass emergence rate increased by 18.4% to 31.7%, and the total yield of fresh grass increased by 35.24% to 113.52%. Overall, 3 improvement measures are conducive to Dingbian Shaanxi saline soil, and the application of desulfurization gypsum had the best effect.

When using to improve saline soil, the harmful elements contained in desulfurization gypsum had no direct pollution, but the content of harmful elements in different parts of the ryegrass have been relatively increased, forming a potential harm to the food chain. The results showed that the application of desulphurization gypsum to improve saline alkali soil significantly increased the content of heavy metals in cultivated soil and ryegrass, but the contents of heavy metals such as Cd, Hg, Pb, As and Cr in Ryegrass and soil did not exceed Grade-II Soil National Standard for Agricultural Environment Quality and National Standard for Feed Qygienic Quality. Therefore, it is safe to apply the desulfuration gypsum to improve the agricultural soil and the ryegrass produced in the saline alkali land.

Ryegrass is a high quality forage and green manure crop widely cultivated in Northern Shaanxi Province. Enlarging the scale of medium and severe saline alkali soil improvement and ryegrass planting was of great reference value for establishing a circular agricultural economic mode of "improving saline alkali soil forage grass, developing animal husbandry, returning organic fertilizer to saline soil" [22].

The formation of salinized soil is a long and complex process, and the improvement of saline soil is also a complex systematic project. Any single measure has its own applicable conditions, technical advantages and limitations [23]. According to the present situation of Dingbian saline alkaline land, the suggestion is as follows: adopting comprehensive measures to improve and heavily using physiological acid fertilizer which also neutralizes soil alkalinity while supplementing soil nutrients. In addition, straw can be covered or covered with sand to further improve the soil physical and chemical structure and inhibit the soil back to salt [24].

Acknowledgments

This work was financially supported by Key scientific and technological innovation team planning project of Shaanxi Province (2016KCT-23) fund.

References

[1] Gou Q Q, Han Z W and Wang G H. 2011. Research progress on soil salinization in arid irrigated area of northwestern china. *Chin. Agri. Sci. Bull.*, **27**(29): 246-50

[2] Ning S R, Han J C, Hao Q L, Li G, Yan B and Zhao X. 2017. Survey and discuss of saline-alkali new farmland and soil improvement measures. *Northern Horticul.*, (8): 172-8

[3] Bezdicek D F, Beaver T and Granatstein D. 2003. Subsoil ridge tillage and lime effects on soil microbial activity, soil ph, erosion, and wheat and pea yield in the pacific northwest, USA. *Soil Tillage Res.*, **74**(1), 55-63

[4] Li F R, Yang J S, Wu Y K, Yao R J, Yu S P and Zhang J B. 2013. Effects of straw mulch at different depths on water-Salt dynamic changes of coastal saline soil in north Jiangsu Province. *Soils*, **45**(6): 1101-7

[5] Xiao G J, Qin P, Luo C K and Zhang F J. 2011. Study on effects of plowing and rotary tillage on improved solonetzic soil with desulphurization gypsum. *Ecol. Envir. Sci.*, **19**(2): 433-7

[6] Wang H J, Shi J C, Zhang H L,Sheng J D and Ma X W. 2014. Soil salinity dynamic change and desalting effect under different improvement measures in severe salinity soil in Xinjiang.

Trans. Chin. Soc..Agri. Engin., **30**(22):102-11
[7] Brown T T, Koenig R T, Huggins D R, Harsh J B and Rossi R. 2008. Lime effects on soil acidity, crop yield, and aluminum chemistry in direct-seeded cropping systems. *Soil Sci. Soc. Ame. J.,* **72**(3): 634-40
[8] Clark R B. 2003. Growth of forages legumes and grasses in acidic soil amended with flue gas desulfurization products. *Commun. Soil Sci. Plant Analysis*, **34**(2): 157-180
[9] Stamford N P, Silva A J N, Freitas A D S and Filho J A T. 2002. Effect of sulphur inoculated with Thiobacillus on soil salinity and growth of tropical tree legumes. *Biores. Tech.*, **81**(1): 53-9
[10] Li B, Wang Z C, Sun Z G, Chen Y and Yang F. 2005. Resources and sustainable resource exploitation of salinized land in China. *Agric. Res. Arid Areas*, **23**(2): 154-8
[11] Hussain I, Olson K R and Siemens J C. 1998. Long-term tillage effects on physical properties of eroded soil. *Soil Sci.*, **163**(12): 970-81
[12] Fan F, Zhang Q G, Tai J C, Hou J H and Sun D Z. 2015. Biological traits on corn straw interlayer in improving saline-alkali soil. *Trans. Chin. Society Agri. Engin.*, **31**(8):133-9
[13] Ke Y and Li F. 2014. Study on microbial flora and diversity in saline-alkali soil improved for different years in Yinchuan Plain. *J. Henan Agri. Sci.s*, **43**(6):75-9
[14] Teakle N L, Snell A and Real D. 2010. Variation in salinity tolerance, early shoot mass and shoot ion concentrations within Lotus tenuis: Towards a perennial pasture legume for saline land. *Crop Pasture Sci.*, **61**:379-88
[15] Li F X, Wang X Q, Guo Y Z, Xu X and Yang J G. 2012. Effect of soil properties and soil enzyme measures activity in different improvement of saline-alkali soil in Yinchuan Plain. Res. Soil Water Conser., **19**(6): 13-8
[16] Qin D L, Wang S L and Liu Y H. 2017. Effects of cotton stalk returning on soil physical and chemical properties and cotton yield in coastal saline-alkali soil. *Acta Agronomica Sinica*, **43**(7): 1030-42
[17] Colmer T D, Munns R and Flowers T J. 2005. Improving salt tolerance of wheat and barley: Future prospects. *Austra. J. Experim. Agri.*, **45**:1425-43
[18] Guo R, Shi L and Yang Y. 2009. Germination, growth, osmotic adjustment and ionic balance of wheat in response to saline and alkaline stresses. *Soil Sci. Plant Nutri.*, **55**: 667-79.
[19] Ruben V, Chen P, Ishibashi T and Matthew C. 2008. A rapid and effective method for screening salt tolerance in soybean. Crop Sci., **48**(5):1773-1779
[20] Baumhardt R L, Wendt C W and Moore J. 1992. Infiltration in response to water quality, tillage, and gypsum. *Soil Sci. Soc. Amer. J.*, **56**(1): 261-6
[21] Zheng P S, Hao B P, Feng Y C, Ding Y C and Li Y F T. 2012. Effects of different saline-alkali land amendments on soil physicochemical properties and alfalfa growth and yield. *Chin. J. Eco-Agri.*, **20**(9): 1216-21
[22] Hajlaoui H, Denden M and El A N. 2009. Differential responses of two maize (Zea mays L.) varieties to salt stress: Changes on polyphenols composition of foliage and oxidative damages . *Industrial Crops Prod.*, **30**:144-51.
[23] Han J C, Xie J C, Wang T, Luo L T and Zhu M L. 2009. Experimental observation of saline alkali of soil in saline land after changing drainage to impoundment in Lubotan of Shaanxi Province. *Trans. Chin. Soc. Agri.Engin.*, **25**(6):59-65.
[24] Zhang J F, Zhang X D, Zhou J X, Liu G H and Li D X. 2005. World Resources of Saline Soil and Main Amelioration Measures. *Res. Soil Water Conser.*, **12**(6):28-30.

IOP Conf. Series: Earth and Environmental Science **185** (2018) 012021 doi:10.1088/1755-1315/185/1/012021

Soil productivity and its relation to the environment in the Czech Republic

V Voltr[1, 3], J Klír[2] and M Hruška[1]

[1] Institute of Agricultural Economy and Information, Mánesova 75, Praha 2, The Czech Republic

[2] Crop Research Institute, Drnovská 507/73, Praha 6 – Ruzyně, The Czech Republic

E-mail: voltr.vaclav@uzei.cz

Abstract. Based on the evaluation of data from agricultural operations of 60 enterprises in the period 2012–2016 on 339 516 ha, frequent increases in nutrient dosages for production are mainly in marginal areas where higher production than is equivalent to optimal production on the soil is required. In the production of crops, it is most manifest in the production of feed crops, which are often used for the continuous flow of livestock production and biogas stations. Because of the above-standard utilization of fertilizers, yields rise above the level corresponding to the standard conditions for soil–climatic conditions, but also to decrease the efficiency of fertilizer utilization and thus to overload the optimal soil productivity. In contrast, the standard use of fertilizers is reflected in winter wheat, grain corn, triticale, potatoes and rye. Because of the lower strength of the humus horizon in marginal areas, it can be assumed that by increased fertilization, the agricultural enterprises solve the lower sorption capacity of the soil. The overall finding is also a warning to the occupation of quality land for non-agricultural purposes because their intensification cannot be transferred into marginal areas with no environmental impact.

1. Introduction

The economic effect of crop production determines the value of the agricultural land fund. This effect, however, is not only a result of soil quality as such but also depends on fertilization. Fertilizers P_2O_5, K_2O, as well as N are limiting crop yields at all sites [1]. In addition, fertilization increases efficiency and produces a better quality of product recovery in agricultural activities; it is one of the most critical techniques for crops [2]. Plant nutrition is one of the key factors in the intensification of production, which is limited by soil–climatic conditions and environmental constraints due to the protection of water resources and the prevention of soil degradation. Soil fertility is the integration of soil physical, chemical and biological properties, and therefore, there would be more merit to include the biological attributes to quantify soil fertility and predict crop yield in further research [3].

Nitrogen fertilization depends on many factors with varying yields. The options for expressing the value of land are for this purpose mainly based on the categorization of the agricultural land fund in the Czech Republic, which is based on the well-established valuated soil–ecological units (BPEJ). For the broader use of land value, other factors given local conditions can be used. For the selection of suitable indicators and compliance with the indicators described in the BPEJ code, the effect of nitrogen on crop yield as an integrating factor of production intensity is analysed. According to Zhang [4], the fertilization factor explains most of the crop yield variability (42%), while the soil organic carbon variance is primarily determined by the interaction of soil and climate factors (32%). Similar results were obtained by Voltr [5] when fertilization factors explain crop yield variability by 20% (for winter wheat). The effect of technology on soil preparation is also significant [6].

To determine soil fertility and the economic value of land, the basic standardization of soil and climatic conditions and economic evaluation of their relationship is used.

According to Neuberg *et al.* [7], the system of plant nutrition is based on the optimal utilization of the production and ecological conditions and the biological potential of the crops, taking into account to the maximum possible extent the economics of the manure measures and applies measures to avoid adverse environmental effects. In the recommended plant nutrition methodology, it was already stated in 1985 that it is necessary to optimize fertilization to the level of production and ecological conditions concerning the required quality of the harvest and the protection of the environment. Practical optimization of nitrogen fertilization includes not only doses but also forms, term and method of application of nitrogen fertilizers.

The need for nitrogen for plant nutrition changes throughout the year. For example, Zimolka *et al.* [8] reported about winter wheat that the proportion of nitrogen taken in the autumn is not more than 12% of total consumption and therefore applying high nitrogen doses before sowing is unnecessary and non-organic. Nitrogen pickup increases in the spring when plants have to regenerate biomass after winter. The growth of its pumping increases until the end of flowering. After flowering, plant nitrogen requirements are relatively low. At the end of the vegetation, up to 75% of nitrogen is accumulated in the grain. Converting to one tonne of grain and the corresponding amount of straw and wheat roots drains on average 25 kg nitrogen, Liu [9] concluded that a considerable amount of residual soil nitrate accumulated in the 0–200 cm soil profile was observed after crop harvest under 240 kg N ha^{-1} treatment, indicating a sizeable environmental risk of NO_3–N leaching loss, while the opposite was true of 120 kg N ha^{-1}. In addition, the current fertilizer management only NP fertilizers applied could lead to an imbalance in soil nutrients, and managers in this region should pay more attention to balanced fertilization.

The excess availability of reactive N has resulted in diverse environmental problems [10]. Excessive use of manure and fertilizers can increase the amount of nitrates in the soil and therefore increase the risk of N leaching and N_2O and NO volatilization. Depending on the amount of nitrate in the soil, the type of soil and the amount of rainfall and use of water and nitrate by plants, nitrate can leach into surface and groundwater, contributing to pollution of drinking water and eutrophication of surface waters. Denitrification depends on the amount of organic matter, soil water content, soil oxygen supply, soil temperature, soil nitrate levels and soil pH. N_2O is a potent greenhouse gas and contributes to climate change. Nitric oxide (NO) also contributes to smog. A part of N in fertilizers and manure applied to the soil, and in a lesser degree in decaying plants, is transformed into ammonia (NH_3) and emitted into the air.

Soil productivity can contribute to different levels of crop production in this process. It is essential to what extent the farmers' approaches to the soil affect the above process. For this reason, the relationship of farmers' level of fertilization with soil quality was evaluated.

2. Materials and methods

To compare the fertilization of individual crops according to the soil production capacity, a database of fertilizer recordings was used at the enterprise level, according to the Crop Research Institute. A total overview of the evaluated crops is given in Table 1. A total of 1366 crop yields and fertilizers were evaluated, with a total area of 339 516 ha, which was evaluated for a total of 60 holdings.

The basis for comparing the level of fertilization with the soil's production capacity is the assessment of the size of the valuated soil–ecological units (BPEJ). The average production capacity of the land, according to the economic evaluation of the BPEJ, ranged from 10 to 26 CZK/m^2 (approximately 0.5 to 1.3 USD/m^2).

2.1. Valuation of the land

The primary indicators for determining soil fertility, productivity and profitability are defined by the physical characteristics of the soil and the climate supplemented with technical data related to crop production. The analysed soil characteristics are those of topsoil and subsoil texture, pH, chemical composition, humus content, soil absorption complex and soil moisture during the vegetation period [10]. The analysed climatic data relate to the average values of precipitation and soil temperature for a specific month at any given location, as collected by the Czech Hydrometeorological Institute. The analysed technological data relate to fertilizers, plant protection and tillage as well as the penetrometric resistance of the soil.

In the Czech Republic, the soil fertility and soil productivity are evaluated based on a soil system broken down by the genesis, moisture conditions and soil texture. It is composed of a total of 78 groups, the so-called main soil units (HPJ) [11]. These main soil units (HPJ) are based on the climate further classified into a total of 557 main soil-climatic units (HPKJ) and concerning land configuration, soil thickness and skeleton altogether 2 199 valuated soil-ecological units (BPEJ) are defined in the Czech Republic. Their characteristics allow for the quantification of the underlying physical properties of soil and climate and the follow-up soil productivity modelling also about the basic soil fertility factors.

The categorization and evaluation of agricultural land in the Czech Republic is used in some Acts and related legislation. It is also used to calculate payments associated with production potential or to limit the value of inputs. The BPEJ system is based consistently on the natural conditions and characteristics of the given soil and site. The BPEJ system is thus associated with soil assessment capabilities within the categories given by the characteristics of each unit of the valuation system.

All BPEJ are in evidence of the cadastral system, see Figure 1, marked with a five-digit number:

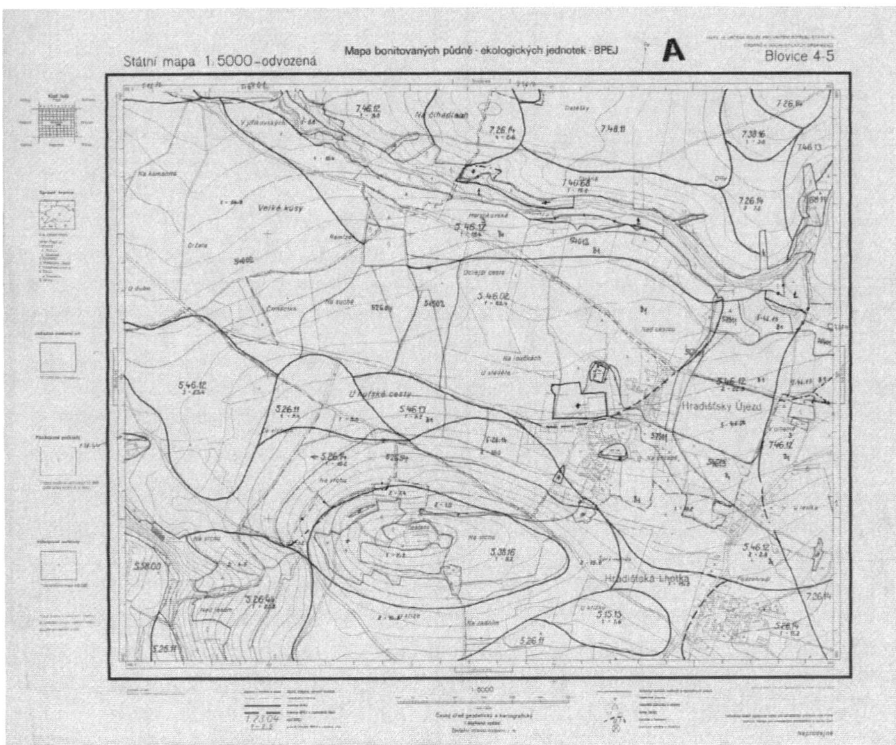

Figure 1. A sample of a BPEJ map on the terrain.

2.2. *Determination of gross annual rental effect (HRRE)*

Gross annual rental effect enables the definition of profitability of land in soil–climatic conditions [12, 13]. The procedures for calculating HRRE are designed to be used when calculating the land price as the main means for determining the profitability of land for a particular farming regime given by the selection of crops and the focus of production. The final land valuation depends on the choice of production orientation and the size of the price support.

Mathematically, the relation of the gross annual rental effect to a particular location can be determined according to relationship 1.

$$HRRE_{ipoz} = HRRE_i * k_{poz}, \quad (1)$$

where: HRRE is the gross annual rent effect of ith BPEJ, $HRRE_{ipoz}$ is the gross annual rental effect of ith BPEJ on the given plot, k_{poz} the HRRE adjustment factor for a particular BPEJ plot

$$HRRE_{i,p} = (CPP_{i,p} - NPP_{i,p}) * K_{i,p}, \quad (2)$$

where: $CPP_{i,p}$ is the price of parametrized production of pth crop on i^{th} BPEJ, $NPP_{i,p}$ the normative cost of parameterized production of pth crop on i^{th} BPEJ, normative costs are described in more detail for all technical operations, $K_{i,p}$ is a dimensionless number resulting from the percentage representation of the pth crop in a given valuation type structure on i^{th} BPEJ (%).

2.3. Calculation of the BPEJ price

Calculation of the official BPEJ price (3) is derived from the adjusted relationship for the calculation of the perpetual rent.

$$UCZPi = BCZP + \frac{(HRREi + P) * (1 - \frac{DP}{100})}{U/100}, \quad (3)$$

where: UCZP is the official price of agricultural land (CZK/ha), BCZP is the basic price of agricultural land (CZK/ha), HRRE gross annual rental effect on BPEJ (CZK/ha), P is the amount to derive BPEJ (CZK/ha), Corporate income tax in %. The calculated rate is 21%, valid for 2015, U is the interest rate for capitalization of HRRE in %.

The dependency of yield on nitrogen dose is derived by equation 4:

$$Y = K + K1 * Ypred + k_2 * difN + sign(difN) * abs(k_3 difN^2), \quad (4)$$

where: K, k_1, k_2 and k_3 are constants, difN is the difference between the actual and the predicted nitrogen dose conditions, Ypred is the predicted yield value for BPEJ under conditions without the effect of actual nitrogen doses [5].

The dependence of predicted yield and nitrogen dose is derived from the statistical survey (2002–2010). Figure 2 shows the rates of yield dependence in different soil–climatic conditions in the percentage of optimal nitrogen consumption relative to the optimal nitrogen dose under given conditions, which is marked above the curve of 0. The standardized dose of N in the next calculation is developed from the optimal point for all crops and valuated soil–ecological units.

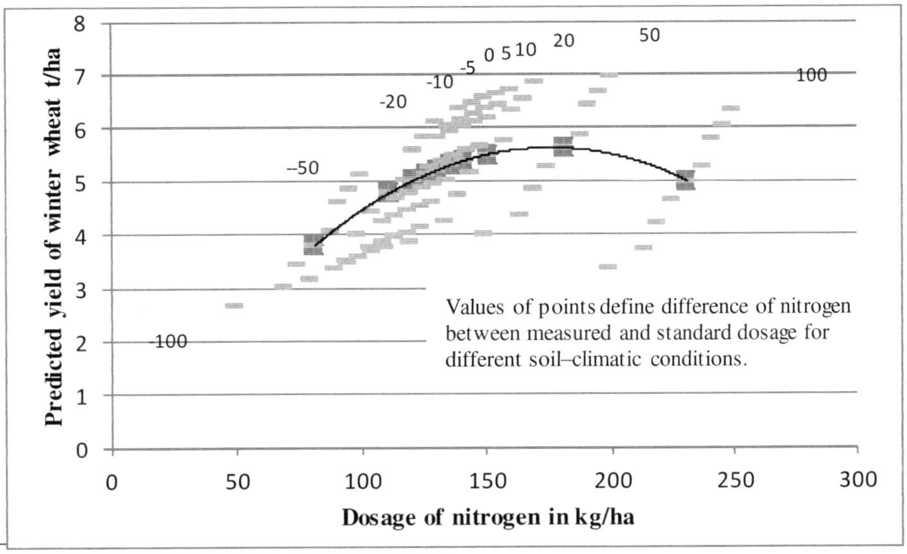

Figure 2. Rates of yield dependence in different soil–climatic conditions.

2.4. Fertilizing valuation

Fertilizing valuations for individual crops were evaluated by the balance of fertilizers with the OECD methodology [10], which are based on the so-called budgets, and the main emphasis is laid on nitrogen budget. The term 'nitrogen budget' is based on statistical dependencies for the major crops.

Table 1. Basic data of the evaluated crops.

Crop	Area (ha)				Yield (t/ha)						
	N	Sum	Minimum	Maximum	N	Mean	Minimum	Maximum	Std. Deviation	Geometric Mean	Variance
Potatoes	36	1330	0.38	188	36	25.97	22.40	30.48	2.50	25.85	6.27
Sugar beet	68	11252	12.00	458	68	63.88	57.75	69.73	2.84	63.82	8.05
Spring barley	170	34896	5.71	857	170	4.72	3.52	5.97	0.52	4.69	0.27
Winter barley	213	24415	3.00	289	213	5.45	4.21	6.34	0.50	5.42	0.25
Corn for silage	196	53100	19.00	1158	196	39.77	25.59	47.37	3.61	39.60	13.02
Corn for grain	113	28871	2.00	1988	113	7.65	6.26	8.82	0.55	7.63	0.31
Poppy	30	2918	4.00	274	30	0.88	0.69	1.03	0.08	0.88	0.01
Oat	55	2586	4.00	252	55	3.99	3.74	4.40	0.17	3.98	0.03
Winter wheat	208	118404	74.00	2674	208	6.11	4.52	7.37	0.54	6.08	0.29
Winter rape	189	54402	19.00	1058	189	3.27	2.74	3.73	0.19	3.27	0.04
Triticale	56	4819	8.00	259	56	5.28	4.46	5.72	0.39	5.26	0.15
Winter rye	32	2523	1.10	211	32	3.69	3.28	4.34	0.30	3.68	0.09
Total	1366	339516	152	9666	1366	171	139	195	12	170	29

The basis for the nitrogen account is built on the work of Leip *et al.* [14]. In this work, the nitrogen budget definition is used for the crop and the farm for a period of five years. A 'balance' is defined as: 'Ideally, the balance of a pool, a sub-pool, or a full Nitrogen Budget is closed, i.e., all nitrogen flows can be explained as input, output or stock changes. The balance equation is then Noutput +

Nstock_change – Ninput = 0. Such a closed N-balance is theoretically possible for each pool defined and for a full Nitrogen Budget. In practice, however, a closed balance is not a requirement of a Nitrogen Budget'.

Normative values for calculation of fertilizers' budget of crops and nutrient content of plants were used for conditions in the Czech Republic from plant nutrition methodology and fertilization and other papers [15, 16]. A global overview of the evaluated crops is given in Table 1.

Fertilizers' budget is given as follows:

Outputs:

Total nutrient consumption (main + by-product). Harvest nutrients are described with the average nutrient consumption of nitrogen, phosphorus, potassium, based on normative nutrient content on the base of kg of nutrients per tonne of crops and summarized according to the area of crops.

Inputs:

N, P_2O_5, K_2O for decomposition of straw, supply N, P_2O_5, K_2O by symbiotic fixation, a quick-release fertilizer with N, (inclusive with P_2O_5, K_2O), slow releasing fertilizers with N (inclusive with P_2O_5, K_2O), mineral fertilizers N, P_2O_5, K_2O

For each crop, the total nutrient supply according to the nutrient balance in the years 2012–2016 was evaluated by the nutrient dosages by the farm and the nutrients collected by the yield of the crop. The balance of nutrients collected by yield is based on the normative intake of fertilizers of VÚRV, calculated by Klír [15]. In the balance, nutrient outflows were also included in the by-product. The balance was determined using the following equations 5, 6:

$$Ny_{N,P,K} = \frac{X_{N,P,K}}{Y_p}; \tag{5}$$

$$Inp_{N,P,K} = d_{p\,N,P,K} * Y_p; \tag{6}$$

where $Ny_{N,P,K}$ is the nutrient dose per tonne of yield (kg/t), $X_{N,P,K}$ the nutrient dose kg/ha N, P_2O_5, K_2O, Inpp the nutrient intake by crops, dp the specific nutrient uptake by harvest, Yp the yield of pth crop.

The total input and output balance are expressed by the nutrient input and crop intake for N, P_2O_5, and K_2O for each crop, enterprise and year of monitoring.

Comparison of the main final results for individual elements and applications of effective nitrogen fertilizers for selected crops is given in Table 2.

For evaluation of specific consumption of nitrogen according to the group of yield according to the division into 10 groups with percentiles was constructed as a graph (Figure 3). The graph illustrates, in the example of maize silage, that a low nitrogen yield per tonne of production is used on agricultural holdings at lower yield levels in percentiles than in other fertile areas. This link may be due to the greater need for silage maize in marginal areas due to the installation of biogas stations on farms that require increased doses of silage maize. The nitrogen dose may also be related to the P_2O_5 and K_2O dose and may also be dependent on the course of climatic conditions in the harvesting year. For this reason, a statistical evaluation of the data yields, N, P_2O_5 and K_2O doses by years using the linear models in the IBM-SPSS version 17 program was used in the following procedure. Because of the same database for different crops, the price of land that characterizes land profitability was used to compare nitrogen doses. These data also included dates for other crops to assess the rate of intensification of nitrogen fertilization. For the unification of the data, the economic value of the soil solvency—yield of the land according to relations 1–5—which are subject to the update of the input values of the income and costs, was used and the derived yields of the soil based on the average of years 2011–2015 were used in this model.

Table 2. Applications of effective nitrogen fertilizers for selected crops.

Soil quality (CZ crowns/m^2)	Winter barley				Maize silage				Winter wheat			
	Balance N per ton of output	Balance P_2O_5 per ton of output	Balance K_2O per ton of output	Dosage kg of N per ton of output	Balance N per ton of output	Balance P_2O_5 per ton of output	Balance K_2O per ton of output	Dosage kg of N per ton of output	Balance N per ton of output	Balance P_2O_5 per ton of output	Balance K_2O per ton of output	Dosage kg of N per ton of output
10	13.31	3.19	9.92	22.24	4.86	1.05	0.48	6.17	1.48	-5.86	-10.21	24.54
11	26.45	3.83	13.72	34.82	4.27	1.24	0.76	5.55	1.19	-6.58	-9.63	23.88
12	10.74	-7.52	-12.08	30.68	10.76	6.01	6.88	9.22	5.85	-9.19	-13.80	30.11
14	5.33	-5.99	-9.21	24.48	5.69	1.95	1.22	6.74	5.72	-3.43	-6.06	27.66
15	13.64	-4.20	-2.45	29.13	3.29	1.28	0.42	5.14	7.78	-5.76	-4.79	28.98
16	6.64	-8.14	-8.16	24.54	3.07	1.64	-2.05	5.59	9.45	-5.13	-1.44	29.35
17	7.80	-2.62	-0.51	23.14	4.64	0.83	-0.33	6.53	3.09	-4.51	-4.71	24.67
18	4.63	-7.09	-9.48	24.36	2.09	0.60	-1.47	4.43	3.17	-5.68	-6.76	25.28
19	5.93	-3.87	-4.24	23.31	2.20	-0.03	-2.57	4.97	5.86	-4.22	-3.26	26.85
20	-1.30	-4.74	-9.12	16.84	0.30	-0.09	-2.69	3.31	10.53	-0.29	-2.70	31.38
21					0.00	-0.63	-3.74	3.50	-1.01	-4.63	-6.03	20.44
22	28.73	-7.80	-6.00	45.73	1.25	-0.18	-2.36	4.34	7.05	-0.78	-2.87	29.61
23					3.20	0.33	-2.12	6.11	19.45	-6.43	-5.15	40.96
24					1.69	-1.09	-4.60	5.39	37.99	-8.72	-10.72	61.16
26					-1.01	-0.87	-3.11	2.69	-4.54	-4.50	2.42	16.63
Average	8.11	-3.81	-3.70	24.46	2.81	0.66	-1.21	5.10	7.42	-3.80	-4.38	28.91

The graph illustrates, in the example of maize silage, that a low nitrogen yield per tonne of production is used on agricultural holdings at lower yield levels in percentiles than in other fertile areas. This link may be due to the greater need for silage maize in marginal areas due to the installation of biogas stations on farms that require increased doses of silage maize. The nitrogen dose may also be related to the P_2O_5 and K_2O dose and may also be dependent on the course of climatic conditions in the harvesting year. For this reason, a statistical evaluation of the data yields, N, P_2O_5 and K_2O doses by years using the linear models in the IBM-SPSS version 17 program was used in the following procedure. These data also included dates for other crops to assess the rate of intensification of nitrogen fertilization. For the unification of the data, the economic value of the soil solvency—yield of the land according to relations 1–5—are subject to the update of the input values of the income and costs, was used and the derived yields of the soil based on the average of years 2011–2015 were used in this model.

Nitrogen dose Ns as an independent variable in linear models related to the intensity of production based on the relationship 7:

$$Ns = \frac{Ntot}{Ytot}, \tag{7}$$

where Ntot: total nitrogen per hectare, Ytot: total crop yield per hectare.

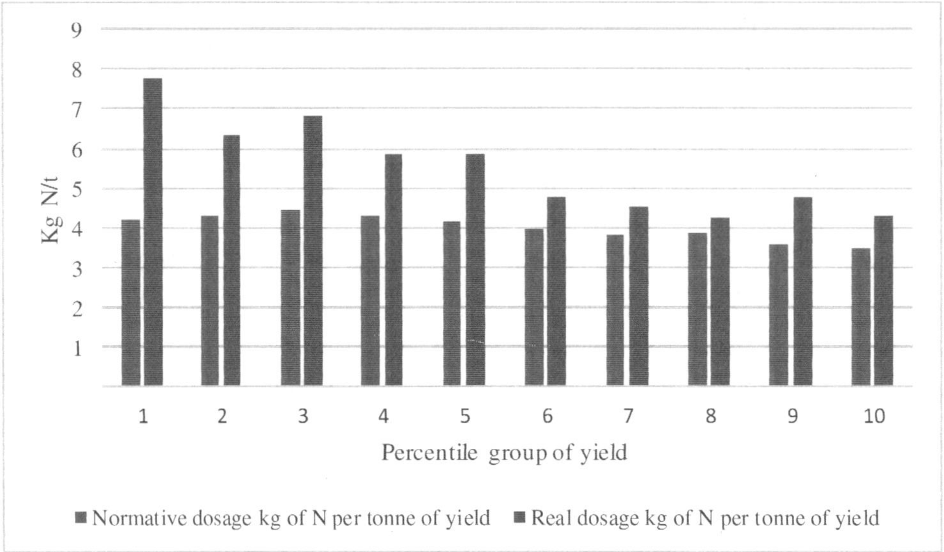

Figure 3. Specific consumptions of nitrogen per tonne of the yield of maize silage.

3. Results

3.1. Results of linear models

Linear models depend on the data of agricultural farms according to the area of each crop. For this reason, linear models were calculated using weighted values of acreage (Table 3).

Table 3. Model Summary[a].

Crop	R[a]	R Square	Adjusted R Square	Std. Error of the Estimate
Grain corn	0.166	0.028	0.027	7.20116
Oat	0.554	0.306	0.305	6.27548
Poppy	0.779	0.607	0.606	34.82108
Potatoes	0.866	0.750	0.750	0.92781
Winter rape	0.324	0.105	0.105	17.81356
Silage maize	0.321	0.103	0.103	2.78136
Spring barley	0.552	0.305	0.305	5.86287
Winter triticale	0.122	0.015	0.014	7.77036
Winter barley	0.526	0.276	0.276	7.49721
Winter rye	0.451	0.203	0.202	8.12836
Winter wheat	0.441	0.194	0.194	9.56603

[a]Predictors: (Constant), year, soil productivity level, K_2O dose, P_2O_5 dose

Coefficients show a relatively low value, which is mainly justified by the individual conditions of agricultural holdings, but relatively few explanatory variables are presented in the models, and the models are significant at the 1% level (Table 4). The number of variables robustly achieved by the coefficient of determination is mostly unchanged, see adjusted R square.

IOP Conf. Series: Earth and Environmental Science **185** (2018) 012021 doi:10.1088/1755-1315/185/1/012021

Table 4. ANOVA[a,b] statistics of models.

	Model	Sum of Squares	df	Mean Square	F	Sig.
Grain corn	Regression	41998.648	4	10499.662	202.475	0.000[a]
	Residual	1484917.900	28635	51.857		
	Total	1526916.548	28639			
Oat	Regression	43428.496	4	10857.124	275.690	0.000[a]
	Residual	98303.743	2496	39.382		
	Total	141732.239	2500			
Poppy	Regression	4481668.492	4	1120417.123	924.049	0.000[a]
	Residual	2900779.479	2392	1212.508		
	Total	7382447.971	2396			
Potatoes	Regression	3421.433	4	855.358	993.631	0.000[a]
	Residual	1137.558	1321	.861		
	Total	4558.991	1325			
Winter rape	Regression	1983095.740	4	495773.935	1562.365	0.000[a]
	Residual	16939847.916	53384	317.323		
	Total	18922943.656	53388			
Silage maize	Regression	44803.710	4	11200.928	1447.901	0.000[a]
	Residual	390181.414	50437	7.736		
	Total	434985.124	50441			
Spring barley	Regression	509577.678	4	127394.419	3706.207	0.000[a]
	Residual	1162422.366	33818	34.373		
	Total	1672000.044	33822			
Winter triticale	Regression	4424.427	4	1106.107	18.320	0.000[a]
	Residual	290644.313	4814	60.378		
	Total	295068.740	4818			
Winter barley	Regression	513626.347	4	128406.587	2284.485	0.000[a]
	Residual	1346066.179	23948	56.208		
	Total	1859692.526	23952			
Winter rye	Regression	41649.213	4	10412.303	157.594	0.000[a]
	Residual	163373.333	2473	66.070		
	Total	205022.546	2477			
Winter wheat	Regression	2491436.518	4	622859.130	6806.544	0.000[a]
	Residual	10328676.636	112871	91.509		
	Total	12820113.155	112875			

[a]. Predictors: (Constant), year, soil productivity level, K_2O dose, P_2O_5 dose
[b]. Dependent Variable: Dose of nitrogen per tonne of yield (kg/t).

Independent variables for all models for dependent variables are soil productivity level, dose of P_2O_5 and K_2O and year.

The results show that soil productivity as an independent variable is significant at 1% level for all used crops. The level in Table 5 in one part of crops is increasing—blue colour for grain corn, potatoes, winter rape, triticale, rye, winter wheat, but with red colour for decreasing Ns for soil productivity level is marked oat, poppy, silage maize, spring barley, sugar beet, spring and winter barley.

Table 5. Linear models.

Crop	Variables	Unstandardized Coefficients B	Std. Error	Standardized Coefficients Beta	t	Sig.

Grain corn	(Constant)	169.809	78.625		2.160	0.031
	Soil productivity level	0.236	0.020	0.071	12.039	0.000
	P_2O_5 dose	-0.019	0.002	-0.114	-11.220	0.000
	K_2O dose	-0.004	0.001	-0.033	-3.241	0.001
	year	-0.075	0.039	-0.011	-1.915	0.055
Oat	(Constant)	1 607.632	213.700		7.523	0.000
	Soil productivity level	-0.379	0.042	-0.159	-9.037	0.000
	P_2O_5 dose	0.199	0.009	0.820	22.897	0.000
	K_2O dose	-0.055	0.005	-0.395	-10.837	0.000
	year	-0.787	0.106	-0.128	-7.420	0.000
Poppy	(Constant)	-10 123.107	1 182.255		-8.563	0.000
	Soil productivity level	-8.920	0.522	-0.263	-17.080	0.000
	P_2O_5 dose	0.813	0.020	0.631	41.110	0.000
	K_2O dose	-0.049	0.015	-0.045	-3.321	0.001
	year	5.151	0.588	0.117	8.753	0.000
Potatoes	(Constant)	-568.974	55.408		-10.269	0.000
	Soil productivity level	0.032	0.011	0.045	2.952	0.003
	P_2O_5 dose	0.006	0.001	0.155	4.953	0.000
	K_2O dose	0.013	0.001	0.670	23.725	0.000
	year	0.283	0.028	0.170	10.290	0.000
Winter rape	(Constant)	9 352.982	132.702		70.481	0.000
	Soil productivity level	0.509	0.025	0.086	20.136	0.000
	P_2O_5 dose	0.075	0.003	0.140	21.420	0.000
	K_2O dose	0.014	0.002	0.046	6.890	0.000
	year	-4.623	0.066	-0.290	-70.134	0.000
silage maize	(Constant)	-470.886	21.107		-22.309	0.000
	Soil productivity level	-0.158	0.004	-0.188	-39.756	0.000
	P_2O_5 dose	0.014	0.000	0.221	29.787	0.000
	K_2O dose	-0.002	0.000	-0.041	-5.269	0.000
	year	0.237	0.010	0.096	22.644	0.000
Spring barley	(Constant)	1 572.429	53.328		29.486	0.000
	Soil productivity level	-0.874	0.010	-0.409	-88.836	0.000
	P_2O_5 dose	0.090	0.001	0.385	60.436	0.000
	K_2O dose	-0.011	0.001	-0.065	-10.036	0.000
	year	-0.766	0.026	-0.135	-28.931	0.000
Winter triticale	(Constant)	-649.771	215.164		-3.020	0.003
	Soil productivity level	0.018	0.046	0.006	0.394	0.693
	P_2O_5 dose	-0.088	0.012	-0.509	-7.213	0.000
	K_2O dose	0.040	0.006	0.468	6.602	0.000
	year	0.334	0.107	0.049	3.130	0.002
Winter barley	(Constant)	6 121.631	80.294		76.241	0.000
	Soil productivity level	-0.758	0.017	-0.272	-44.454	0.000
	P_2O_5dose	-0.022	0.003	-0.084	-7.636	0.000
	K_2O dose	0.027	0.002	0.180	15.551	0.000
	year	-3.023	0.040	-0.422	-75.787	0.000
Winter rye	(Constant)	-553.262	310.317		-1.783	0.075
	Soil productivity level	0.396	0.059	0.129	6.695	0.000
	P_2O_5 dose	0.232	0.016	0.419	14.876	0.000
	K_2O dose	-0.011	0.010	-0.031	-1.119	0.263
	year	0.281	0.154	0.033	1.823	0.068
Winter wheat	(Constant)	8 196.931	49.764		164.717	0.000
	Soil productivity level	0.100	0.010	0.028	10.393	0.000
	P_2O_5 dose	0.027	0.002	0.065	17.223	0.000
	K_2O dose	-0.004	0.001	-0.010	-2.702	0.007
	year	-4.059	0.025	-0.443	-164.208	0.000

IOP Conf. Series: Earth and Environmental Science **185** (2018) 012021 doi:10.1088/1755-1315/185/1/012021

The results show that agricultural holdings behave selectively for crop fertilization. Some crops, in marginal areas, are over-fertilized with nitrogen and are marked in red, these crops being typically fed in the production areas in a standard manner. These are mostly fed crops, but they are grown from more fertile areas with a profitability of CZK 15/m^2 and more and spring barley. For other crops labelled blue, these are mainly market crops, but these crops do not normally have an increase in N consumption above the normative limit. The role of P_2O_5 and K_2O in models is somewhat individual. It can be inferred from these results that increased doses of fertilizers for crop production in marginal areas are mainly based on the need to ensure higher production of livestock feed and to ensure the operation of biogas stations, which are used in the peripheral parts of Czech Republic for supply in Germany and Austria.

3.2. Summary of nitrogen use according to the profitability of land

The results are particularly interesting from profitability where sufficient funds in marginal territories due to subsidy titles allow the use of inefficient nitrogen doses for crop production.

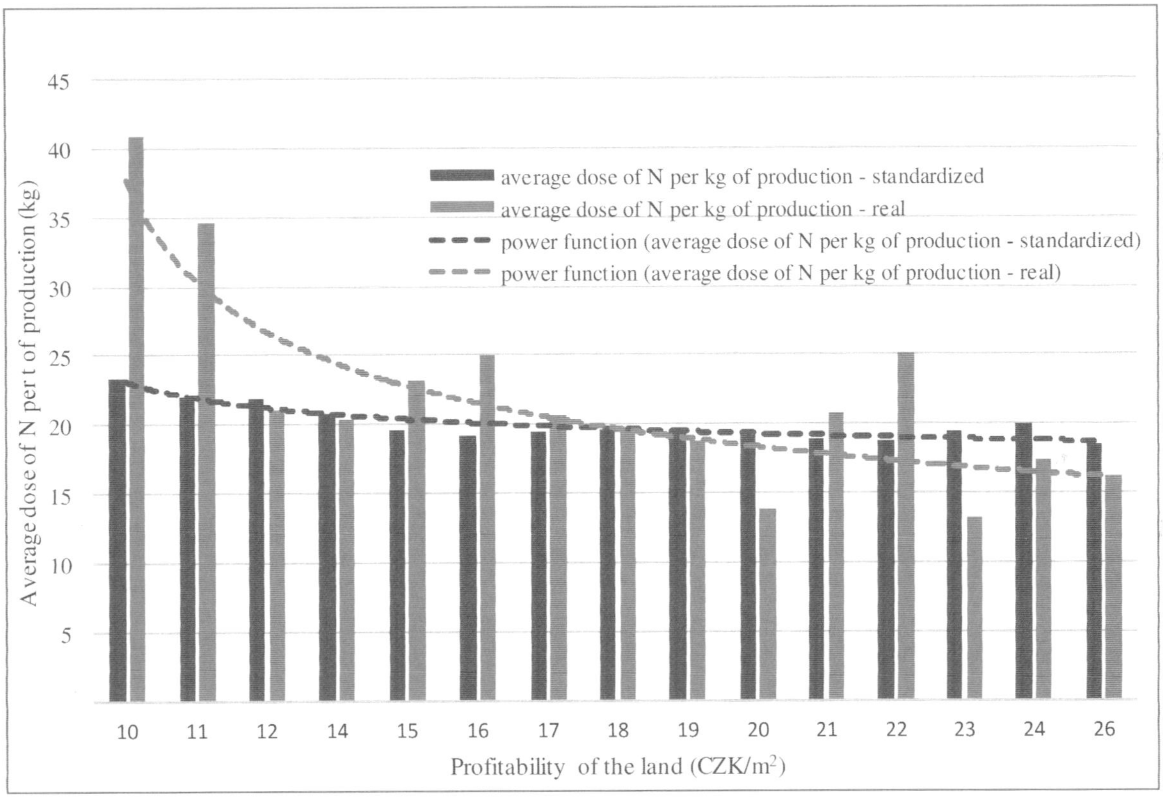

Figure 4. The weighted average of a dose of N per tonne of production for all selected crops.

The result of the findings mainly affects the deteriorating ecological parameters of the soil in marginal areas that have not yet been subject to fertilizer management restrictions under the Nitrate Directive. The increased nutrient load has an impact on their crop resorption (Figure 4) and leads to the leaching of fertilizers into the water.

The overall average consumption per tonne of production, depending on the yield of the soil, shows that the trends described are the predominant average of the total of enterprises and crops, and in marginal areas, there is a higher nutrient supply than plants need for their growth.

4. Conclusion

Data on nutrient intake for crops from 60 agricultural holdings were obtained in 2012–2016. Data evaluation shows that in marginal areas there is increased fertilization with nitrogen fertilizers, especially for feed crops. Linear regression models for the main crops were developed, taking into account both the nitrogen fertilizer dose per tonne of production and the P_2O_5 and K_2O doses and the crop year of the crop that confirmed this trend. These conclusions are partly related to climate change, where farms may expect higher temperatures at higher elevations to support plant growth. Because of the lower strength of the humus horizon in the marginal areas, it can be assumed that by increased fertilization the agricultural enterprises solve the lower sorption capacity of the soil. This is also related to the need to provide feed for livestock production, along with the supply of biogas stations that often occur in marginal areas. At the same time, there is a demand for silage maize for foreign biogas stations.

The overall finding is also a warning to the occupation of quality land for non-agricultural purposes because their intensification cannot be transferred into marginal areas with no environmental impact.

Acknowledgement

We would like to express our gratitude towards the financing of the research by the Ministry of Agriculture directly and also with NAZV – project of MoA QK1710307 ‘Economic support for strategic and decision-making processes at national and regional level leading to the optimal use of renewable energy sources, especially biomass, while respecting food self-sufficiency and soil protection’.

References

[1] Zhao B Q, Li X Y, Liu H, Wang B R, Zhu P, Huang S M, Bao D J, Li Y T and So H B 2011 Results from long-term fertilizer experiments in China: The risk of groundwater pollution by nitrate. *NJAS - Wageningen J. Life Sci.* **58** 177-83. https://doi.org/10.1016/j.njas.2011.09.004

[2] Savci S 2012 Investigation of effect of chemical fertilizers on environment. ICESD 5-7 January, Hong Kong. *APCBEE Procedia* **1** 287–292. https://doi.org/10.1016/j.apcbee.2012.03.047

[3] Zhang J, Zhu A, Xin X, Yang W, Zhang J and Ding S 2018 Tillage and residue management for long-term wheat-maize cropping in the North China Plain: I. Crop yield and integrated soil fertility index. *Field Crops Res.* **221** 157–165. https://doi.org/10.1016/j.fcr.2018.02.025

[4] Zhang J, Balkovič J, Azevedo L B, Skalský R, Bouwman A F, Xu G, Wang J, Xu M and Yu C 2018 Analysing and modelling the effect of long-term fertilizer management on crop yield and soil organic carbon in China. *Sci. Total Enviro.* **627** 361–72. https://doi.org/10.1016/j.scitotenv.2018.01.090

[5] Voltr, V *et al.* 2011 *Hodnocení půdy v podmínkách ochrany životního prostředí (Evaluation of land in environmental conditions).* (Prague: Ústav zemědělské ekonomiky a informací), ISBN 978-80-86671-86-4, 480

[6] Voltr V, Froněk P and Hruška M. 2013 Impact assessment of trim levels of mechanization in company on yield of winter wheat. ČZU *Trends Agri. Engin.*, **5**, 651–656, ISBN: 978-80-213-2388-9

[7] Neuberg J 1989 Komplexní metodika výživy rostlin. (Comprehensive plant nutrition methodology) *Metodiky zavádění výsledků výzkumu do praxe. (Methodology for implementing research results into practice)* (Prague: Ústav vědeckotechnických informací pro zemědělství, Knihovna Antonína Švehly E29092/1990/1b, 327)

[8] Zimolka J 2005 *Wheat, cultivation, grading and grain use* (Prague: Profi Press, s.r.o.) ISBN 80 8672609-6

[9] Liu Z, Chen Y, Ma P, Meng Y and Zhou J 2017 Effects of tillage, mulching and N management on yield, water productivity, N uptake and residual soil nitrate in a long-term wheat-summer maize cropping system. *Field Crops Res.* **213** 154–164.

https://doi.org/10.1016/j.fcr.2017.08.006
[10] Kremer A K 2013 *Methodology and Handbook Eurostat/OECD. Nutrient Budgets.* (Luxemburg: OECD), 112. http://ec.europa.eu/eurostat/cache/metadata/Annexes/aei_pr_gnb_esms_an1.pdf
[11] Voltr V 2012 Concept of soil fertility and soil productivity: evaluation of agricultural sites in the Czech Republic.*Archives Agron. Soil Sci.* **58** S243-S251 DOI: 10.1080/03650340.2012.700511
[12] Voltr V, Hruška M, Šařec P, Leština J and Froněk P 2012 *Land Valuation Methodology for Valuated Soil–Ecological Units (BPEJ).* (Prague: ÚZEI), 282. http://www.uzei.cz/data/usr_001_cz_soubory/metodika_oceneni_bpej.pdf
[13] Voltr V and Hruška M 2013 Effect of crop selection on the economy of crop production and quality of the environment. *Agrarian Perspect.*, **22**, ČZU, 389–402, ISBN 978-80-213-2419-0 https://ap.pef.czu.cz/en/r-12193-conference-proceedings
[14] Leip A, Brity W, Weiss F and Vries W D 2011 Farm, land, and soil nitrogen budgets for agriculture in Europe calculated with Capri, Envir. Pollut. 159 pp 3243–3253. https://ac.els-cdn.com/S0269749111000625/1-s2.0-S0269749111000625-main.pdf?tid=4036e28f-35f8-4230-921e-42d9c6e32260&acdnat=1524219435_13b02fef0399a383e6feb019f58fefdf
[15] Klír J, Kunzová E and Čermák P 2008 *Framework methodology of plant nutrition and fertilization. Metodika pro praxi.* (Prague: VÚRV) p 48, ISBN 978-80-87011-61-4 https://www.vurv.cz/sites/File/Publications/ISBN978-80-87011-61-4.pdf
[16] Csathó P *et al.* 2007 Agriculture as a source of phosphorus causing eutrophication in Central and Eastern Europe. *Soil Use Manag.*, **23** (1 suppl.): 36–56 DOI: 10.1111/j.1475-2743.2007.00109.x

IOP Conf. Series: Earth and Environmental Science **185** (2018) 012022 doi:10.1088/1755-1315/185/1/012022

Contemporary condition and conservation of wetland resource in Pingshan, Hebei Province

Z C Wang, P Y Ma, S Y Wang, Q Li, S Zhang and Y J Zhao[1]

College of Resources and Environmental Sciences, Hebei Normal University, Hebei, China

E-mail:ecoenvir@163.com

Abstract. The current work envisages the use of Remote sensing,Geography information systems and Global positioning systems.literature analysis of precipitation, field investigation and laboratory analysis to perform a comprehensive investigation on the present situation of wetland resources of Pingshan Wetland in Hebei Province. The results showed that there were 5 types of wetlands, including the riverine wetland, flood plain, reservoir and pond, water-conveyance canal, and paddy field, with a total area of 162.61 km^2. There were 311 species of wetland higher plants in Pingshan County, including 10 species of bryophytes, 296 angiosperm species, 5 species of ferns, and 81 phytoplankton species. There were 186 wetland birds, 35 species of fish, 5 species of amphibians, 14 species of reptiles, and 19 species of mammals in Pingshan County. In view of the present situation and the threat of wetland in Pingshan County, certain suggestions have been proposed to strengthen the protection and utilization of the wetland.

1. Introduction

Wetlands are the most unique ecosystems with the highest biodiversity and ecological function resulting from the interaction of the earth's surface water and land [1-3]. They not only provide a variety of resources for the survival of mankind, but also play an important role in the conservation of biodiversity, regulation of the regional climate, water conservation, reserve for migrating birds, and maintaining the food chain and the carbon cycle. Wetlands, forests and oceans are known to be the world's three big ecosystems, and also known as the "kidney of the earth" [1, 4-11]. For a long time, due to the lack of understanding of the importance of wetlands, wetlands have suffered from invasion and damage, resulting in the shrinkage of the wetland area, the decrease of biodiversity, and the decline of wetland function and benefit [12-15]. The problem of ecological environment is becoming more and more prominent. The protection of wetland ecosystem, rational utilization and recovery of resources are currently hot topics in the world [2]. Since China joined the "Convention on Wetlands of International Importance Especially as Waterfowl Habitat" in 1992, two large-scale wetland surveys were conducted successively. However, due to the limitation of the administrative regions, it is still not possible to provide a comprehensive scientific basis for the protection, utilization and management of wetlands. In order to solve this problem, it is necessary to conduct a scientific investigation, especially on some important area of wetlands [16]. In this case, using the county as a unit for wetland resource protection, the efficiency of protection can be improved.

Wetland is an important part of Pingshan County regional, and this research was carried out in Pinshan County. Pingshan County is located in the west of the Hebei Province (113°31'~114°51', 38°09'~8°45'), the eastern foot of Taihang Mountain, and at the upper reaches of the Hutuo River [17], with a total land area of 2415.58 km2. The total wetland area of Pingshan County is 162.61 km 2, accounting for 6.73% of the total land area. The wetland includes river wetland and artificial wetland, which can be divided into five types such as riverine wetland, flood plain, reservoir and pond, water-conveyance canal, and paddy field, accounting for about 14.7% of the country's 34 wetland types.

2. Scope, content and method of wetland resources investigation

2.1. The scope investigation

2.1.1. Riverine wetland. The boundary is defined according to the area submerged in the highest annual maximum water level during the investigation period, and fine-tuning according to the second national land survey results.

2.1.2. Flood plain. This includes the flooded river banks, river valleys, and seasonal overflow of grasslands during the flood season.

2.1.3. Artificial wetland. The area is not less than 5 hm^2 reservoir and pond, water-conveyance canal, paddy field.

2.1.4. Reservoir and pond. These water catchment areas include reservoirs, agricultural ponds, urban park landscape etc., caused by water storage, power generation, agricultural irrigation, urban landscape and rural life.

2.1.5. Water-conveyance canal. These include the artificial river wetlands built for water transmission for irrigation purpose.

2.1.6. Paddy field. This includes a paddy field or winter water storage or a wet farmland.

2.2. Survey content

With reference to the National Technical Regulations for Wetland Resources Survey (the trial), we choose research contents included the wetland type, area, distribution (administrative region, coordinate range), average altitude, vegetation type, the main plant species, etc. The survey of wild wetland animals include assessing the type, distribution and living environment of main aquatic and terrestrial vertebrates in wetland, such as aquatic birds, mammals, reptiles, amphibians and fish; wetland management and protection, wetland utilization, and threatened status.

2.3. Survey method

The investigation of wetland resources in Pingshan County was carried out by means of "3s" technology as well as combining the literature analysis of precipitation, field investigation and laboratory analysis. According to the National Technical Regulations for Wetland Resources Survey (the trial) issued by the State Forestry Administration (December 2008), we collected the original data from the 0.5m resolution aerial photography image, contour vector data, 1:50000 land use status map. The other raw data was sorted using ARCGIS10.2 to make a field investigation map, and a hand-held GPS to conduct field and spot visit. Finally, the wetland type, distribution, area, average altitude, main vegetation type, wetland wildlife, and the protection management status were obtained.

3. Results of the survey

3.1. Types and distribution of wetlands

3.1.1. Identification of wetland patches. The wetland in Pingshan County is classified as a wet area because of its close hydrological relationship. According to the difference of wetland type, the wet area of Pingshan County is divided into five wetland patches including the Huangbizhuang reservoir wetland patch, Gangnan reservoir wetland patch, Yehe wetland patch, Hutuo upstream wetland patch and other wetland patches (Figure 1).

3.1.2. Type, area and distribution of wetland. Pingshan Wetland is located in the upper and middle reaches of the Hutuo River, and belongs to an inland wetland and the transitional zone from Shanxi platform to North China Plain [18]. It consisted of 2 wetland kinds and 5 wetland types, with a total area of 162.61 km^2 at an altitude of about 115~710 m. The area of various types of wetlands and their proportion are shown in Table 1.

The permanent rivers of Pingshan County include the Hutuohe River mainstream and its 12 branches, which are fan-shaped, with a total area of 43.96 km^2. The flood plain that includes the flood waters and seasonal grasslands in flood season, are mainly distributed along the banks of permanent rivers with a total area of 19.78 km^2. The storage area of the reservoir and pond distribute in the whole Pingshan County, the area of which is less than 5 hm^2. The total reservoir and pond area was 94.74 km^2. The water area of the two large reservoirs is 72.10 km^2, and the water surface area of the small and medium-sized reservoirs is about 22.64 km^2. The water-conveyance canal is distributed in the low mountains and hills and plain areas of Pingshan County with a total area of 1.33 km^2, of which the paddy fields consisted of a total area of 2.80 km^2.

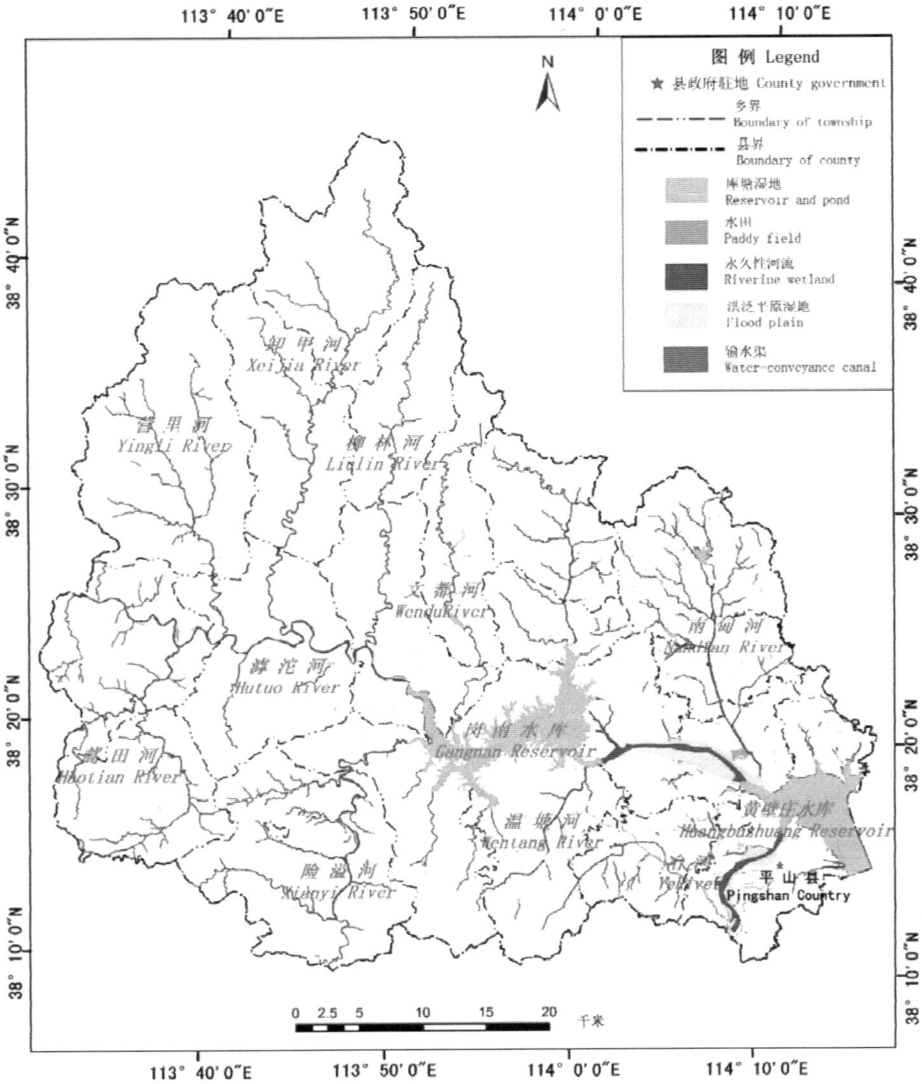

Figure 1. Spatial distribution map of wetlands in Pingshan County.

Table 1. Statistics of types of wetlands in Pingshan County.

Wetland types	Area/km²	The proportion of the total area /%
Riverine wetland	43.96	27.04%
Flood plain	19.78	12.16%
Reservoir and pond	94.74	58.26%
Water-conveyance canal	1.33	0.82%
Paddy field	2.80	1.72%

3.2. Plant resources in wetland

3.2.1. Composition of plant in wetland. Pingshan County has a diverse variety of wetlands. There are 311 species of wetland higher plants, accounting for 10.13% of the 3071 species in Hebei province. These include 10 species of bryophytes accounting for 3.2%, 5 species of fern (1.6%), and 296 species of angiosperms (95.2%). Among the angiosperms, there are 42 species of gramineous plant, accounting for 14.2%. Of them, 40 species of composite (13.5%), 22 species of cyperaceae (7.4%). There are 20 kinds of legumes, accounting for 6.8%; there were 20 species of polygonaceae, accounting for 6.8%. No gymnosperms are found in this investigation. Among them, Glycine soja is the national secondary protection plant. The survey area is dominated by the herbaceous plants, and the dominant species included the *Echinochloa crusgalli*, *Setaria viridis*, *Polygonum maackianum*, *Typha orientalis*, *Phragmites australis*, *Carex spp.*, *Cyperus rotundus*, *Potentilla chinensis*, *Artemisia annua*, *Salsola collina*, *Glycine soja*, as well as a variety of algae, etc.

3.2.2. Types of plant communities. According to the sample lines and the sample sites, there are 19 species of the main plant populations in Pingshan County wetland, as shown in Table 2.

Table 2. Major types of plant communities

Order Number	Community Name	The Main Plants
1	*Echinochloa crusgalli* community	Echinochloa crusgalli, *Cyperus rotundus, lisma plantago-aquatica, Sagittaria trifolia* L. var. *Sinensis*
2	*Echinochloa crusgalli~Rumex acetosa* community	*Rumex acetosa, Echinochloa crusgalli, Polygonum aviculare, Potentilla chinensis*
3	*Echinochloa crusgalli~Cyperus rotundus* community	*Echinochloa crusgalli, Cyperus rotundus, Typha orientalis, Salix babylonica, lisma plantago-aquatica, Rumex acetosa, Potentilla chinensis*
4	*Cnidium monnieri* community	*Cnidium monnieri, Potentilla chinensis, Artemisia argyri, Conyza canadensis, Echinochloa crusgalli*
5	*Ixeridium sonchifolium~Humulus scandens* community	*Ixeridium sonchifolium, Humulus scandens, Salsola collina, Lagopsis supina, Salix babylonica, Ulmus pumila, Dendranthema indicum*
6	*Equisetum arvense* community	*Equisetum arvense, Setaria viridis, Artemisia annua, Tribulus terrester, Chenopodium glaucum, Humulus scandens, Lespedeza bicolor*
7	*Vicia sepium~Artemisia*	*Vicia sepium, Glycine soja, Artemisia argyri, Potentilla supina, Chenopodium glaucum, Xanthium sibiricum,*

	argyri community	*Humulus scandens, Plantago asiatica*
8	*Artemisia argyri* community	*Artemisia argyri, Bidens pilosa, Medicago sativa*
9	*Zoysia japonica* community	*Zoysia japonica, Carex spp., Polygonum aviculare, Amaranthus lividus, Atriplex patens, Cyperus nipponicus, Portulaca oleracea, Rorippa indica, Eleusine indica*
10	*Abutilon theophrasti* community	*Abutilon theophrasti, Zoysia japonica, Amaranthus tricolor, Tribulus terrester, Chenopodium glaucum*
11	*Scirpus triqueter* community	*Scirpus triqueter, Bidens pilosa, Potentilla supina, Polygonum hydropiper*
12	*Chloris virgata* community	*Chloris virgata, Cyperus rotundus, Carex rigescens, lisma plantago-aquatica, Rumex acetosa, Rorippa indica*
13	*Heleocharis dulcis~Polygonum aviculare* community	*Heleocharis dulcis, Polygonum aviculare, Scirpus triqueter, Plantago asiatica, Bidens pilosa, Sparganium stoloniferum, Salvinia natans, Artemisia annua, Phragmites australis*
14	*Halerpestes ruthenica* community	*Halerpestes ruthenica, Arthraxon hispidus, Echinochloa crusgalli, Setaria viridis, Bidens pilosa*
15	*Calamagrostis pseudophragmites* community	*Calamagrostis pseudophragmites, Miscanthus floridulus, Phragmites australis, Dendranthema indicum, Leonurus artemisia, Humulus scandens, Artemisia lavandulaefolia*
16	*Phragmites australis* community	*Phragmites australis, Humulus scandens, Artemisia scoparia, Dendranthema indicum, Artemisia mongolica, Chenopodium glaucum*
17	*Typha orientalis* community	*Typha orientalis, Humulus scandens, Scirpus triqueter, Bidens pilosa, Commelina communis, Draba nemorosa, lisma plantago-aquatica, Glycine soja, Polygonum orientale, Phragmites australis*
18	*Sagittaria trifolia* L. var. *Sinensis~Potamogeton distinctus* community	*Sagittaria trifolia* L. var. *Sinensis, Potamogeton distinctus, lisma plantago-aquatica, Typha orientalis, Polygonum hydropiper, Scirpus triqueter, Cyperus rotundus, Myriophyllum verticillatum*
19	*Eleusine indica~Typha orientalis* community	*Eleusine indica, Typha orientalis, Plantago asiatica, Oenanthe javanica, Scirpus triqueter, Artemisia argyri*

3.3. Wetland resources of phytoplankton

According to a case study of the investigation on the aquatic biological resources of the national aquatic germplasm resources reservation at the Gangnan reservoir and referring to the results of the survey conducted in the case of Gangnan reservoir, we find 81 types of phytoplankton. Among them, there were 46 species of *Chlorophyta*, 14 species of *Bacillariophyta*, 5 species of *Cyanophyta*, 5 species of *Xanthophyta*, 5 species of *Cryptophyta*, 3 species of *Euglenophyta*, 2 kinds of *Pyrrophyta*, and 1 species of *Chrysophyta*.

The results of this survey are quite different from those of Liang Chunguang *et al.* (2014) [19]. Our survey has found an increased number of species of *Xanthophyta* by 39, including 24 species of *Chlorophyta*, 6 species of *Diatoms*, 5 species of *Bacillariophyta*, 4 species of *Cryptophyta*, and 1 of *Euglenophyta*, *Pyrrophyta*, and *Chrysophyta*, respectively.

3.4. Wetland wildlife resources

3.4.1. Species of wetland birds. According to the results of the field investigation combined with the research on the scientific investigation and biodiversity of the Hebei Tuoliang reserve and China Vertebrate Bibliography [20-21], we have found 186 species of birds in Pingshan wetland belonging to 17 orders, 52 families, and 105 genus. The species of wetland birds in Pingshan wetland accounted for 13.96% of the 1332 species of known birds in China, accounting for 39.74% of the 468 species of the known birds in Hebei Province.

Among the wetland birds in Pingshan County, there are 74 species of 23 families, 42 genus, and 44.23%, 40% and 39.78% of the total families, genus, and species respectively of the passeriformes. There are 112 species of non passeriformes, 63 genus, which accounted respectively for 55.77%, 60.00% and 60.22% of a total family number, genus number and species number. In level of the family, genus, species, the non passeriformes birds was more than that of the passeriformes.

Among the 238 species of birds in the wetland and *Ciconia nigra* protection village in Pingshan County, 20 species of national key protected birds accounted for 10.75% of the total number of birds in the region. Among them, three species of national II level protected birds are *Ciconia nigraa*, *Grus japonensis* and *Otis tardaa*, which accounted for 1.7% of the total number of birds. The 17 species accounted for 9.14% of the total number of birds belongs to the national II level, including *Platalea leucorodia, Cygnus Cygnus, Cygnus columbianus, Anser albifrons, Aix galericulata, Milvus korschun, Accipiter nisus, Pandion haliaetus, Grus Grus and Grus vipio*.

3.4.2. Major fishes. According to the preliminary investigation and statistics, combined with the relevant data records [22],there are 35 species of fishes in Pingshan wetland, belonging to 5 orders, 11 families and 34 genus, and accounted for 41.2% of freshwater fishes in Hebei Province. The *Cyprinidae*, a total of 23 species, accounted for 65.7% of the total area of fish. *Cobitidae*, a total of 3 species, accounted for 8.6% of the total fishes. 5 families and 5 species of *Perciformes*, 2 families and 2 species of *Siluriforms*, and one species each of *Cyprinodontiformes* and *Synbranchiformes*, accounted for 14.3%, 5.7% and 2.9%, respectively, of the total number of fishes in the each investigation area. Among the 32 monotypic species, a total of 20 species belong to *Cyprinidae* accounting for 62.5% of the monotypic genus of fish; 3 *Cobitidae*, accounting for 9.6% of the monotypic genus of fish, and one species each of *Bagridae*, *Siluridae*, *Oryziatidae*, *Eleotridae*, *Giurinus*, *Anabantidae*, *Channidae*, *Mastacembelidae*, *Synbranchidae*, accounting for 3.1% of fishes of the monotypic genus.

3.4.3. Major amphibians and reptiles. Through the combination of field visits and the wild animal survey data, 5 species of amphibians are found in Pingshan County, which belong to 3 families. There are 14 species of reptiles that belong to 9 genus and 6 families.

Rana chensinensis was the dominant species of amphibians in the wetland area of Pingshan County accounting for 84.7% of the total number found in amphibians. *Bufo raddei* and *Bufo gargarizans* are found often in a humid environment full of luxuriant trees and abundant grasses that accounted for 8.6% and 5.1%, respectively. *Rana nigromaculata* and *Kaloula boreals* are found in the pit ponds and stream waters and accounted for 1.6%.

The amphibian family, genus, and species of wetland in Pingshan County accounted for 60%, 60% and 50% of the Hebei Province respectively, and the reptile family, genus and species accounted for 75%, 69.2% and 58.3%, respectively; thus it can be seen that the amphibians and reptiles in Pingshan wetland are relatively abundant in the whole province (Table 3).

Table 3. Composition and proportion of amphibians and reptiles in Pingshan wetland.

Group of Animals	Pingshan wetland	Hebei Province	The proportion of the total number of Hebei/%	China	The proportion of the total number of China/%
	Family: Genus: Species	Family: Genus: Species		Family: Genus: Species	
Amphibian	3:3:5	5:5:10	60.0:60.0:50.0	11:51:321	27.3:5.9:1.6
Reptiles	6:9:14	8:13:24	75.0:69.2:58.3	24:125:407	25.0:7.2:3.4

There are 19 species of amphibians and reptiles in the Pingshan wetland. Among them, *Gekko swinhonis*, *Eumeces elegans* and *Eumeces capito* are the endemic in China, while *Pelodiscus sinensis* is included in IUCNS and VU in the red list of Chinese species. *Elaphe anomala*, *Elaphe davidii*, *Elaphe taeniura*, *Gloydius brevicaudusus* and *Gloydius intermedius* are also listed as vulnerable VU in the red list of Chinese species. *Rana nigromaculata* is listed in the red list of Chinese species (NT).

3.4.4. Major mammals. According to investigation, there are 19 species of mammals belonging to 6 orders, 13 families, 18 genus in the wetland and surrounding areas of Pingshan County. There are 2 species of 2 families of *Insectivora*, 2 species of 2 families of *Chiroptera*, 1 speciy of 1 familiy of *Lagomorpha*, 3 species of 3 families of *Lagomorpha*, 7 species of 4 families of *Carnivora*, 1 speciy of 1 familiy of *Artiodactyla*.

Due to Pingshan wetland are located in warm temperate zone, belonging to semi-humid to semi-arid monsoon type continental climate, the geographical environment is complicated. There are many different kinds of vegetation and abundant resources and Pingshan County offering a variety of different types of good animals breeding environment. This is the result of long-term evolution and animals' adaptation to the environment. The species composition of wetland in Pingshan County is relatively rich. The number of the order and the genus in Pingshan County is at the middle level of all protection areas. However, for animal species, the number and proportion of wetland species are all at a low level.

4. Problems in the utilization and protection of wetlands

4.1. Water pollution in wetlands
The main sources of pollution in Pingshan County are the industrial wastes from mining and the household waste. In addition, agricultural production caused by pesticide and fertilizer runoff, also pollutes the water, in flood season. When flowing into the reservoir of high concentrations of nitrogen, phosphorus and other nutrients [23], a large number of pollutants can't realize self-purification, the wetland system of wetland water quality declines, the biomass decreases, ecological system is damaged.

4.2. Unreasonable development and utilization reduces sustainability of wetlands
The unreasonable exploitation and utilization of wetland is a serious of threats to the sustainable development of wetland. In 2006, the total area of wetlands in Pingshan County was 195.50 km^2 [24]. In 2016, the total area of wetland in Pingshan County was 162.61 km^2. The main reason is reclamation that reduced the wetland area and declined the plant community, especially the distribution of reed and shrub forest. The number of animals is reduced and the function of wetlands is degraded. Overgrazing and irrational farming have damaged the vegetation in the wetlands, making the habitat of birds smaller and fewer. Overfishing has led to a huge decline in the number of fish in the reservoir and the rivers, and the variety has reduced.

4.3. Poor planning and low protection efficiency for wetlands
Due to the lack of coordination, utilization and protection of wetland resources, the wetland management departments have no unified scientific planning. In addition, the laws and regulations are not stringent enough to protect a relatively perfect system due to lack of specification and basis of the wetland protection. Although some departments have launched some research and taken measures to protect the wetland, the effect is not significant, and the destruction of the wetland resources and waste are still growing.

5. Countermeasures and suggestions for wetland utilization and protection

5.1. Strengthening the protection of wetland water quality
The mining factories and enterprises should be prohibited to discharge sewage into the wetland, and instead the municipal solid waste and domestic sewage treatment shall be used. By controlling the species and quantity of fish and plants in the wetland, the ecological stability and the ability of purifying water can be controlled. The quality inspection of water bodies should be strengthened.

The sampling points should be reasonably arranged in the water body, and real-time comprehensive data should be obtained to control the dynamic changes of wetland water quality in a timely manner.

5.2. Wetland restoration and ecological restoration mechanisms
According to the actual situation of Pingshan County, we propose the measures to transfer farmland into wetland and fishery. We propose that by prohibiting overfishing, illegal and unreasonable grazing in farming, reclamation activity and increase the penalties, the wetlands can be duly protected. Besides, we can appropriately adjust the compensation, and provide economic benefits to local residents, which will reduce the interference of human activities on the wetland, and the sustainable development and utilization of wetlands can be harnessed.

5.3. Refining wetland reserve planning and wetland science management
We propose to perfect the wetland protection management agencies and administration department that will lead to strengthen the departments' linkage, form resource sharing mechanism, and implement the relevant work. Hierarchical planning for nature reserves in wetland areas should be undertaken, such as rational utilization of areas, conservation areas, restoration and reconstruction areas, etc. Improvement in the construction of wetland laws and regulations, rational exploration of the current situation of wetland resources, and provision for the basis of the protection and management of wetlands can help in managing the wetlands successfully.

Acknowledgment
The authors sincerely appreciate and acknowledge the field investigation work done by Li Jianping, Li Wei, and Luo Vanping, Pingshan County Forest Service.

References
[1] Liao Y J and Song C C 2009 Research advances for the degradation of wetland ecosystem *Chin. J. Soil Sci.* **40(5)**1199-1203
[2] Zhang L Z, Zhang X L, Bai X F, Sun Y J 2014 Ecological effects and protection measures of Yimin river upstream wetland *Protection Forest Sci. Tech.*.**(4)**90-2
[3] Quan Q Z 2017 Evaluation of heavy metal pollution in wetland:Research progress in China *Chin. Agric. Sci. Bull.* **33(8)**60-7
[4] Wang Y Y, Sheng L X and He C G 2005 Outlook for international wetland ecology research:the 7th INTECOL international wetlands conference and its inspirations on China *Geo. Geo-Infor. Sci.*(**6**)56-60
[5] Zhang F W, Liu A H, Li L N, Li L N, Zhao L and Wang Q X 2008 CO_2 flux in alpine wetland ecosystem on the Qinghai-Tibetan plateau *Acta Ecol. Sin.* **28(2)**453-62
[6] Martha A S, Brian C P and Enrique R 2003 Factors affecting spatial and temporal variability in material exchange between the Southern Everglades wetlands and Florida Bay(USA) Estuarine,*Coastal Shelf Sci.* **(57)**757-81
[7] Sovikak K 2007 Emission of N_2O and CH_4 from a constructed wetland in southeastern Norway *Sci. Total Environ.* (**380**)28-37
[8] Ren G H, Deng B and Hou Y 2015 Changes of plant community characteristics in the degradation process of the alpine swamp wetland in the Yellow River source area *Pratac.Sci.* **32(8)**1222-29

[9] Wang Y, Li Y H and Sun X Y 2015 Impact of climate change on the eco-environment in the Yellow River source *Pratac .Sci.* **32(4)**539-57
[10] Zhang Y, Wei Y M, Wei M H and Li L 2016 The resources survey report on the Gansu Yellow River Three Gorges wetland nature protection area *Pratac. Sci.* **33(8)**1509-17
[11] Li Y M and Li Z Q 2013 Progress and prospect of research on wetland at home and abroad *Yunnan Geographic Environ. Res.* **25(1)**36-43
[12] Mitsch W J and Gosselink J G 2000 The value of wetlands:importance of scale and landscape setting *Ecol. Wconomics* **35(1)**25-33
[13] Cole C A 2002 The Assessment of herbaceous plant cover in wetlands as an indicator of Function *Ecol. Indic* **2(3)**287-293
[14] M Quinton W L and Pietroniro A H 2004 Hydrologic functions of wetlands in a discontinuous permafrost basin indicated by isotopic and chemical signatures *J. Hydrol.***296(4)**81-97
[15] Zhang F 2004 Ecological environment degradation characteristics of wetland in Shanxi and restoration strategies *J. Soil Water Conserv* **18(1)**151-4
[16] Wu F Z, Wang J L, Zhong X Y and Fan T 2012 A study on the survey and protection of wetland at Dashanbao Black-Necked crane national nature reserve *Yunnan Geographic Environ. Res.* **24(2)**78-89
[17] Wang H, Zhang S, Zheng Q, Hou F, Chen L and Zheng *Y et al.* 2015 Investigation and evaluation of Chinese herbal resources in Pingshan county *Modern Chin. Medicine* **17(10)**1051-6
[18] Guan W L 2008 Evaluation and protection countermeasures of wetland resources in Pingshan county *Hebei Fore. Sci. Tech.* **(3)**52-4
[19] Liang C G, Liu Z W and Wang L P 2014 Assessment on community structure of phytoplankton and water quality in Gangnan Reservoir *Water Resour. Hydr. Eng.***45(9)**27-9
[20] Zhao J C, Wu Y F and Guan W L 2008 *Area of scientific investigation and study on the biodiversity of Hebei Tuoliang nature reserve*, Beijing:Science Press, 254
[21] Liu M Y, Jie Y H and Ji D M 2000 *Encyclopedia of vertebrate animals in China,* Shenyang:Liaoning University Press, 681
[22] Wang S A, Wang Z M and Li G L 2001 *Hebei fauna:fishes*, Shijiazhuang:Hebei Science and Technology Press, 201
[23] Wang L 2016 Watershed pollution control and ecological rehabilitation of Gangnan Reservoir and Huangbizhuang Reservoir *Value Engineering* **35(9)**180-2
[24] Sun J Y, Zhao X Y and Lu Z 2006 Evaluation and development of wetland sightseeing agricultural resources in Pingshan County, Hebei Province *J. Shanxi Normal Uni. (Natural Science Edition)* **(3)**88-91

Poverty alleviation of "smart countryside": an empirical study on mechanism, innovation model and guarantee systems

L Chen[1]

Business School of the University of International Business and Economics, Beijing, China

E-mail: chenlian0208@126.com

Abstract: Since twenty-first century, the third revolution of science and technology as the core of the internet information technology makes society become more and more "smart". Under the guidance of the strategy of "smart planet", the concept and practice of "smart countryside" emerged. Using of energy and information & communication technology access to rural society to build a new "Smart countryside" in the 3I situations (Instrumented, Interconnected, Intelligent), which helps rural and poor areas out of poverty with a new form of poverty alleviation under the background of the internet. The paper analyses the mechanism of "smart countryside" poverty alleviation from the four perspectives: sustainable development of energy, social capital and knowledge, bridging the digital divide and promoting the information rights. Then, the paper investigates and summarizes seven kinds of innovation model of "smart countryside" poverty alleviation. The conclusion shows that the innovation model is conducive to the "multi-level" ecological interaction of poverty alleviation, which is "government-led, social participation, industry development and the poor benefit". It also brings environment, economy, service, the humanities, and the people's livelihood performance for poverty alleviation. Finally, the paper puts forward policies supporting system for "smart countryside" poverty alleviation from the aspects of technology, space, capital and people.

1. The Proposal of the problem

Since twenty-first Century, in the development of the Internet era led by the third scientific and technological revolution, information differentiation has formed a clear technology partition system [1] in today's society, which causes the information gap to aggravate the opportunity poverty, the efficiency poverty and the evasion of the risk of poverty [2]. Therefore, it is necessary to pay high attention to the problem of information poverty in the Internet environment, because information poverty is not only an important cause of income poverty and human poverty, but also the result of their poverty [3]. In China, the level of information poverty in national poverty counties is significantly lower than the national average level of [4]. Luo Chuanling puts forward the use of rural informatization to improve the information literacy of farmers in an all-round way and take the way of developing poverty alleviation [5]. As an important work of developing rural informatization, "village to village project" can effectively narrow the digital gap between urban and rural areas and eliminate the information poverty phenomenon in rural areas [6]. Empirical studies show that the level of information service in rural communities has positive significance for increasing farmers' income and reducing the incidence of poverty [7]. Therefore, the application of Internet information technology to strengthen the rural informatization construction, accelerate the "Internet+" poverty alleviation projects is the inevitable choice of rural anti-poverty. Nowadays, the third technological revolution centered on Internet information technology has made society become more and more intelligent. Under the guidance of "smart planet" [8] strategy, the concept and practice of "smart countryside" emerge as the times require.

In recent years, the "smart tourism to help the poor" in Tibet Lu Langzhen Zha Xigang village, "intelligent e-commerce poverty alleviation' in Henan Xi Ying village, "intelligent energy poverty alleviation" in Hunnan District of Shenyang, "wisdom culture support poverty" in Hebei Luanshuihe village, and "wisdom information support poverty" in Qinghai's Daijia village are constantly emerging. From the grass-roots practice of "smart countryside", a series of urgent problems need to be explored: what is the mechanism of "smart countryside" to help the poor? In practice, what are the innovative models for poverty alleviation? How to ensure its smooth operation in the rural area where the Internet information society is in the "fringe"?

2. The mechanism of poverty alleviation in the "smart countryside"

"Smart Countryside" is originated from the concept of "Smart Planet" proposed by Sam Palmisano, CEO of IBM, in 2008 [9]. The strategy of "Smart Planet" indicates that the full application of the new generation of information technologies in all walks of life is essentially the embedment and equipping of sensors in various articles for universal connections to form "Internet of Things" and then the Internet of Things is integrated with the Internet to conduct real-time management and control of personnel, machines, equipment and infrastructure in the network, so as to realize refined and dynamic human production and living and lead to a "smart" instrumented, interconnected and intelligent status of the planet [10]. Therefore, "Smart Countryside" is defined in this paper as "a digitalized, networked and intelligent brand-new rural form constructed through connection to sustainable energies and application of information and communication technologies".

The paper expounds the mechanism of "smart countryside" from the following four aspects: sustainable development of energy, social capital and knowledge, bridging the digital divide and enhancing information rights.

2.1. Energy sustainable development

Around the world, nearly 130 million people still do not have access to electricity. These groups are usually located in remote areas, far away from the urban centers, and are placed outside the national grid. In many areas far away from power grid, renewable energy is gradually regarded as the most practical and economic choice. Green poverty alleviation is based on the protection of the ecological environment in the poverty-stricken areas, the rational exploitation of natural resources, the promotion of green low-carbon economy and sustainable development in the poor areas, and further help the poor people in the fragile areas to get out of poverty and become rich [11]. For example, "Internet + tourism poverty alleviation", increasing "cost-utility" photovoltaic technology applications for the poor areas (green energy) force for poverty reduction. "Internet + tourism" poverty alleviation is in the information age in the new poverty reduction by green, using the Internet to solve the information barriers between poverty areas and non-poverty areas and to improve the fairness of the application of information technology.

2.2. Increasing social capital and knowledge

Information and communication technology promotes human resource and social capital growth by promoting information flow, building endogenous knowledge and reducing human poverty [12]. The core contents of social capital and knowledge include social network resources, cooperative behaviors, and behavioral norms and information to promote reciprocal cooperation. The World Bank [13] believes that access to income and the ability to cope with poverty are the key factors to solve poverty, and social capital and knowledge can play a role in reducing poverty by increasing farmers' income, providing them with various opportunities and enhancing their capacity [14]. Information and communication technology can be regarded as a "correct" social capital to promote community economic and social progress. The weaker the social capital is, the harder the growth of knowledge and human capital is in a community, so poverty will continue to exist.

2.3. Plugging the digital gap

The so-called digital divide refers to the gap between the people who are exposed to the latest information technology and those who are not exposed to the latest technology [15]. The digital divide means the widening gap between the "owners" and "non-owners" of information, communication and technology all over the world. In developing countries, the digital divide may occur in places where infrastructure (such as electricity supply) is lacking or access to modern technology (Internet, computer or mobile phone). The World Bank (1998) pointed out that the new technology greatly facilitated the acquisition of knowledge and provided unprecedented educational opportunities for developing countries to bring greater business opportunities for the poor. [16] Moreover, information and communication technology can provide effective information for villagers 'economic welfare and social welfare [17].

2.4. Promotion of information rights

In the Internet information society, information right is a basic right of citizens. In the information age of the Internet, the lack of information rights of "information disadvantaged groups" has evolved into an important resistance to economic development and social progress. The inadequacy of the information resource acquisition ability of the disadvantaged groups inevitably leads to the lack of the opportunity to enjoy the information rights fully and equally, thus damaging the information rights. Amartya Sen viewed poverty from the perspective of deprivation of rights and deprivation of "feasibility" [18]. Poverty must be regarded as a deprivation of basic practical ability, not just a low income. Therefore, in the Internet information society, the anti-poverty management of rural and poverty-stricken areas can be carried out from the perspective of the promotion of information rights and the "feasible ability" recharge of information access to the rural and poor areas [19].

3. Innovative poverty alleviation models of "Smart Countryside"

With the significant improvement of rural network infrastructure and the flourishing of the Internet economy, the construction of a "Smart Countryside" has been implemented all over the China. Through summarizing the novel experience from the construction of "Smart Countryside", the poverty alleviation models are concluded into 7 forms in this paper and described in the following sections.

3.1. Smart tourism poverty alleviation model

Smart tourism refers to a systematic and intensified management providing tourists with high-quality and individualized services and social resources based on information and communication technologies [20]. As a new poverty alleviation model, the smart tourism poverty alleviation model boasts extensive participation of the poverty-stricken population, quick effect of poverty alleviation and low poverty-returning rate. Smart tourism is an intelligent service management mode centering on provision of tourism information. For example, the City of Linzhi, Tibet initiated the implementation of a "Smart Countryside" Tourism Information Engineering Construction Project and established a platform comprising rural information construction data to make it convenient for tourists to intelligently sense and conveniently utilize various kinds of tourism information, thus improving tourists' independent experience awareness. In the meantime, as an advanced stage of tourism informatization development, smart tourism demands technical supports from information, network and artificial intelligence.

3.2. Smart E-commerce poverty alleviation model

The national poverty alleviation work meeting of 2014 included e-commerce into the Chinese poverty alleviation policy system for the first time. Given that e-commerce has the market expansion effect, market flattening effect and information feedback mechanism, the development of e-commerce in rural poverty-stricken regions benefits the effective connection between the poverty-stricken population and big market, so that the poverty-stricken population can acquire opportunities to participate in the market and the effect of poverty reduction and elimination can be realized [21].

The e-businessman is a core subject of Smart E-commerce. The so-called e-businessman refers to any enterprise or individual making deals through the Internet. In rural areas, some villagers are lacking basic e-commerce operation skills, which give birth to the e-commerce "intermediary". For example, in the strategy of "Smart E-commerce" of Xinxiang County, Henan Province, "Rural Taobao Partners" helped villagers "vicariously buy and sell", thus opening the purchase and sales channels between the villagers and the outside world. Besides the e-businessman, platform enterprises are also a key subject in Smart E-commerce. The connection of the local market and national big market can be better realized only by fully utilizing the resources and strengths of the platform enterprises.

3.3. Smart information poverty alleviation model

Within the scope of "Smart Countryside", smart information poverty alleviation mainly refers to the application effect of "Smart Governmental Services" in the rural basic-level society on improvement of the information rights of poverty-stricken population [22]. Smart Governmental Services refer to a public service paradigm providing the public with seamless connection based on reality network and through comprehensive application of information and communication technologies. This kind of "ubiquitous" governmental service management model "centering on the public" and "benefiting all" presents certain convenient, transparent, autonomous, real-time and intelligent features, so as to provide enormous conveniences for the general public to get to know and handle governmental affairs. For example, the "Rural Digitalized Social Management Platform" of Daijia Village, Qinghai Province and "Smart Governmental Services" platform of Jiangzhu Township, Sichuan Province vigorously promoted the information construction of rural basic-level governments.

3.4. Smart energy poverty alleviation model

Smart Energy is a new energy system utilizing Internet thinking and technologies to reconstruct the traditional energy industry and realizing a high degree of integration between energies and information [23]. Smart Energy relying on the Internet information technology finally forms a high energy market and realizes full, transparent and free configuration of energy flow through the erection of a distributed energy transaction market. In this way, Smart Energy benefits the breaking of isolation and monopoly of traditional energy markets and ensures that the energy efficiency can benefit ordinary people in remote regions. For example, Hunnan District of Shenyang created an "Internet + New Energy" smart countryside construction model called "Aiyi Sheepherder" Internet + Rural Household Photovoltaic Power Station Project. The application of photovoltaic roof power stations cannot only solve the problem of energy shortage of peasant households, but also enable peasant households to acquire stable power generation benefits and promote the development of low-carbon and environment-friendly green economy.

3.5. Smart network poverty alleviation model

The urban-rural "binary" economic and social structure formed for a long time results in the serious lack of balance of China's information construction and the relatively severe situation of "great information gap" between urban and rural areas [24]. The establishment of Smart Network facilities is a key step to solve the problem of weak rural infrastructure, make the network system unimpeded and overcome the phenomenon of "Information Isolated Island". By accessing the Internet, peasants are able to get to know the cutting-edge agricultural development and improve their agricultural science quality. Besides, the formation of the Internet thinking of the peasant groups can be promoted and the good effect of information poverty alleviation can be achieved. For example, the integration of "Communication Network, Broadcast Television Network and Internet" has been successively realized in Tankun Village of Guangxi Zhuang Autonomous Region, Huizu Village of Shanxi Province and Dazhai Village of Shanxi Province, thus improving the local historical puzzle of information poverty.

3.6. Smart culture poverty alleviation model

It is clearly pointed out in Outline of Construction and Planning of the Public Culture Service System in Poverty-stricken Regions during "13th Five-year" Period that an important part of Smart Culture poverty alleviation is to implement "public digitalized culture construction project in poverty-stricken

IOP Conf. Series: Earth and Environmental Science **185** (2018) 012023 doi:10.1088/1755-1315/185/1/012023

regions". Library digital comprehensive service platforms for rural poverty-stricken regions shall be erected to realize one-stop acquisition of massive resources, provide diversified service modes and satisfy the users' individualized demands through integration of digital resources, design of unified exhibition interfaces and utilization of new media approaches [25]. For example, China Association of Poverty Alleviation & Development organized the commencement of the "Accurate Poverty Alleviation Smart Countryside Multimedia Reading Room" Public Welfare Project in Luanshuihe Village, Dongxiaobaiqi Township, Hebei Province, which was an innovative practice of "Internet + Public Culture Service". It provided people in the poverty-stricken regions with abundant and accurate culture information services, so as to promote the development of poverty-stricken regions.

3.7. Smart rural area poverty alleviation model

The Smart Rural Area Poverty Alleviation Model is an advanced model of "Smart Countryside" poverty alleviation, aiming at constructing agricultural informatization of rural areas and promoting the development of "Agriculture, Rural Areas and Peasants". To be specific, this model fully utilizes information and communication technologies, improves peasants' comprehensive quality and realizes their personal development, promotes the transformation of the agricultural development mode to reliance on modern technologies, realizes comprehensive and coordinated development of each public service undertaking of rural areas and finally forms a new pattern of integrated social development of urban and rural areas [26]. Such kind of "Smart Countryside" poverty alleviation model has been widely applied in relatively developed regions in China. For example, Xibaidian Village, Pinggu District, Beijing constructed "Smart Countryside" through a series of agricultural and rural informatization application engineering to realize the comprehensive integration of e-agriculture affairs, e-government affairs, e-commerce and information services with rural production, living and ecology. The construction path is mainly about the application of IOT technology to transform the "fuzzy" processing of agricultural production to "accurate control", so as to realize scientific management and improve production efficiency. In addition, emphasis is placed on the embedment of Internet information technology into financial and tourist services.

4. Establishing a policy supporting system for poverty alleviation of "Smart Countryside"

4.1. Technical level

To build the information industry ecology as the main body, information infrastructure planning and information resources development and utilization as the two wings, with green, intelligent and ubiquitous "one body, two wings, three dimensional" technical policy support system. We should enhance and improve ubiquitous and advanced information infrastructure, expand 4G network coverage, and promote the next generation Internet evolution and upgrading. Deploy cloud computing data center and Internet of things facilities in advance, so as to achieve optimal matching and effective collaboration between application technology and broadband network. It should vigorously support the construction of green cloud computing data center in rural and poverty-stricken areas. We will fully implement broadband rural areas and increase network investment and network coverage for remote and poverty-stricken areas.

We should accelerate the sharing mechanism of government, society and Internet data resources, and establish a unified and open big data system. Based on government information resources and relying on government data sharing platform, we accelerate cross layer and cross department data resource interaction. We should establish laws and regulations related to data opening, property rights protection and privacy protection as soon as possible, and strengthen data security protection.

4.2. Space level

To build a deep and wide coverage information service system, so as to achieve unified planning, systematic deployment and collaborative promotion.

Firstly, it needs to build a three level service platform system. At the national level, a unified big data platform for poverty alleviation and development is built to provide information collection and analysis support for poverty alleviation departments at all levels. At the county level, we should build an e-commerce platform in poverty-stricken counties, and vigorously develop the rural e-commerce. At the township level, we should establish an online-offline and interactive service provider for the poor towns, and give full play to the key role of the township node network in helping the poor.

Secondly, it needs work deployment with in the full factors. We should set up a network poverty alleviation team with high quality and ability, train Internet business leaders, help poor households improve their information skills, and achieve rich incomes. A simple and easy to operate network terminal is designed to solve the problem of poor households accessing the Internet.

Lastly, it needs to build a comprehensive information service system. In order to serve the people’s livelihood as the value orientation, we should set up the information service system of network poverty alleviation. To speed up the implementation of Internet plus government services, livelihood information systems and services integration and utilization of resources, build a comprehensive poverty alleviation service system, make public services inclusive of poor people.

4.3. Capital level

Establish a market-oriented and diversified investment and financing mechanism. Taking financial capital as the leading factor, attracting social capital such as corporate capital and crowd raising funds and social donation, we will build a new path to investment and financing of smart countryside for poverty alleviation.

To make full use of the "leveraged" guidance of financial funds, make full use of the funds in the rural self-financing, the central and local budget, temporary or one-time financial aid, so that it will become the basic capital for the construction and implementation of the “smart country”.

We should fully mobilize the enthusiasm of private capital investment in building smart countryside. From the international experience, telecom operators and high-tech enterprises are important investment entities in smart countryside construction. From the point of view of the practice of helping the poor in rural areas, Tennent Inc., Alibaba group, Qinghai giant Thailand Technology Company, China Telecom and other telecom operators and high-tech enterprises have actively invested in the construction of smart countryside, which shows that private capital has become one of the main financing sources of “smart country” construction.

4.4. Human level

To establish skills training mechanism, we should set up a skill training system for poor groups to take the initiative to help them out, so as to help the poor groups change from passive participation to voluntary participation.

First, from the stage of compulsory education to higher education, we should carry out all field pair assistance to poor students and accurately target poor groups. Secondly, we should infiltrate, enlighten, nourish and educate the poor people in culture and wisdom, so that they can widen their horizons and enhance their inner driving force. In the end, we should increase the education and publicity of the poor areas and invest in the funds of education, so that the children in the poverty-stricken areas are exposed to rich educational resources and are well educated to avoid the vicious circle of "intergenerational poverty".

References

[1] Xie J G, 2003, Information differentiation and information poverty alleviation in the process of social informatization, *Inform. Sci.*, **11**,1138-41.

[2] Wang W L, Mao Q Q and Yu J, 2008, Redefinition of poverty based on risk and opportunity perspective,*China Popul., Resour. Envir.*, **12**, 147-53.

[3] Zhou X H, 2016, From digital divide to digital poverty: basic concepts and research framework, *Academia Bimestrie*, **4**, 154-57.

[4] Li G and Joe H C, 2017, Under the background of rural poverty alleviation evaluation index system research, *J. Agrotech. Econ.*, **5**, 2120-128.

[5] Li S L, 2009, Commentary on the research of rural informatization and development oriented poverty alleviation, *Econ. Geogra.*, **10**, 1759-60.

[6] Zhang W L, 2011, Information and communication technology makes rural life better--- from “digital city” to “digital countryside”,*China New Telecommun.*,**10**,26-9.

[7] Gao M T, He Y and Shi H L, 2008, Information service and farmers' Income: China's empirical evidence, *J. World Econ.*, 6, 50-8.

[8] Zhang Z C and Lu G, 2015, The wisdom of the Earth concept, *Studies in Dialectics Nat.*, **11**, 117-22.

[9] Wang X T, 2018, China's smart city construction: partial “leading”, *China economic herald*, 2018-01-09, edition 07.

[10] Li H, 2014, The foundation of the Internet of things and its business mode research, *Beijing: China Fortune press*, 23.

[11] Zhang Q and Feng D M, 2016, Poverty alleviation practice and theoretical innovation in China: 1978~2016 years, *Reform*, **4**, 27-42.

[12] Cathy U, Shantha L and Muhammadoum K, 2008, ICTs and poverty reduction: a social capital and knowledge perspective, *J. Inform. Tech.*, **23**, 203-13.

[13] World Bank, 1990 World Development Report,*Beijing: China financial and Economic Publishing House*, 35.

[14] Guo J Y, 2011, Poverty reduction analysis of poor farmers under the perspective of social capital, *Comm. Res.*, **3**, 193-7.

[15] Compaine B M, 2001, The digital divide-Facing a crisis or creating a myth? *Cambridge: MIT press*,XI.

[16] Dalman D, 1998, A world bank expert says -- knowledge can narrow the gap between rich and poor, *Leadership decision-making information,* 16.

[17] Anwer M, 2006, ICTS for Better Society,*Dhaka: International Conference on Computer and Information Technology*, 56-70.

[18] Zhou M H, 2009, Farmers' rights poverty and its governance -- an analysis based on amartya sen's "feasible ability", *Gansu theoretical J*, **5**, 78-81.

[19] Xiang L L and Niu L H, 2016, The cause of information poverty based on Amartya Sen rights method, *Inform. Sci.*, **34**, 47-51.

[20] Li Y P, Hu Z, Huang C and Duer L, 2014, The concept of Intelligent Tourism in the perspective of tourism information service is discussed, *Tourism Tribun*, **29**,106-15.

[21] Lin G Y, 2016, The function mechanism and the mechanism of poverty alleviation in rural areas, *Chinese Academy of Social Sciences*, 118.

[22] Zhao D, 2013, From e-government to smart Government: paradigm shift, key issues and government coping strategies, *J. Intelligence*, **32**, 204-07+197.

[23] Guo Y W and Cheng A N, 2015, "Internet +" smart energy: Future direction of energy development, On Economic Problems, **11**,61-4.

[24] Li X W, 2011, Problems and causes of rural informatization operation mechanism and performance in western minority areas, *Social Scie. Ningxia*, **3**, 145-7.

[25] Jiang X X, 2017, Construction of digital integrated service platform for Library in poor county in the"Internet +"context, *Library Tribune*, **37**, 18-22+76.

[26] Hu Y M, 2016, Research on rural informatization construction under the background of urban and rural co-ordination development, *Jiangxi Social Sci.*, **36**, 208-213.

IOP Conf. Series: Earth and Environmental Science **185** (2018) 012024 doi:10.1088/1755-1315/185/1/012024

Evaluation method of new cultivated land quality in Guanzhong Plain of Shaanxi Province: Taking Longting Town land consolidation project as an example

J Li[1, 2, 3, 4, 5], H Guo[1, 2, 3, 4] and C S Shi[1, 2, 3, 4]

[1] Institute of Land Engineering & Technology, Shaanxi Land Construction Group Co. Ltd., Xi'an, Shaanxi 710075, China

[2] Shaanxi Provincial Land Engineering Construction Group Co., Ltd., Xin Group Co. Ltd., Xi'an, Shaanxi, China

[3] Key Laboratory of Degraded and Unused Land Consolidation Engineering, Ministry of Land and Resource, Xi'an, Shaanxi 710075, China

[4] Shaanxi Provincial Land Reclamation Engineering Technology Research Center, Xi'an, Shaanxi 710075, China

E-mail: lijuan8136@163.com

Abstract. In order to evaluate the quality of newly cultivated land as for the land reclamation project of Guanzhong Plain in Shaanxi Province, and to meet the requirements of seeking a simple and quick, convenient and accurate and universal method, this paper adopted the Analytic Hierarchy Process. The main evaluation index was selected to construct the judgment matrix. The weight coefficient of each index was determined. Each index was scored and the total score of new cultivated land was calculated. The quality of the new cultivated land was evaluated and the improvement suggestions were given. An analysis on quality of the new cultivated land - Longting Town in the land remediation project of Longting Town was conducted. After evaluation, the score of the new cultivated land was 62.8, which was a qualified level, and measures were proposed to improve the quality of the land and fertility. This method will provide a practical application for the evaluation of new cultivated land quality in other land reclamation project in temperate semi-arid plains.

1. Introduction

With the occupation of arable land because of economic development and the increasing demand for food due to population growth, it is imminent to solve the problem of maintaining the quantity and improving the quality of cultivated land. In order to achieve the dynamic balance of the total arable land, an important way is land consolidation and reclamation. China implemented the policy of "balancing the quantity of cultivated land being taken up and made up" so as to stick to the red line of 1.8 billion acres arable land [1]. While ensuring that the total area of arable land does not decrease, maintaining quantity and improving quality of arable land are important measures to promote food production and solve the contradiction between people and land. In this case, carrying out new farmland quality evaluation after finishing land reclamation project can reach the goal of project acceptance and give directional guidance on how to improve the quality of cultivated land.

The specific implementation of the quality evaluation of newly cultivated land includes the selection of indicators affecting the quality of cultivated land, site observation of cultivated land and post-sampling laboratory tests. Then, the evaluation of the quality of cultivated land can be carried out basing on the observational results and measurement results. Evaluation can be achieved through programming, mathematical methods, GIS software and other means. Hu *et al* [2] developed an application system that is compatible with the acceptance scheme by visual programming language Visual Basic 6.0, which can evaluate the cultivated land pollution index and the productivity of cultivated land. Luo [3] quantitatively compared the quality of newly cultivated land and the surrounding high-yield and middle-yield fields by using fuzzy mathematics comprehensive evaluation method. Li *et al.* [4] proved the feasibility of GIS-based comprehensive evaluation of the quality of newly cultivated farmland in small-scale in Karst hilly areas, which provided ideas and methods for the application of GIS in the evaluation of the quality of newly cultivated farmland at small scales. The ultimate goal of using various evaluation methods is to achieve a comprehensive, objective and accurate evaluation of the quality of arable land. The current study mainly focus on the hilly areas [4-6] and the valley areas [7], and researches on the farmland quality evaluation resembling that of the Guanzhong Plain in Shaanxi Province are rare.

This paper takes the newly cultivated land in Guanzhong Plain of Shaanxi Province as the object for quality evaluation.. Tomographic analysis is used to select evaluation indicators suitable for the region, and scores and calculations are made. According to the comprehensive scores, comprehensive assessment of cultivated land quality can be carried out. Taking the Longting Town Land Renovation Effort Compensation and Balance Project as an example, the new farmland quality evaluation process in the plain area is given to provide reference for the new cultivated land quality evaluation of the acceptance and compensation of farmland compensation and balance projects in temperate semi-arid plain areas.

2. Method for evaluation of the quality of newly cultivated land

Completion and acceptance of land consolidation and reclamation projects include the following seven parts: technical files, engineering tasks, newly cultivated land quality, newly cultivated land, land ownership, fund management, and management and maintenance measures [2]. The acceptance of newly cultivated land acting as an important and complicated content of acceptance requires comprehensive on-site measurement and combination with laboratory testing and literature research.

In order to complete the evaluation of the land consolidation and reclamation project in Guanzhong Plain, this paper focuses on how to carry on the new farmland quality evaluation.

2.1. Evaluation process

The procedure of newly cultivated farmland evaluation is shown in Figure 1. According to the quality requirement of farmland basing on Classification Rules for Agricultural Land and Regulations for Acceptance of Land Development and Consolidation, main indexes are selected to perform the evaluation. The indexes are shown in Table 1, and the index can be added or removed according to actual situation. Then, the weight of the indexes is calculated. The evaluation standards are given and experts would mark each index. According to the weight and score of each index, the total score is obtained. Then, the quality evaluation of newly cultivated land finishes.

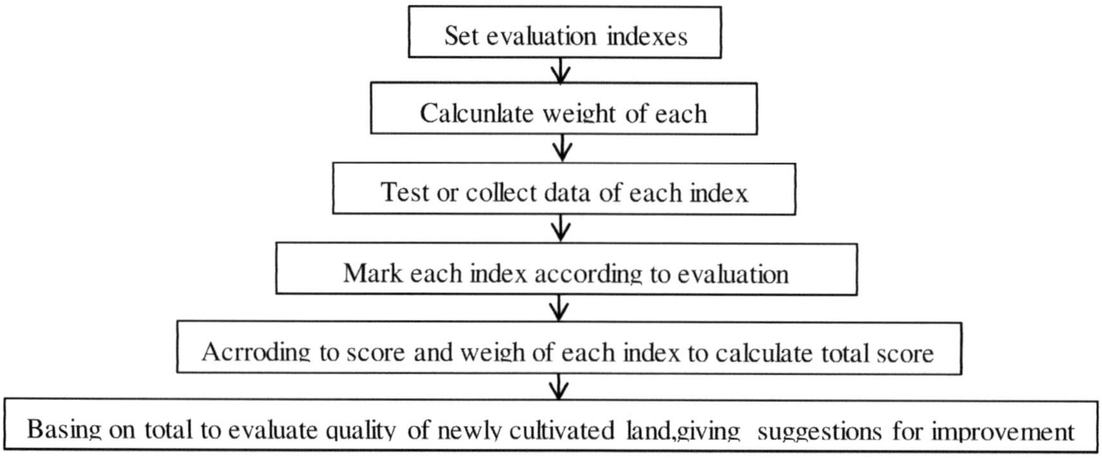

Figure 1. The procedure of quality evaluation of the new cultivated farmland.

Table 1. Evaluation criteria for quality evaluation of new cultivated land.

Evaluation index	80~100 (excellent)	60~80 (qualified)	<60 (unqualified)
Flatness [2]A_1	≤3°	3°～6°	>6°
Contiguous degree[2]A_2	field plots contiguous, easy for mechanical farming	field plots are relatively concentrated and contiguous, easy for mechanical farming	field plots much scattered, not easy for mechanical farming
Mastery[5]A_3	Field Road and Production Road form into a network, near settlements, transportation is convenient	Field Road is less, Production Road form into a network, Production Road can reach all tillage fields	No Field Road, Production Road is less and its distribution is uneven, traffic conditions is quite inconvenient
Macro layout[2]A_4	The proportion of various types of land is	The proportion of various types of land is less	The proportion of various types of land is

	coordinated and the layout is reasonable	coordinated and the layout is less reasonable	not coordinated, the layout is not reasonable
Soil texture[4] A_5	loam	powder soil	sandy soil
Soil pH[4] A_6	7.0~8.0	8.1~9.0	9.1~10.0
All salt A_7	<0.5%	0.5-1.0	>1.0
Topsoil organic matter content A_8	>30 g/kg	10-30 g/kg	<10 g/kg
Total nitrogen A_9	>1.5 g/kg	0.75-1.5 g/kg	<0.75 g/kg
Available phosphorus A_{10}	>20 g/kg	5-20 g/kg	<5 g/kg
Available potassium A_{11}	>150 g/kg	50-150 g/kg	<50 g/kg
Cultivated land pollution index [8] A_{12}	Cd ≤0.2 mg/kg	Cd 0.2-1.0 mg/kg	Cd —
	As ≤15 mg/kg	As 15-20 mg/kg	As 20-30 mg/kg
	Cu ≤35 mg/kg	Cu 35-100 mg/kg	Cu 100-400 mg/kg
	Pb ≤35 mg/kg	Pb 35-350 mg/kg	Pb 350-500 mg/kg
	Cr ≤90 mg/kg	Cr 90-350 mg/kg	Cr 350-400 mg/kg
	Zn ≤100 mg/kg	Zn 100-300 mg/kg	Zn 300-500 mg/kg
	Ni ≤40 mg/kg	Ni 40-60 mg/kg	Ni 60-200 mg/kg

Note: A_8: topsoil organic matter content; A_9: total nitrogen; A_{10}: available phosphorus, A_{11}: available potassium. Those were divided according to the national second soil census classification criteria for standard division

2.2. Determination of the weight index by using the judgment matrix

Analytic Hierarchy Process method (AHP method) is used to determine the weights of the indicators [2]. After making an objective judgment on the multiple indicators of the system, the corresponding quantitative values of the relative importance of each indicator in comparison with other indicators are given. Then, the judgment matrix is established and the weight of the relative importance of all the indicators is calculated. The sum of the weight of all indicators is 1. In practical application, we can sort the weights of each index, and finally plan and make decisions and choose the solutions to the problems according to the sorting results.

2.2.1. Establishment of a judgment matrix. Experts are asked to compare the importance of indicators Ai and Aj in Table 1 to the quality of newly cultivated land. If Ai is more important than Aj, take bij=K. K is in the range of 1 to 9, and the greater the degree of importance, the K value is larger. If Ai and Aj are equally important, then take bij=bji=1. If Aj is more important than Ai, then take bij=1/K, as shown in Table 2. The judgment matrix B is obtained by traversing all 12 indicators. The number of indicators can be increased or decreased according to the actual situation.

$$B = \begin{matrix} b_{11} & b_{12} & L & b_{1,12} \\ b_{21} & b_{22} & L & b_{2,12} \\ L & L & L & L \\ b_{12,1} & b_{12,2} & L & b_{12,12} \end{matrix}$$

Table 2. The criteria for the determination of the matrix elements of judging matrix.

A_i compared to A_j	Quiet unimportant	Unimportant	Very Unimportant	Most Unimportant	Less Important	Import-ant	Most important	Quiet Important	Equal
Value of b_{ij}	1/3	1/5	1/7	1/9	3	5	9	7	1

Note: Degree is between the two degrees next to each other in Table 2, the value of b_{ij} can be 8, 6, 4, 2, 1/2, 1/4, 1/6, 1/8.

2.2.2. *Weight coefficient calculation.* Calculate the eigenvectors of the judgment matrix, which is the weight coefficient of each index

$$\mathrm{M_i} = \prod_{j=1}^{n} \mathrm{b_{ij}} \tag{1}$$

where n is the number of evaluation indicators, and then M_i' is calculated according to the equation

$$M_i' = \sqrt[n]{M_i} \tag{2}$$

Finally, W_i as the weight coefficient is calculated basing on the equation

$$W_i = \frac{M_i'}{\sum_{i=1}^{n} M_i'} \tag{3}$$

[5]

Finally, it is necessary to test the consistency of the weighting coefficients obtained, which can verify whether the judgment matrix is successfully established and accurately reflect the importance of each evaluation index in affecting the quality of arable land. The calculations involved are calculated using Excel.

3. Application

Taking the project of balancing land occupation and land reclamation in Longting Town, Hancheng City as an example to evaluate the quality of newly cultivated land. First, the evaluation indicators are determined, which are flatness A_1, contiguous degree A_2, accessibility A_3, macro layout A_4, soil texture A_5, soil pH A_6, total salt A_7, tillage organic matter content A_8, total nitrogen A_9, available phosphorus A_{10}, available potassium A_{11}, cultivated land pollution index A_{12}. Then, the weight coefficient of 12 indicators is determined.

3.1. *Determination of the judgment matrix*

After 12 indicators being selected, the value of matrix elements of the judgment matrix is determined according to the relative importance of A_i to A_j. The cell content of the A_i column in Table 3 is the value of the matrix element of judgment matrix i.e. B_{ij}.

Table 3. The relative importance level of Ai in comparison with Aj.

	A_1	A_2	A_3	A_4	A_5	A_6	A_7	A_8	A_9	A_{10}	A_{11}	A_{12}
A_1	1	1/2	1/3	1/4	1/8	1/7	1/7	1/9	1/8	1/8	1/6	1/5
A_2	2	1	1/2	1/3	1/9	1/6	1/6	1/8	1/7	1/6	1/5	1/4
A_3	3	2	1	1/2	1/8	1/5	1/5	1/7	1/6	1/5	1/4	1/3
A_4	4	3	2	1	1/7	1/5	1/5	1/6	1/5	1/4	1/3	1/2
A_5	8	6	5	4	1	4	4	1/2	4	4	4	7
A_6	6	5	4	3	1/4	1	2	1/4	1/3	1/3	1/3	2
A_7	6	5	4	3	1/4	1/2	1	1/5	1/4	1/4	1/4	1/2
A_8	9	7	6	5	2	4	5	1	2	2	2	6
A_9	7	5	4	3	1/3	3	3	1/2	1	2	2	3
A_{10}	7	5	4	3	1/4	2	2	1/2	1/2	1	2	3
A_{11}	6	4	3	2	1/5	2	2	1/2	1/2	1/2	1	3
A_{12}	4	3	3	2	1/7	1/2	1/2	1/6	1/3	1/3	1/3	1

3.2. *Calculation of the weight coefficient*

Calculate the weight coefficient from the judgment matrix.. First, calculate Mi according to $M_i = \prod_{j=1}^{12} b_{ij}$as shown in Table 4. Then M_i is calculated basing on equation (1) As shown in Table 4. M_i' is calculated according to on equation (2), and finally, the weight coefficient W_i is calculated based on equation (3).

In order to judge whether these two indexes are objectively compared with each other, the consistency of the results is tested. The consistency index C.I. and the random consistency ratio C.R. are calculated according to the method of Ref. [9]. The consistency index C.I.=0.0027. For C.R.=C.I./R.I. and R.I.=1.54 in the 12th order matrix according to Ref. [9], the random consistency ratio C.R. is calculated to be 0.0018. The C.I. and C.R. are both less than 0.01. Therefore, the judgment matrix has satisfactory consistency.

Table 4. Weight coefficient.

	1	2	3	4	5	6	7	8	9	10	11	12
M_i	6.1512×10^{-9}	1.5310×10^{-7}	5.9524×10^{-6}	1.9048×10^{-4}	34406×10^{6}	3.3333	0.0703	3.6288×10^{6}	7560	630	43.2	0.0159

M_i'	0.2069	0.2704	0.3669	0.4898	3.5053	1.1055	0.8015	3.5208	2.1048	1.7111	1.3686	0.7080
W_i	0.0128	0.0167	0.0227	0.0303	0.2169	0.0684	0.0496	0.2179	0.1302	0.1059	7	0.0438

3.3. Score according to monitoring results of newly cultivated land's quality in Longting Town

The actual assessment of the newly cultivated land's quality includes the actual field observation and taking sample back to the laboratory for testing. The two parts are combined for evaluation.

3.3.1. Score of field road, layout and so on. The field observations show that the plots were of high flatness. There is almost no gradient. The plots are concentrated and contiguous. All types of land are reasonable distributed, which is close to village and convenient in transportation. The layout, quantity and quality of the road in the field are all good. Therefore, the scores of the four evaluation index are as followed, which are 95, 93, 95, 95 points for flatness A_1, contiguity A_2, accessibility A_3, macroscopic layout A_4, respectively.

3.3.2. Score of testing results of soil samples.

A. Score of soil texture etc.

The testing results of soil texture A_5, soil pH A_6, total salt A_7, topsoil organic matter content A_8, total nitrogen A_9, available phosphorus A_{10}, available potassium A_{11} and their evaluation criteria (from Table 1 in this paper) are shown in Table 5.

Table 5. Testing results and scores of the indicators.

Index	Testing results	Score
Soil texture A_5	Silty sand	65
Soil pH A_6	8.35	72
Total salt A_7	1.6	55
Topsoil organic matter content A_8	5.195417	56
Total nitrogen A_9	0.186042	50
Available phosphorus A_{10}	2	52
Available potassium A_{11}	48.02083	58

B. Pollution index

In order to ensure that the edible part of the crop planted on newly cultivated land is not polluted, the situation of land pollution needs to be evaluated. According to the relevant national and local food hygiene standards, the soil heavy metal contents of newly cultivated land in Longting Town were evaluated. In detail, soil samples are collected and the content of heavy metal elements is determined. The heavy metal elements in newly cultivated land being tested include Cd, As, Cu, Pb, Cr, Zn and Ni. The test results are compared with those of Table 1 to get the score of each individual item (Table 6). Then the N.C.Nemerow index method is used to calculate multi-factor integrated pollution index, and to get the comprehensive score of cultivated land pollution index [10-11]. The formula used is,

$$F = \sqrt{\frac{(F_{iave})^2 + (F_{imax})^2}{2}} \quad (4)$$

where F is the comprehensive score of the cultivated land pollution index, and F_{iave} is the average value of the individual pollution items of the soil, and F_{imax} is the maximum score of the individual pollution items. The project's land pollution index score is calculated to be 94 points.

Table 6. Test results and scores of heavy metal content.

	Testing results (g/kg)	Score
Cd	0.118571	88
As	10.55143	88
Cu	14.55786	92
Pb	13.63571	92
Cr	25.11214	95
Zn	28.77143	96
Ni	18.03357	93

3.3.3. Final total score. The scores of the above 12 evaluation indexes and their respective weight coefficients are summarized in Table 7. According to the formula, the total score of new farmland quality evaluation of Longting Town Project is 62.8, reaching the qualified level. It can be seen that the overall macro-layout of cultivated land in this project is good. The soil is not polluted and of high cleanliness. The reason for its low quality is that the soil texture and the fertility need to be improved. Problem of slight salinization needs to be solved. Rational fertilization and ploughing system needs to be set. As time goes by, after time of baptism, soil reaches maturation and ultimately becomes fertile farmland.

Table 7. Scores and weights of quality evaluation indicators for Longting Town New Cultivated Land Project.

Index	Score	Weight Coefficient
Flatness A_1	95	0.0128
Contiguous degree A_2	93	0.0167
Mastery A_3	95	0.0227
Macro layout A_4	95	0.0303
Soil texture A_5	65	0.2169
Soil pH A_6	72	0.0684
All salt A_7	55	0.0496
Topsoil organic matter content A_8	56	0.2179
Total nitrogen A_9	50	0.1302
Available phosphorus A_{10}	52	0.1059
Available potassium A_{11}	58	0.0847

Cultivated land pollution index A_{12}	94	0.0438

4. Conclusion

In this paper, the acceptance of newly cultivated land in the land consolidation and reclamation project in China is investigated. Taking the Guanzhong Plain in Shaanxi Province as the research object and the whole evaluation process is illustrated by the Longting Town Land Improvement Project.

(1) The steps of the method are clear and the key steps are the determination of the main evaluation index, the construction of the judgment matrix, the calculation of the weight coefficients and the consistency test.

(2) The difficulty of this evaluation method lies in the construction of a judgment matrix that objectively and accurately reflects the importance of each evaluation index. That is to say, the determination of the matrix elements of the judgment matrix should try to avoid the subjective influence of man and satisfy the consistency test.

(3) The method has the advantages of simple and quick, convenient in calculation and with strong accuracy. Based on this, it is possible to evaluate the quality of newly cultivated land in the temperate semi-arid plains. Based on the evaluation results, a constructive suggestion is given on how to enhance the ground fertility.

Acknowledgments

This research was supported by the Scientific Research Item of Shaanxi Provincial Land Engineering Construction Group (DJNY2018-14). We are grateful to researchers at the Fuping pilot base of Shaanxi Provincial Land Engineering Construction Group, for their help with the field experiments.

References

[1] Sun J, Bai X H and Yan G Q 2009,To save intensive land for the main line to protect the development of the red line - written in the 19th national "land day". *China Land Res. Econ.*, **46**:10-2.

[2] Hu Z Q, Zhao Y L, Jiang J and Wang J F 2005, Scheme and criteria for checkup and acceptance of land consolidation and reclamation projects. *Transact. CSAE*, **21**:59-63.

[3] Lv X J and Luo L T 2004, Application of fuzzy mathematics to the quality evaluation of the added cultivated field through exploitation and upgrading. *J. Xi'an Uni. Sci. Tech.*, **24**(1):65-8.

[4] Li Y Q and Zhao C W 2011, Quality of new farm land in Karst Hilly Area: a case study of Dingqi town of Zhenning county in Guizhou Province. *Guizhou Agri. Sci.*, **39**(7):204-8.

[5] Zhou J S, Liu X H, Liao X Y and Liu Y 2004, Research of the appraisal of the quality of newly-increased cultivated land in hill regions of south China. *J. Southwest Agri. Uni. (Social Science Edition)*, **2**(1):30-3.

[6] Hu Y Q and Luo Z. A 2007, study of quality evaluation methods for newly-increased cultivated land in hill regions of southeastern China——a case study of Yongqi town of Chongqing’s Dazu county. *J. Southwest Agri. Uni. (Social Science Edition)*, **5**(3):1-4.

[7] Zhou X Q and Shao Z L 2016, YEERNAZI Yushan’aili. New arable land quality of Xinyuan in Ili river valley. *Chin. Agri. Sci. Bull.*, **32**(23):59-64.

[8] Xia J Q 1996. *Soil environmental quality standards.* China Environmental Science Press, 8.

[9] Guo F M and Cheng J H 1997. Study on inspection to determine unanimity of the matrix in a level analysis method and deviation correction method. *China Geolog. Mining Econ.*, (7):19-24.

[10] Li T J 1995. *Soil Environmental Science*. Beijing: Higher Education Press, 304-9.

[11] Luo L T, Tong W, Han J C, Wang H Y, Xie J C and Ma Z H 2014, Analysis and evaluation of heavy metal content in the arsenic rock, sand and compound soil of Mu Us Desert. *J. Safety Envir.*, **14**(1):258-62.

IOP Conf. Series: Earth and Environmental Science **185** (2018) 012025 doi:10.1088/1755-1315/185/1/012025

Oreganum vulgare; techno – economic analysis for the alternative uses and the role of soil – climate conditions

C Tsanaktsidis[1,6], E Zafeiriou[2], K Spinthiropoulos[3] and G Kasapidis[4] and A Itziou[5]

[1]Department of Environmental Engineering and Pollution Control and Technologies Laboratory of Qualitative Fuel Control, Technological Education Institute of Western Macedonia, 501 00, Kila, Kozani, Greece

[2]Department of Agricultural Development, Democritus University of Thrace, 193, Pantazidou str., Orestiada, Greece

[3]Department of Accounting and Finance, Technological Education Institute of Western Macedonia, Kila, Kozani, Greece

[4]School of Agriculture, Aristotle University of Thessaloniki, Thessaloniki 54124, Greece, Thessaloniki 54124, Greece

[5]Department of Genetics, Development and Molecular Biology, School of Biology, Aristotle University of Thessaloniki, 54124, Greece, Thessaloniki 54124, Greece

E-mail: prof.tsanaktsidis@gmail.com

Abstract. The connection between research activity of higher educational institutions and productive institutions plays a special role in the economic development of local communities. The exploitation of research results in areas such as the primary sector enables the development of innovative projects in the rural economy, as well as in the energy sector. In the current study, the research activity in the field of medicinal plants as a dynamic form of cultivation is presented. Our research refers to the qualitative and quantitative composition of distillate aromatic plants (flowers).The results can be reflected in the primary sector of Western Macedonia, in order to apply appropriate soil and climatic conditions in the region and to develop new crops adjusted to the specific conditions, especially in the field of medicinal plants. The proposal in cooperation with the Local Cooperative of Aromatic and Medicinal Plants has been presented in a series of workshops and has been utilized appropriately to strengthen rural culture in the region. The need to deal with the particular crop stems from the fact that the final products (oil and water) are products that may have direct application in the food sector, cosmetics, aromatherapy, human and animal health.

1. Introduction

Origanumvulgare ssp. hirtum is internationally known as Greek oregano (English Greek oregano), which is native to our country. It is a plant with a very flexible developmental character in terms of soil requirements. It appears in a wide variety of soils and climates, from coastal to mountainous areas, from islands to mainland Greece, in rich and poor soils. Oregano with stands drought and can be grown arid. However, in prolonged drought, especially during the period of spring, watering it once or twice (30-40 mm each time, corresponding to 30-40 n.M./Ha.) can be beneficial, increasing the efficiency without particularly reducing the quality.

Oregano is multiplied by seed or by division of suckers and plants taken from old plantations. Plantation may be performed during the fall as well as during spring. In order to obtain seedlings for autumn plantation installation, seedbeds get prepared during late July to early August, while for spring crops the preparation is done during January, according to the local climatic conditions, and even better in specially designed lawns under cover in fare.

The planting of seedlings or cuttings can be fulfilled during two periods: in October or in March. If it is possible, the planting drip irrigation can be performed until mid-May. The field should be properly prepared for the growing. Seedlings or shoots are planted at a distance of 60 to 80m between the rows and 30 to 40m between plants in line. In arid and weak (infertile) fields, it is planted densely while in loudly (fertile) fields or when the crop can be helped with some watering, less so. The plantation should be kept clean of weeds until summer. Therefore, treads are required between the lines, carved with a special machine (as in tobacco culture) and between plants, made with hoe by hand.

Harvesting is done once at full flowering by cutting all the exposed parts at a height of 8-1cm from the ground. The plantation reaches full production in its third year and continues with a stable rate for another 3-4 years.

The cutting is done with a trimmer. Oregano of the best quality can be acquired when the cut mass is transported loose loaded in special drying machines or when it is dried in the shade.

The mass can also with adjust to the combine harvester threshing on the spot in the field, by being already drained after cutting and by being kept in sacks. Then depending on the direction of the product to the market, it is either sieved and left with the leaves and flowers, destined for medical consumption (drug) or distilled threshed as it is intended for the production of essential oil of oregano. The utility of this product is either commercial, i.e. pure for direct consumption in the population or for medical use in the manufacture of medicaments.

In poorly drained farmland during the cultivation of oregano, rot of roots may be present, caused by soil fungi; for this reason, cultivation of oregano in such farmland should be avoided.

Figure 1. Oreganumvulgare, a vibrant rural culture, which finds application in the region of Kozani, Greece.

In a dry form (leaves, flowers),the oregano is used in cooking, as well as in the essential oil and water, but in small portions. Especially the essential oil finds applications in feed for pigs, poultry and lamb serving as a natural antibiotic. Also strong antioxidant, anticancer, antibacterial and antiseptic properties of oregano (dry, medical drug, essential oil, oregano water) make these products precious good health advice and taste improvement in human nutrition [1-4]. In our research, a complete production management system and a dynamic rural culture of the oregano is presented, which finds application in the region of Kozani, Greece (Figure 1).

An attempt to present the cultivation, harvesting, production process and analysis of plant distillates properties is made, through specific soil and weather parameters that make it environmentally friendly.

This type of crop is presented in this survey taking into account all the features such as soil analysis, distillation and the process through which oregano essential oil is produced. With regard to the essential oil of oregano, it was necessary to present its chemical composition and the flavorings contained therein. But in addition the oregano essential oil, the process of oregano water production is presented, followed by the analysis of physicochemical and microbial parameters. The above research is presented by Constantinos Tsanaktsidis, Eleni Zafeiriou, Konstantinos Spinthiropoulos, George Kasapidis, and Aikaterini Itziou.

Based on the above, conclusions are drawn about the type of the proposed cultivation, as environmentally friendly, and the quality characteristics of the products (oregano water and oregano oil) that make them competitive for potential use and applications in everyday human life.

2. Materials and methods

In order to move on to our empirical research, a Soil analysis is presented. The Soil Analysis was performed in a representative sample of one field. We chose a sample of soil from the whole area, since the soil is similar in the whole area. The Soil analysis was determined by the methods [5-6], Qualitative method modified [7-8], Practical Handbook [9].

Figure 2. Presentation of the distillation process for the production of oregano essential oil.

The dried oregano that came from the flower of oregano is put in the retort in a dry form (the whole shaft and the ground). Then, it is cut with a specific machine and left for two days in the field to dry, in order to then be gathered in large sacks (we can keep them in storage for several days). Then the oregano is put in the cauldron (a cauldron of 4000 liters holds around 500-600 kg of dry oregano and about 700 -750 kg of milled oregano). The distillation process lasts for about 3 hours. The distilled product (water and oil) is fed into a specific container where two tiers and based on the difference in specific gravity between the oil and the water, the oil is separated from the water. Therefore, the top tier is occupied by the water and the lower one by the oil. The capacity of the tank is 100 liters and usually about 70 liters of the distilled product are pumped in as we can observe on Figure 2.

In our effort to determine both the chemical composition and the composition of the existing aromatic substances we used a specific sample. This sample of diluted essential oil (1/10 in n-heptane, v/v) for the GC-MS type assays. It is worth mentioning that these analyses were done by using the Agilent 5975C Series GC-MSD (7890A GC and 5975C dormant MSD) system. This system operated in an EI state at 70 eV, equipped with a capillary column HP-5M5 (30 m x 0.25 mm, film thickness 0.50 μm). In the process, 2 ml of dilute essential oil was introduced (separately in a split and splitting ratio 50:1). Throughout the experiment, the inlet temperature remained constant at 250 °C and Helium was used as a carrier gas at a constant flow rate 1.2 ml / min It should be noted that the oven temperature was as follows: 40 °C for 10 minutes, raised to 220 °C (4 °C / min) for 10 minutes and raised to 320 °C (5 °C / min) for 20 minutes The MSD operates in a scanning operation in the 50-700 m / z range.

Continuing our analysis, mass spectra and the known RI index were compared to NIST (National Institute of Standards and Technology, Gaithersburg, MD, USA, NIST 05 and NIST 05 s). Based on the peak areas of the known GC integrator, the individual components were calculated. At this point, we have to mention that the whole process proceeded without the use of corrective factors.

The comparison process is generally known. The relative retention index (PH) and mass spectra were decided to compare with those of the NIST commercial spectral library. In order to be able to calculate the relative amounts, we have taken into account the peak areas of the well-known GI integrator. The whole procedure was done without the use of corrective factors

The process gave a complex range of organic compounds mainly terpenes. This analysis did not detect any components present at very low concentrations.

The microbiological and physicochemical analyses in a sample of oregano water were determined with the methods: "ISO 11290-2, ISO 16649-2:2001, ISO 21528-2:2004 ,ISO 6888-2, ISO 6222:1999".

3. Results and discussion

3.1. Soil analysis

In Table 1 we can see that the soil analysis was done in a sample (representative) of a field.

Table 1. Soil analysis by using a representative sample.

a/a	Parameters of our analysis		Methodology used
1	pH	7,26	ISO 10390:2005[5]
2	Conductivity (μS/cm)	145,3	ISO 11265:1994
3	Mechanic constitution	Characterization: ammopilodes of sand	Qualitative method modified from The, 1979
	of soil	50-70%. Sandy loam (SL)	
4	Nitrogen (%w/w)	0.074	ISO 11261:1995[7]

5	Phosphate(mg/kg)	22.03	ISO 11263:1994[8]
6	Potassium (mg/kg)	88.67	Practical Handbook N.T. Faithfull 2002, CABI Publishing
7	Cuprate(mg/kg)	1.02	ISO 14870:2001[9]
8	Zinc (mg/kg)	3.18	ISO 14870:2001[9]
9	Maganese (mg/kg)	7.09	ISO 14870:2001[9]
10	Ferrate (mg/kg)	13.61	ISO 14870:2001[9]
11	Boron (mg/kg)	0.18	METHODS OF ANALYSIS FOR SOILS, FAO, 2007

3.2. Oregano water composition analysis via GC/MS

We can observe the results of the physicochemical properties analysis of oregano water in Table 2. In each case we noticed that there is no concern about these parameters.

Table 2. Analyses and Results of physicochemical attributes in a sample of oregano water.

Analysis Of Physicochemical Properties	Result	Units	Ref. Values
Pη	4.42		
Conductivity	62.8	μS/cm	<2000
Total Hardness	0.5	d0	
Nitrates	3.8	Mg/lt	50
Nitrate	0.03	Mg/lt	0.5
Ammonia	0.11	Mg/lt	0.4

The results of the microbial content for the first sample of 25ml are showed in the Table 3 below. Taking into account these results, we could mention that there is no problem, as far as these parameters are concerned. At this point, it should be stressed that in order to avoid possible errors in our measurements, we repeated the experiment in a subsequent period and the results were the same. Taking into account the results, it can be ascertained that there is no possibility for microbiological change. Chances remain extremely low even during storage of the oregano oil (small bottles of 25ml).

Table 3. Microbiological results and analyses in a sample of oregano water.

Sample	16/4-6-2015/364	Units	Ref. values	Method
Salmonella spp	-	-	-	ISO 6579:2002
Listeria monocytogenes	-	-	-	ISO 11290-2
Count for positive to β-glycuronidase Escherichia coli	0	Cfu/ml	0	ISO 16649-2:2001
Enterobacter count	0	Cfu/ml	0	ISO 21528-2:2004

Staphylococcus aureus	0	Cfu/ml	0	ISO 6888-2
O.M.X. 30 0 C	0	Cfu/ml	100	ISO 6222:1999

3.3. Essential oil composition analysis via GC/MS

The results obtained from the analysis of oregano oil gave spectrums of all the detectable compounds listed below. Using the Agilent 7890A Gas Chromatograph / 5975C Mass Selective Detector System we received the following results (Figures 3-5) and (Table 4).

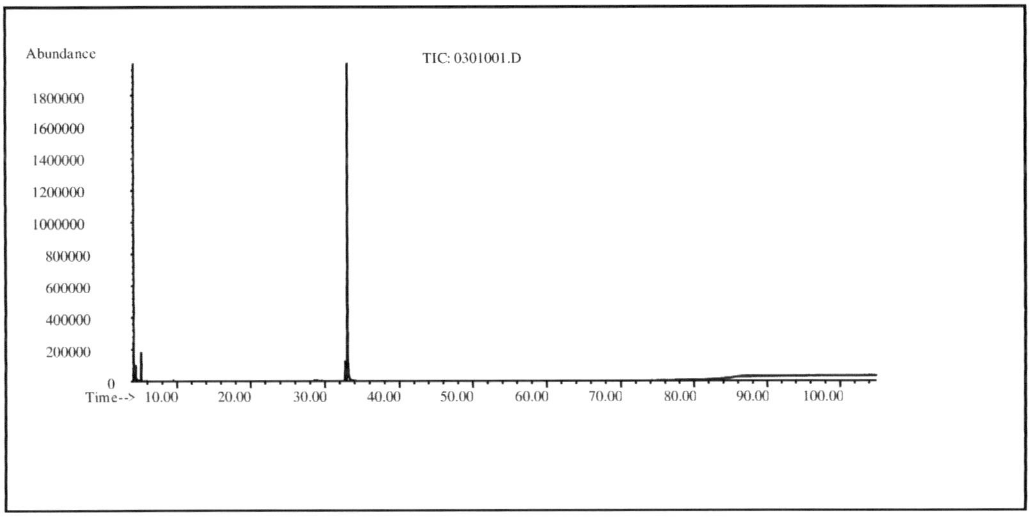

Figure 3. TIC (total ion chromatogram): Aqueous solution containing carvacrol. TIC organic extract (dilution 1/500).

The use of the GC-MC method lead to the detection of carvacrol, which is the major compound (67.34%) and thymol (6.94%), p-cymene (5.72%) and γ-terpinene (5.21%) , β-caryophyllene (RT: 37.12) 1.97% and β-bisabolene (RT.: 39.76) 1.95%. .They are listed below (Figure 4).

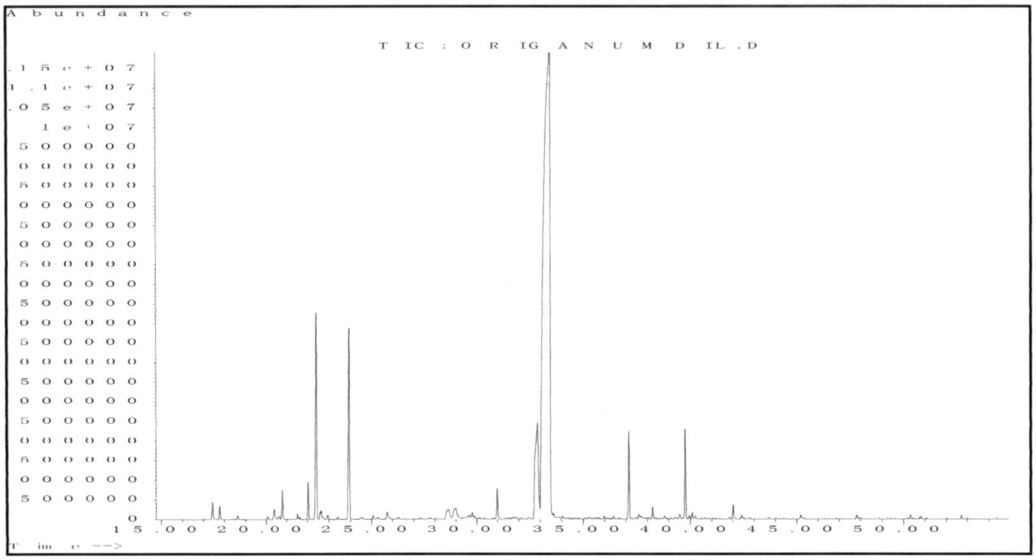

Figure 4. The GC-MS chromatogram for the sample Spec oregano essential oil.

It is known that when we have Lower essential oil concentrations then we can gain better peak shapes. It should also be mentioned that in cases such as the one described above it is more efficient to separate the components, but compounds present in a small percentage may not be detected. (Figure 5)

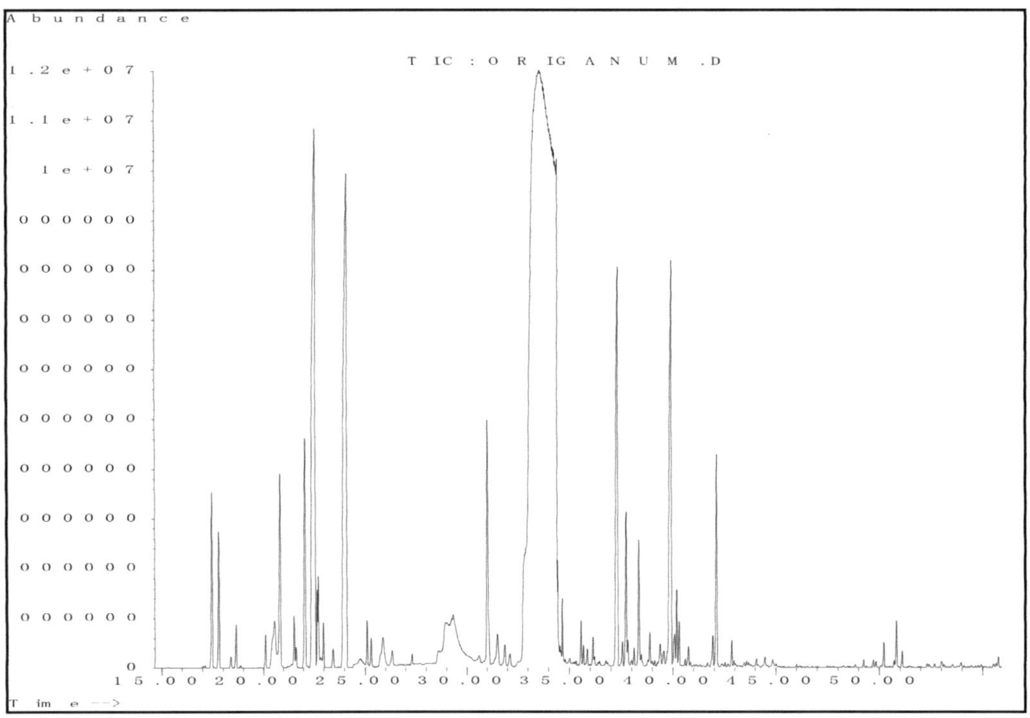

Figure 5. Spectrum analysis with oil sample gas chromatograph from oregano distillation.

Table 4. The major compounds of oregano oil as analysed with Gas Chromatograph/5975C Mass Selective Detector.

A/A	RT	Compounds	%	A/A	RT	Compounds	%
1	17.44	α-Thujene	0.85	20	29.33	cis-β-Terpineol + Borneol	3.64
2	17.79	α-Pinene	0.62	21	31.02	Thymol methyl ether	1.54
3	18.62	Camphene	0.21	22	31.50	Dihydrocarvone	0.39
4	20.07	β-Pinene	0.16	23	32.87	Thymol*	*
5	20.52	1-Octen-3-ol	0.52	24	33.44	Carvacrol + Thymol	60.10
6	20.82	Myrcene	1.20	25	35.53	Thymolacetate	0.40
7	21.52	à-Phellandrene	0.38	26	36.12	β-Bourbonene	0.27
8	21.62	3-carene	tr	27	37.29	β-caryophyllene	2.98
9	22.06	à-terpinen	1.30	28	37.52	Germacrene	tr
10	22.51	p-cymene	7.27	29	37.73	unknown	1.02
11	22.66	Limonene	tr	30	38.34	α-Humelene	0.82
12	22.74	β-Phellandrene	tr	31	38.87	ç-Muurolene	0.28
13	22.97	trans-β-ocimene	0.20	32	39.36	Varidiflorene	tr
14	23.44	cis-β-ocimene	0.13	33	39.87	β-Bisabolene	3.87
15	24.04	γ-Terpinene	5.84	34	40.04	τ-Cadinene	tr
16	25.07	Terpinolen	0.36	35	40.17	δ-Cadinene, (+)-	0.53
17	25.27	à,p-Dimethylstyrene	Tr	36	40.29	β-Sesquiphellandrene	tr
18	25.84	β-Linalool	0.49	37	41.90	Spathulenol	tr
19	28.90	Borneol *	*	38	42.08	Caryophyllene oxide	1.20

Environmental conditions, such as soil quality, sun exposure, temperature etc, may also contribute to the determination of phenolic concentrations [10-13]. The empirical analysis of this work has taken into account the international literature in order to ensure its validity. Older empirical studies [4, 10-11] have shown that the presence of a carbamole / thymol chemist of Origanum vulgare ssp. hirtum was positive. The qualitative composition of oregano has been the subject of several studies, however, it has not been connected either with the composition of the soil where it is grown or with the production of a secondary product, the oregano water. In the present work, these parameters and indeed presented, and, what is more, a high percentage of carvacrol as a component of oregano oil is found, which has not been reported in the previous literature in such high concentrations.

3.4. Economic analysis of oregano cultivation

The oregano belongs to the most profitable variety of aromatic plants while its use is not limited to the traditional one but also serves as medicinal syrup. The two alternative uses are profitable with different production costs while in case they are certified as organics, they provide the greatest revenue. A major part of the total production is exported to major destination countries such as the USA and Germany.

In order for an investment in Oreganum vulgare to be promoted, the classic method of net present value may well be employed while the destination of its cultivation should also be taken into consideration. Furthermore, the profitability of the particular cultivation, as mentioned above, does not only depend on annual income but it takes two or three years to reach its maximum profit, in the third year and, therefore, a ten-year time period should be selected in order for a real perception of profitability to be formed.

First of all, the settlement cost reaches 220 euros per acre, with all the expenditures included, while the price per kg ranges from 1.8 to 2.3 euros for a conventional crop. On the other hand, in case of an organic crop, the price can even reach six euros per kg. As an alternative use, the production of oregano oil can be selected with a proportion of 7% of the total production. The role of irrigation conditions is substantial since it may well lead to a productivity increase per acre. Furthermore, the income generated by oregano oil reaches 70 euros per liter or 300 euros per acre. As far as revenues per acre are concerned, they can reach 400 euros per acre (for the classic use) while a farmer's revenue may become 1000 euros in case of an organic certified cultivation. Sustainability for the particular cultivation asks for a minimum price of 0.7 euros in order for the production costs to be satisfied.

For the aforementioned reasons we employ the methodology of net present value for a time period of ten years. In year 0 the settlement of the crop occurs, while the profit increases gradually and its greatest value is reached in the third year. The formula used in the particular methodology is the following;

$$NVP = \sum_{t=1}^{10} \frac{\Pi_t}{(1+r)^t}$$

where t denotes the year whose cash flow is being discounted, Π denotes the cash flow for each year,R denotes *Discount rate*.

As mentioned above and according to the production costs and the prices per kg used, the cash flows are formed as follows for the case of the dry oregano;

$$\Pi_0 = -220,\ \Pi_1 = 200(2{,}5 - 07) = 360,\ \Pi_2 = 250(2{,}5 - 0{,}7) = 450,$$
$$\Pi_3 = 300(2{,}5 - 0{,}7) = 540,\ \Pi_{4-10} = 540$$

For our case r is equal to 5%.

According to our findings NVP=4665,714 (per acre).

The investment for the oregano crop if the production is for oregano oil is good, because the cash flows associated with the net income obtained reaches EUR 300 per acre. The calculation of the NVP in this case gave us the following results.

$$\Pi_0 = -220, \Pi_1 = 300,\ \Pi_{2-10} = 300,$$

For the same discount rate the NVP =2637,43. What is evident is that the net income that comes from oregano oil for the producer is almost half despite the difference in price per kilo.

4. Conclusions

The essential oil of the Greek oregano consists of individual associations, while two of the most important are thymol and carvacrol. Thymol (2-isopropyl-5-methylphenol) is an oxygenated monopropene, and a phenol derivative of cymene. The carvacrol ($C_6H_3CH_3$ (OH) (C_3H_7)) is a monoterpenoid phenol.

The thymol, due to antifungal, antioxidant and antibacterial properties, is a component of many commercial preparations such as mouthwashes, toothpastes, soaps, shampoos, deodorants and medications for cough and cold. In fact, laboratory experiments performed on mice showed anti-inflammation and a better healing process. Also, plants containing thymol are used both as aromatic and in cooking for the characteristic smell of thyme.

The carvacrol may inhibit the growth of several strains of bacteria. It is often used as a food additive for the prevention of bacterial contamination because of its low toxicity and its pleasant taste and smell. The antimicrobial properties are due to the breakdown of the bacterial membrane. The carvacrol is easily metabolized and excreted from the body. It could be supported that the properties of carvacrol, in combination with the low PH levels of the oregano oil (4,42) may contribute to its use as a preservative in food chemistry. Moreover, this high concentration of carvacrol in oregano oil is considered the most important characteristic of the high quality of the specific oregano type.

It has been supported that Oreganum vulgare ssp. hirtum may direct its biosynthesis to the formation of a certain compound that is considered preferable. More specifically, the metabolic pathway of phenolic compounds is through the autooxidative conversion of γ-terpinene to *p-cymene* followed by hydroxylation of *p-cymene* to thymol or carvacrol [10].Compared to the essential oil produced by Rosa damascene, thymol and carvacrol are detected in both oils [14]. However, the two oils are very different, a fact that may well be attributed to the different concentrations of these substances that are found in the two oils.

Finally regarding the economic issues of the two alternative uses in terms of producer profit and for a ten year time period, the dry oregano should be preferred over the oil given the high extraction cost. An issue left unanswered is the case of organic oregano that is much more profitable than the conventional crop.

Acknowledgments
The authors would like to thank Dr. N. C. Kokkinos and Prof. Dr. S. Mitkidou for providing the GC-MS analysis.

References

[1] Panizzi L, Flamini G, and Cioni P and Morelli I 1993. Composition and antimicrobial propertiesof essential oils of four Mediterranean Lamiaceae. *J. Ethnopharmacol*, **39**:167-70
[2] Sivropoulou A, Papanikolaou E, Nikolaou C, Kokkini S, Lanaras T and Arsenakis M 1996 Antimicrobial and cytotoxic activities of Origanum essential oils. *J. Agri. Food Chem.*, **44**:1202-5
[3] Aligiannis N, Kalpoutzakis E, Mitaku S and Chinou I 2001. Composition and antimicrobial activity of the essential oils two Origanum species. *J. Agri. Food Chem.*, **49**:4168-70
[4] De Martino L, De Feo V, Formisano C, Mignola E and Senatore F 2009. Chemical composition and antimicrobial activity of the essential oils from three chemotypes of *Origanum vulgare* L. ssp. hirtum (Link) Ietswaart growing wild in campania (Southern Italy) *Molecules*, **14**:2735-46.
[5] ISO 10390:2005 Soil quality -- Determination of pH
[6] ISO 11265:1994 Soil quality -- Determination of the specific electrical conductivityQualitative method modified from the 1979
[7] ISO 11261:1995 Soil quality -- Determination of total nitrogen -- Modified Kjeldahl method
[8] ISO 11263:1994 Soil quality -- Determination of phosphorus -- Spectrometric determination of phosphorus soluble in sodium hydrogen carbonates solutionPractical Handbook N.T. Faithfull 2002, CABI Publishing
[9] ISO 14870: 2001 Soil quality --Extraction of trace elements by buffered DTPA solution. Methods of analysis for soils, FAO, 2007
[10] Poulose A and Croteau R 1978. Biosynthesis of aromatic monoterpenes. Conversion of γ-terpinene to p-cymene and thymol in *Thymus vulgare* L. *Arch. J. Biochemi. Biophys.*, **187**:307-14
[11] Russo M, Galletti G, Bocchini P and Carnacini A 1998. Essential oil chemical composition of wild populations of Italian oregano spice (*Origanum vulgare* ssp. hirtum (Link) Ietswaart): A preliminary evaluation of their use in chemotaxonomy by cluster analysisInflorescences. *J. Agri. Food Chem.*, **46**:3741-6
[12] Kokkini S, Karousou R, Dardioti A, Krigas N and Lanaras T 1997. Autumn essential oils of Greek oregano. *Photochemistry*, **44**:883-886
[13] Jerković I, Mastelić J and Milos M 2001. The impact of both the season of collection and drying

on the volatile constituents of *Origanum vulgare* L. ssp. hirtum grown wild in Croatia. *Interna. J. Food Sci. Tech.,* **36**: 649-54

[14] Tsanaktsidis C, Tamoutsidis E, Kasapidis G, Itziou A and Ntina E 2012. Preliminary results on attributes of distillation products of the rose Rosa damascene as a dynamic and friendly to the environment rural crop, ICESD (Hong Kong)

IOP Conf. Series: Earth and Environmental Science **185** (2018) 012026 doi:10.1088/1755-1315/185/1/012026

Impact of soil chemical properties on rice yield in 116 paddy fields sampled from a large-scale farm in Kinki Region, Japan

D Li[1], T Nanseki[1,3], Y Chomei[1] and Y Fukuhara[2]

[1] Faculty of Agriculture, Kyushu University, 819-0395, Morooka 744, Fukuoka, Japan
[2] Fukuhara Farm Co. Ltd., 521-1147, Satsuma 339-3, Hikone, Shiga, Japan

E-mail: nanseki@agr.kyushu-u.ac.jp

Abstract. Japan is in urgent need to reduce the production costs through increasing rice yield, which depends highly on soil fertility. This study aimed to investigate determinants of rice yield, from the perspectives of fertilizer nitrogen and soil chemical properties. The data were sampled in 2014 and 2015, from 116 paddy fields, on a large-scale farm located in the Kinki Region of Japan. The rice included Koshihikari and other seven varieties, cultivated in conventional, special and organic regimes. The nine soil chemical properties included pH, cation exchange capacity, ammonium nitrogen, effective phosphoric and silicic acid, saturation of base elements, exchangeable potassium, lime and magnesia. Multiple regression analysis indicated that positive effects were identified for silicic acid, exchangeable potassium, and ammonium nitrogen; while phosphoric acid affects the yield negatively, controlling the rice variety, cultivation regime, and field area. Finally, countermeasures were put forward to improve soil fertility and rice yield.

1. Introduction

Soil chemical properties are critical to the growth, yield, quality and market competitiveness of crop and their degradation will result in decreased soil fertility, nutrients and thus productivity [1-4]. Relating to sustainable agro-ecosystem as well, soil chemical properties can be improved through fertilization, cropping adjustment, and other farm managerial practices [2, 5-8]. Therefore, many researchers have studied variability and impact on rice yield of soil chemical properties. Juhos *et al* [1] explored the yield determinants of crops through constructing soil quality index out of more than ten chemical and physical indicators, on a farmland scaled 225 ha in east Hungary. Obade and Lal [4] tested four methods in constructing soil quality index (SQI), and identified the properties determining soil quality and crop yield, in the private fields of Ohio, US.

In Japan, although still be the largest among individual crops, ratio of rice (*Oryza sativa L.*) in gross agriculture output decreased from 27.8% in 1990 to 18.0% in 2016 [9]. Under the acreage reduction policy, the total planted area of rice decreased by 24% in the past two decades [10]. Comparing with other crops, rice yield is determined more by soil fertility. For higher rice production, precise measurement of soil properties and their impact on yield is more needed in Japan, where much soil nutrient is drained by rich, or deposited in the dammed rivers [5]. A large literature has focused on the soil chemical properties of paddy fields in Japan. Katayanagi *et al* [11] analyzed them in a nationwide sample of 986 plots, adopting the individual indicators of pH (H_2O) and total carbon. To estimate the effect of soil chemical properties, Matsumoto *et al* [12] included content of available arsenic, phosphorus and acid ammonium oxalate extractable iron and aluminum. In addition to the content of cadmium, copper, and zinc, Mori *et al* [13] represented the soil chemical properties by pH, cation exchange capacity (CEC), and oxidation-reduction potential. Li *et al* [14] assessed the determinacy of soil chemical properties on rice yield, using on-farm data of individual paddy fields.

This study aimed to investigate the soil chemical properties and their determinacy on rice yield, using on-farm data of individual paddy fields, controlling the rice varieties, cultivation regimes, and field area, following by countermeasures for improved soil chemical properties and higher rice yield.

2. Materials and methods

2.1. *Sample and data*

We sampled the data of 116 paddy fields in 2014 and 2015, from a farm scaled over 170 ha in Kinki Region, Japan. Rice yield was measured using paddy of 15% moisture content. The paddy weight and moisture content were monitored by combine harvester, through a matchbox-sized sensor equipped at the input slot of the grain tank. The data was then conveyed by Global navigation satellite system (GNSS), via the cloud server connecting farms, institutes, and companies. Presented by nine indicators, the soil chemical properties were sampled and analyzed by professional company [15].

2.2. *Analysis framework*

Following the most significant soil chemical properties identified adopting multivariate regression. We estimated effects of the properties on rice yield, incorporating the natural features of paddy soil and rice production in Japan. The analyses were conducted using SPSS 23.0 for Windows, IBM Corp.

3. Results and discussion

3.1. *Rice yield analysis*

In the two years, the average yield decreased from 6347 kg to 5730 kg per ha, with smaller coefficient of variance. Over the two years, 85 paddy fields were planted the same varieties, while 31 fields changed. Within the either rice varieties, Nakateshinsenbon possessed the largest average yield of 6652 kg per ha with least variance. The second highly-yielded variety was Koshihikari, which accounts for the largest share, 36.2% by 2016, of planted area in Japan [16], with good taste and appearance, strong cold resistance and stable yield [17]. Within all the varieties, the conventional and special cultivation regimes yielded 6355 kg and 5910 kg per ha, respectively, which were much higher than that of the organic cultivation, 4734 kg per ha on average (Table 1).

Table 1. Rice yield specified by variety and cultivation regime in 2014-2015.

Specification	N	Paddy yield of 15% moisture (kg ha^{-1})				CV [a] (%)
		Min	Max	Mean	Std. D	
Total in two years	232	4040.60	8452.40	6038.63	1016.92	16.84
Year: 2014	116	4040.60	8452.40	6347.26	1129.74	17.80
2015	116	4070.00	8220.00	5730.00	780.23	13.62
Two-year variety: different	170	4040.60	8452.40	6057.80	1009.74	16.67
same	62	4230.00	7933.70	5986.07	1042.87	17.42
Variety: Koshihikari	9	4862.10	7483.30	6402.18	859.34	13.42
Milky queen	28	4090.00	7449.10	5663.57	745.52	13.16
Kinuhikari	9	4581.00	8452.40	6269.33	1529.22	24.39
Nipponbare	11	4134.90	7564.50	6259.53	1110.70	17.74
Nakateshinsenbon	29	5280.00	8298.30	6651.69	857.27	12.89
Yumeoumi	24	4040.60	8358.80	6189.56	1201.38	19.41
Nikomaru	30	4340.00	7999.90	5999.68	1052.49	17.54
Himenomoti	92	4070.00	7933.70	5848.31	937.95	16.04
Cultivation: conventional	83	4040.60	8358.80	6355.10	1064.85	16.76
special [b]	143	4070.00	8452.40	5909.69	934.54	15.81
organic [c]	6	4090.00	5190.00	4733.80	442.29	9.34

[a] Coefficient of variance.
[b] nitrogen content is reduced by 50% in compost and chemical fertilizers and chemical pesticides according to the national guidelines.
[c] soil fertility is improved by compost fertilizers, not by chemical fertilizers and chemical pesticides.
Data source: on-farm survey by the authors.

3.2. Soil chemical properties

As shown in Table 2, the paddy fields scaled from 634 m^2 to 13562 m^2, with an average value of 4965 m^2. The pH indicates soil acidity (<7) or basicity (>7). CEC showcases soil fertility from the capacity to hold the positive ions of NH_4^+, Ca^{2+}, Mg^{2+}, and K^+, thus protecting the groundwater from cation contamination [5, 13]. Ammonium nitrogen, referring the nitrogen directly absorbable by rice plant, is essential to constituent the major plant components such as proteins, nucleic acids, chlorophyll. It promotes growth by vigorously activating cell division, nutrient absorption, and anabolism [5]. The effective phosphoric acid refers to those can be absorbed by the crops directly. Although essential to ensure grain quality, its excessive content may lead to premature or low yield [18]. Silicic acid is indispensable for rice growth in reducing softened stem and leaf, and decayed root [5, 18]. Within the exchangeable contents of base elements, lime is important for root growth, magnesia is necessary for photosynthesis, while potassium is crucial for anthesis and seed-setting [5].

Correlation analysis is indicated the significant properties, i.e., ammonium nitrogen, effective silicic acid and exchangeable potassium affected the yield positively; while effective phosphoric acid and CEC affected the yield positively (Table 2).

Table 2. Field area and soil chemical properties of 116 paddy fields in 2014-2015.

Variable	N	Min	Max	Mean	Std.D	CV [a] (%)	Corr. [b]
Filed area (m^2)	232	634.00	13562.00	4964.91	2863.40	57.67	-0.105
pH	232	4.98	6.63	5.97	0.29	4.83	-0.102
CEC (meq 100 g^{-1})	232	5.89	18.21	10.67	1.92	18.04	-0.128*
Ammonium nitrogen (mg 100 g^{-1})	232	0.10	2.25	0.51	0.42	81.98	0.135**
Effective phosphoric acid (mg 100 g^{-1})	232	3.10	26.51	11.60	4.32	37.27	-0.254***
Effective silicic acid (mg 100 g^{-1})	232	6.38	69.13	16.14	10.14	62.82	0.147**
Saturation of base elements (%)	232	52.63	123.23	79.70	11.21	14.06	0.037
Exchangeable potassium (mg 100 g^{-1})	232	3.21	23.09	10.43	3.34	32.03	0.182***
Exchangeable lime (mg 100 g^{-1})	232	101.01	297.49	193.50	37.14	19.19	-0.108
Exchangeable magnesia (mg 100 g^{-1})	232	3.80	50.73	26.18	7.18	27.41	-0.009

[a] Coefficient of variance.
[b] Pearson correlation coefficient with paddy yield of 15% moisture; ***, **, * indicate significance at 1%, 5% and 10%, respectively.
Data source: on-farm survey by the authors.

3.3. Significant rice yield determinants

In the multivariate regression analysis, values of the yield, soil properties and field area were taken natural logarithmic transformations, to capture the linearity and interpret the regression coefficients in terms of elasticity. Dummy variables were included to control the effect of year (1=2015, 0=2014), rice varieties (1=each, 0=others) and their changes in the two years (1=same, 0=different), cultivation regimes (1=each, 0=others). Using backward procedure, the significant independent variables were selected. The result indicated that each regressor (t-test) and the whole model (F-test) are significant. All the VIFs less than 10 thus no multicollinearity was identified. The Adj. R^2 indicates that 24.4% of the yield variations were explained. White test eliminates the existence of heteroskedasticity, and hence the bias of excluding variables that may affect the regressors significantly (Table 3).

Among the soil chemical properties, effective silicic acid, exchangeable potassium, and ammonium nitrogen affected the yield positively, significant at 0.05; while effective phosphoric acid and exchangeable lime affected the yield positively, significant at 0.01 and 0.10, respectively. With respect to the control variables, field area and organic cultivation negatively relate to yield, while Nakateshinsenbon is found yielded higher than the other varieties. These findings are in accordance with the correlation analyses. The estimated multiple coefficients indicated the change of yield with respect to each regressor, holding the other variables fixed. For the continuous regressors, a coefficient indicates the elasticity, e.g., 1% increase of ammonium nitrogen increased the yield by 0.034%, while a 1% increase of effective phosphoric acid decreased the yield by 0.099%. For the dummy regressors, average yield of Nakateshinsenbon is 16.6% higher than the other varieties, and organic cultivation yielded 19.8% lower than the other regimes, *ceteris paribus* (Table 3).

Table 3. Result of multivariate regression on the significant rice yield determinants.

Independent variable [a]	Unit	Unstandardized coefficient		t	Sig.	Collinearity statistics	
		B [b]	Std. E			Tolerance	VIF
Ln (Effective silicic acid)	mg 100 g^{-1}	0.060 **	0.024	2.482	0.014	0.678	1.476
Ln (Exchangeable potassium)	mg 100 g^{-1}	0.068 **	0.031	2.178	0.030	0.870	1.149
Ln (Ammonium nitrogen)	mg 100 g^{-1}	0.034 **	0.017	2.050	0.042	0.606	1.650
Ln (Effective phosphoric acid)	mg 100 g^{-1}	-0.099 ***	0.026	-3.816	0.000	0.866	1.155
Ln (Exchangeable lime)	mg 100 g^{-1}	-0.102 *	0.057	-1.806	0.072	0.781	1.280
Ln (Filed area)	m^2	-0.030 **	0.015	-2.092	0.038	0.912	1.097
Nakateshinsenbon	Dummy	0.166 ***	0.031	5.315	0.000	0.869	1.151
Organic cultivation	Dummy	-0.198 ***	0.063	-3.167	0.002	0.938	1.066
(Constant)	—	7.115 ***	0.315	22.588	0.000	—	—
N = 232, R = 0.519, R^2 = 0.270, Adj. R^2 = 0.244, F (8, 223) = 10.294*** (p=0.000); White test: 0.228×232=52.896<χ^2(8)=20.090							

[a] Selected using the Backward procedure, and the excluded variables included year, two-year same/different varieties, rice varieties except for Nakateshinsenbon, conventional and special cultivation, ln(pH), ln(CEC), ln(exchangeable magnesia), and ln(Saturation of base elements). Dependent variable: ln(paddy yield at 15% moisture content per ha).
[b] ***, **, *: significant correlation at 1%, 5% and 10%, respectively.
Data source: on-farm survey by the authors.

3.4. Further discussion

To confirm and supplement the result of the multivariate regression, Figure 1 illustrates the soil properties identified as statistically significant at the level of 0.05 in Table 3. (1) Silicic acid accounts for 10-15% of the dry straw and 2-3 for paddy grain, due to the high silicon accumulation ability of rice [5, 18]. Its deficiency may result in declining growth, delayed heading, and hence worse grain-filling, softened stems and leaves, and increased risks of pest damage and lodging. Although unlikely to occur, excessive contents may cause changed soil pH, alkalization disorder and decreased yield [18]. Comparing with the national minimum criterion of 15 mg $100g^{-1}$ [5], higher content of silicic acid can also increase rice yield. However, significant quadratic relation was identified as well, indicating the optimal amount of 39.8 mg 100 g^{-1}. (2) In Japan, potassium and other base elements easily be drained by rich precipitation [5]. For soils rich in humus, like surface gray lowland soil sampled here, the optimal range of exchangeable potassium is 20-30 mg 100 g^{-1}. It is higher when CEC is lower than 30 meq 100 g^{-1} [5]. In the sampled fields, average exchangeable potassium was 10.43 mg 100 g^{-1}, with a maximum CEC of 18.21 meq 100 g^{-1}, thus its increase undermine the yield. (3) Rice growth relies heavily on nitrogen, 60% of which is supplied by soil [5, 18]. Ammonium nitrogen, directly absorbable by crops, is essential for high rice yield. Here, its highest content was merely 2.25 mg $100g^{-1}$, much lower than the optimal range of 10-20 mg $100g^{-1}$ [18]. Therefore, its increased content benefit higher rice yield. (4) In Japan, soil of the paddy fields is easy to be acidic, as shown by the average pH of 5.97 in this sample, attributing to drained calcium and magnesium [5]. In acidic soil, more iron and aluminum are dissolved and phosphoric acid is easily fixed and less applicable to crops. Moreover, difficult to be observed, as brown streaks on leaf blade, whitened leaf tip [5], phosphoric acid is easily over-supplied, through organic fertilizer and compost, as indicated by the negative correlation with rice yield here.

Table 4. Correlations between the soil properties significant to rice yield.

Variable	Effective silicic acid	Exchangeable potassium	Ammonium nitrogen	Effective phosphoric acid
Effective silicic acid	1.000	—	—	—
Exchangeable potassium	0.117	1.000	—	—
Ammonium nitrogen	0.306***	0.206***	1.000	—
Effective phosphoric acid	0.170***	-0.111	0.146**	1.000

N=232, ***and ** indicate significant correlation at 1% and 5%, respectively.
Data source: on-farm survey by the authors.

In practice, significant correlations exist between some chemical properties (Table 4). Thus it is essential to consider and adjust them collectively through, say, creating soil improvement prescription and deciding proper fertilization rate by professional agencies, in different paddy fields [5].

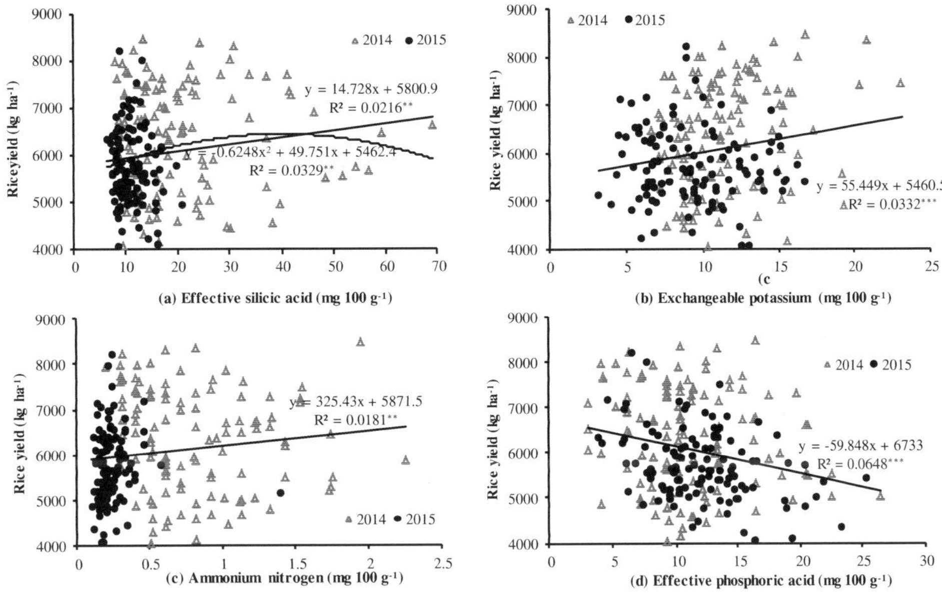

Figure 1. Relationship of rice yield and the significant soil chemical properties.
(N=232, ***and ** indicate significant correlation at 1% and 5%, respectively)
Data source: on-farm survey by the authors.

As a high-yield and lodging-resistant variety, Nakateshinsenbon is widely planted in Kinki Region. With no application of pesticides and chemical fertilizers, although reduces agro-pollutions, organic cultivation tends to suffer lower yield from increased pest damages and insufficient fertility. Over-scaled paddy field may undermine rice yield, in terms of unbalanced fertilization hence soil fertility and irrigation. A quadratic relation indicates the optimal scale of 0.38 ha in this sample.

4. Conclusion and suggestions

This study provided a case study of estimating the soil properties and their impact on rice yield. Based on the nine soil chemical properties and control variables, the regression model explained 24.4% of the yield variation. The significant chemical properties included were generally in accordance with the direct correlation analysis. Positive and significant impacts were identified for the contents of silicic acid, exchangeable potassium and ammonium nitrogen, while phosphoric acid affects the rice yield negatively, *ceteris paribus*. Further discussion indicated that these empirical findings were in accordance with the soil property, and rice production in Japan. In addition, the yield was significantly affected by rice variety and cultivation regime, while smaller paddy field tended to relate to higher yield.

Accordingly, soil fertility is improvable to increase rice yield in the sampled paddy fields. Soil silicic acid can be supplemented through the application of silicate-calcium and other siliceous fertilizers. Potassium can be directly supplemented by chloride or sulfate potassium that do not contain other base elements; or by sulfate potassium magnesia and silicate potassium, considering the balance with other base elements. Ammonium nitrogen can be increased through enhanced nitrogen mineralization, which is promotable by accelerated microbial activities due to irrigating the dried soil after spring-plowing, irrigation after silicate-calcium fertilizers were applied, and when ground temperature rises over 30°C. To decrease the content of phosphoric acid, fertilizer-free farming is adoptable when it is over-supplied, say, over 30 mg 100 g^{-1}.

In future studies, this model can be expanded to accommodate more soil types and rice varieties. For precise analyses on soil chemical properties and soil fertility improvement, regularly soil testing and local technical guidance are necessary as well. Proper selection of rice variety and cultivation regime is of great importance to increase rice yield.

Acknowledgements
This study was supported by the Cabinet Office, Government of Japan, Cross-ministerial Strategic Innovation Promotion Program (SIP), "Technologies for creating next-generation agriculture, forestry, and fisheries" (funding agency: Bio-oriented Technology Research Advancement Institution, NARO).

References

[1] Juhos K, Szabób S and Ladányic M 2016 Explore the influence of soil quality on crop yield using statistically-derived pedological indicators. *Ecolo. Indic.* **63** 366-73
[2] Li Y, Chen X, Shamsi I H, Fang P and Lin X 2012 Effects of irrigation patterns and nitrogen fertilization on rice yield and microbial community structure in paddy soil. *Pedosphere* **22(5)** 661-72
[3] Liu Z, Zhou W, Shen J, He P, Lei Q and Liang G 2014 A simple assessment on spatial variability of rice yield and selected soil chemical properties of paddy fields in South China. *Geoderma*. **235-236** 39-47
[4] Obade V and Pand Lal R 2016 A standardized soil quality index for diverse field conditions. *Sci. Total Environ.* **541** 424-34
[5] DFPJA (Department of fertilizer and pesticide, National federation of agricultural cooperative associations of Japan, JA Zen-Noh) 2011 *Guide to Soil Diagnosis and Fertilizer Calculation* (Tokyo, Japan: Rural Culture Association Press) pp 12-3, 16, 72-7 (in Japanese)
[6] Bouma J 2002 Land quality indicators of sustainable land management across scales. *Agric. Ecosyst. Environ.* **88** 129–36
[7] Qi Y, Darilek J L, Huang B, Zhao Y, Sun W and Gu Z 2009 Evaluating soil quality indices in an agricultural region of Jiangsu Province, China. *Geoderma* **149** 325-34
[8] Gray L C and Morant P 2003 Reconciling indigenous knowledge with scientific assessment of soil fertility changes in southeast Burkina Faso. *Geoderma* **111** 425-437
[9] MAFF 2017a National agricultural gross output and production agricultural income in 2016, http://www.maff.go.jp/j/tokei/kouhyou/nougyou_sansyutu/attach/pdf/index-3.pdf (in Japanese, accessed 2018.Feb.19)
[10] MAFF 2017b Planted area and production of paddy rice, http://www.maff.go.jp/j/tokei/kouhyou/sakumotu/sakkyou_kome/attach/pdf/index-35.pdf (in Japanese, accessed 2018.Mar.2)
[11] Katayanagi N, Fumoto T, Hayano M, Takata Y, Kuwagata T, Shirato Y, Sawano S, Kajiura M, Sudo S, Ishigooka Y and Yagi K 2016 Development of a method for estimating total CH_4 emission from rice paddies in Japan using the DNDC-Rice model. *Sci. Total Environ.* **547** 429-40
[12] Matsumoto S, Kasuga J, Makino T and Arao T 2016 Evaluation of the effects of application of iron materials on the accumulation and speciation of arsenic in rice grain grown on uncontaminated soil with relatively high levels of arsenic. *Environ. Exp. Bot.* **125** 42-51
[13] Mori M, Kotaki K, Gunji F, Kubo N, Kobayashi S, Ito T and Itabashi H 2016 Suppression of cadmium uptake in rice using fermented bark as a soil amendment, *Chemosphere* **148** 487-94
[14] Li D, Nanseki T, Chomei Y and Shuichi Y 2017 Fertilizer nitrogen, soil chemical properties and t heir determinacy on rice yield: evidence from 92 paddy fields of a large-scale farm in the Kan to Region of Japan. *IOP Conf. Ser.: Earth Environ. Sci.*, June. 25
[15] Tsukuba Analysis Center, Katakura Cop-Agri Co., Ltd. http://www.katakuraco-op.com/site_tsuk uba/index.html (in Japanese, accessed 2018. Feb.19)
[16] Komenet 2016 Planting areas of the major Japonica varieties in Japan, http://www.komenet.jp/ju kyuudb/826.html (in Japanese, accessed 2018.Feb.19)
[17] Goto Y, Nitta Y and Nakamura S 2000 *Crops I: Paddy Cultivation* (Tokyo Japan: Agricultural Development and Extension Association Press) 210-1 (in Japanese)
[18] Fujiwara S, Anzai T and Kato T 1996 *Approaches and Application of Soil Diagnosis* (Tokyo, Japan: Rural Culture Association Press) pp 23, 26-7, 88 (in Japanese)

IOP Conf. Series: Earth and Environmental Science **185** (2018) 012027 doi:10.1088/1755-1315/185/1/012027

Research of mining wasteland reclamation and regeneration modes in Shendong Mining Region

K H Chen[1,2,3,4,5]

[1] Shaanxi Provincial Land Engineering Construction Group Co., Ltd, Xi'an 710075, China

[2] Institute of Land Engineering and Technology, Shaanxi Provincial Land Engineering Construction Group Co., Ltd, Xi'an 710075, China

[3] Key Laboratory of Degraded and Unused Land Consolidation Engineering, the Ministry of Land and Resources, Xi'an 710075, China

[4] Shaanxi Provincial Land Consolidation Engineering Technology Research Center, Xi'an 710075, China

E-mail: 383121210@qq.com

Abstract. Mining resource is one of essential fundamental energy resources in China. In recent years, mining exploitation activities has been more frequently, which plays an important role in protecting the economic and social development of mining areas. However, it has led to severely destroyed land resources and environmental pollution. Considerable amount of mining wastelands have been appearing accordingly. Therefore, it is necessary to reclaim the mining wastelands and do the regeneration project. In the study, Shendong Mining Region across Shaanxi Shenmu and Inner Mongolia Erdos is the research site. The paper discussed the ecological restoration engineering measures suitable for the landform reconstruction, soil improvement, vegetation restoration and ecological restoration of the abandoned mining areas in Shendong Mining. Three new modes of development and utilization, including ecological agriculture model, construction land mode and mine park model, are proposed in order to rehabilitate the land resource and make a reference to the economic, social and ecological sustainable development in Shendong Mining Region.

1. Introduction

As one of the basic energy sources for the development of the national economy, the coal industry is an important part of the industrial and agricultural production and the social and economic development [1-2]. According to statistics, more than 95% of the basic energy, 80% of the industrial raw materials and 70% of the raw materials of agricultural production come from the coal industry. The importance of the coal industry is evident [3]. Since the 1980s, the coal industry and other heavy industry has been unprecedented prosperity, but sustained high intensity mining activities caused serious damage to the land (such as surface subsidence, solid waste pollution, geological disasters) and destruction of natural environment (soil pollution, water pollution, air pollution, etc.). Even if the mining operation is over, the destruction of the large area of the land in the mining area still exists.

The development history of the developed countries in the world (including Britain and Germany) has also experienced similar "industrial diseases". Through land consolidation and reclamation, the land that has been polluted and destroyed can rejuvenate the vitality [4-5]. In order to reduce the environmental pollution and protect the ecology of the mining area, many mining areas in China have carried out the land renovation and reclamation work [6-8]. With the development of social economy in Shendong mining area, the demand for construction land is increasing gradually. Actively promoting the abandoned mining land reclamation, improving the local natural and ecological conditions can improve the social, economic and ecological benefits at the same time, and provide protection for the local intensive and efficient and sustainable land resources [9-10].

2. The general situation of the research area and the land use status

2.1. Socioeconomic and natural conditions

Shendong mining area operated by China Coal Enterprises - Shenhua Group, Shenhua Group reached 23 billion 605 million yuan of commercial coal sales revenue in 2009, and its revenues increased significantly to 2013. Shendong mining area as a core area of Shenhua coal industry, there are total reserves of about 200 billion tons of coal resources. It is regarded as one of the eight coal field of the world [11-12]. Shendong mining area built coal production base in the first hundred million tons of modern large, and it included 19 high-yield and high efficiency coal mines, the annual production capacity of is over 2 tons, and the production capacity ranks the first in China [13-14]. With the development and construction of Shendong mining area, it greatly promoted the rapid development of the regional economy, and Shenmu and Yijinhuoluo have become the fastest growing economy in the county, and made great contribution to national economic development.

Shendong mining area is located in border area of the northwest area of Yulin city in Shaanxi Province and the southern Ordos of the Inner Mongolia Autonomous Region City (Figure 1), located in the complex transition zone of the Loess Plateau, Inner Mongolia grassland and the Mu Us Sandland. It is located in the east longitude 109 degrees 30 '30' to 38 degrees north latitude 30 '32 ", above 1000 m [15]. The geomorphic units is the Loess Hilly and gully region, parts of the surface is mostly covered by the quaternary wind sand and sandy loess; It is semi-arid continental monsoon climate, and the annual rainfall is 402.7 mm, however, evaporation is as high as 1753.8 mm~1978.7 mm, and wind erosion and water erosion are very strong [16,17]. Because of the effects of climate and sand area with sparse vegetation, drought tolerance and drought Psammophyte plants mainly showed sparse shrub vegetation landscape; regional ecological environment is very fragile, so it is the national key water and soil erosion monitoring area.

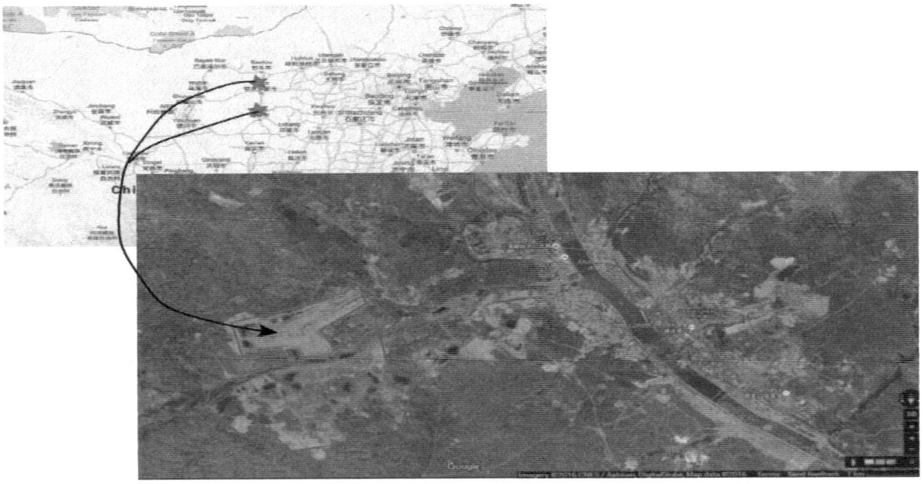

Figure 1. Geological location map of Shendong Mining Region.

2.2. Situation of reclamation project area

Shendong mining area contains large amounts of high quality coal, and it is shallow burial. After the coal mining, it is easy to form the ground collapse pit [18]. With the increase of coal mining intensity, the goaf gradually becomes larger and more, and the collapse is gradually expanded, and a large number of cracks [19] are developed around it. The large-scale exploitation of coal resources not only destroyed the original topography, landforms and natural landscapes, but also left barren mining fields, tailings and collapsed goaf and waste rock sites. In the large coal mine area of Daliuta mining in Shenmu County, the surface area of the goaf and subsidence area is seriously deformed. The surface cracks are as wide as 2m, and the longitudinal depth is 10~18m. The overall subsidence of some parts of the area is 2 to 3m, and the maximum collapse depth is tens of meters [20-21], as shown in Figure 2.

Figure 2. Surface collapse fracture and desert sandy land.

On the other hand, with the large-scale exploitation of the mining area and the widespread disturbance of the surface, the vegetation has been severely damaged, and the accumulation of the waste residue has reached 68 million 580 thousand tons, and the land desertification has intensified.

Due to the shallow coal seam in the mining area, the drainage and mining caused by artificial draining and groundwater cause a large and large scale decline of groundwater. The water supply of farmland is insufficient, and the surface vegetation is withered. At the same time, the discharge of coal chemical wastewater has seriously polluted the water resources. See Figures 3 and 4 in detail.

Figure 3. Scarce plants and destroyed grassland.

Figure 4. Polluted water from industrial and mining sites.

3. Theory and method of ecological restoration of mine wasteland

3.1. Basic principles of land reclamation and ecological restoration

According to the law of mineral resources, there are complete plans for waste disposal and land ecological restoration and corresponding technical measures and evaluation indicators after mining or mining activities. Land reclamation is taking remedial measures for the production and construction activities and land damaged by natural disasters, to restore to the desired state of available activities, and its purpose is to restore the ecological balance of land use, recycling, the production and construction for the healthy development of the land resources at the same time, and also protect the land resources and make it sustainable [22-23].

Ecological restoration is a local ecosystem that restores the ecosystem destroyed by human disturbance and restored into biological diversity and dynamic balance. The essence is to restore or rebuild the damaged regional environment into an ecosystem that is compatible with the local nature. It includes [24] Re-vegetation, Reclamation and Reconstruction. Among them, the key to ecological reconstruction is to restore the necessary structure and function of the ecosystem, so that it can be self-sustaining [25].

3.2. Objectives and principles of ecological restoration

Ecological restoration is an important measure to restore the natural ecological landscape and improve the regional environmental quality in the abandoned mines. The determination of ecological restoration targets is based on the [26] Physicochemical properties of discarded land, natural conditions and social needs. Physicochemical properties is based on the analysis of sampling and investigation, and topography, landforms and geological stability; natural conditions needs to consider the climate, landscape, flora; social needs to consider the subject of wasteland ecological restoration, environmental protection and economic development needs. The ecological restoration in the mining area follows the following basic principles:

(1) The principle of combination of qualitative and quantitative. Methods of combination of qualitative and quantitative evaluation is research on the comprehensive analysis of soil, climate, topography, biological and other natural factors and economic conditions, planting habits and other social factors, considering the extent of damage and the type of garden, and determine the direction of [27] mine ecological restoration reasonable.

(2) The principle adheres to local conditions. According to the different block failure characteristics, ecological environment conditions, it needs economic, reasonable and effective methods, and strictly control the exploitation of mineral resources in the mining area environment disturbance and damage to the protection and keep the utilization of the original vegetation and land reclamation, and minimize the affection of the development activities on the ecological environment [28].

4. Engineering measures for ecological restoration of abandoned in Shendong Mining Area

4.1. Landform remolding and soil improvement

The remolding of landform and soil improvement in mining area is the basic project of ecological reconstruction in mining area. In Shendong mining, the first work is to open pit filling and soil leveling work. Filling and reclamation technology refers to the use of coal mining waste (such as fly ash and coal gangue, etc.) as filling material to repair the damaged land. Among them, burned coal gangue is a strong acid substance, which can neutralize the basic components of the soil, so it is used for filling agriculture and forestry reclamation. The unburned coal gangue can consolidate the foundation, and can build up the capital construction land [29]. Reclamation of fly ash is suitable for dry land. Its use is soil improvement, which can change soil texture, increase soil water holding capacity, improve soil pH value and increase soil fertility. The technology of reclamation and reclamation in wasteland of mining area is shown in Figure 5.

When filling the abandoned area land in Shendong mining reclamation, firstly complete stripping soil (the surface soil thickness of about 30cm and 30 ~ 60cm thick subsurface soil) needs to be taken, collected and placed in the designated area. Coal gangue and fly ash are used as filling materials in order to fill in open pit. According to the shape and size, the large coal gangue is arranged in the bottom layer, and the small pieces of coal gangue are loosely placed on the surface. After filling, the spare soil is covered in the surface of reclamation area, and restores the physicochemical characteristics and nutritional status of the soil, to provide growth conditions for the reconstruction of vegetation in mining area.

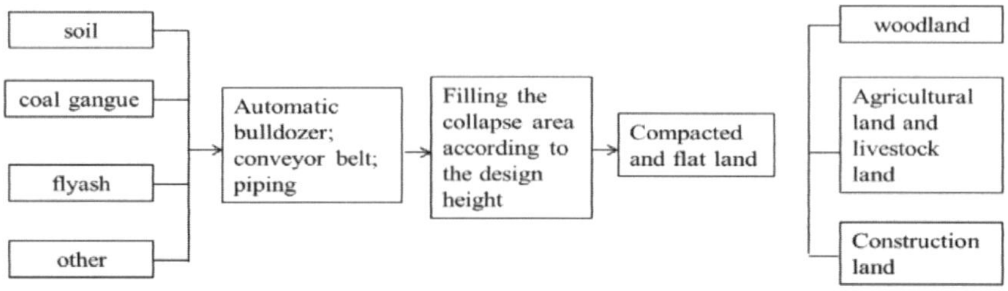

Figure 5. The process of filling reclamation at the opencast mining area.

Soil improvement of Shendong mining wasteland include physical properties improvement, pH improvement and nutrition improvement and other aspects [30-31]. The goal of physical properties improvement is to increase soil porosity, reduce soil bulk density and improve soil structure. It can be used to plough, farmyard manure and other methods [32]. Soils with low pH value can be repeatedly applied to phosphate rock, controlling pH value over a long period of time, and acid soils with low pH value can use lime to regulate acidity, reducing soil pH and promoting microbial activity, and improving soil structure.

4.2. Restoration of vegetation and ecological reconstruction

The ecological environment is vulnerable in Shendong mining area. After exploitation of coal resources, environmental damage and ecological imbalance is more serious, therefore, the vegetation restoration and reconstruction of ecological engineering measures are necessary. Vegetation restoration is not only meant to plant crops on abandoned land, but to create a biological community [33], which is self-sustaining natural environment. Vegetation restoration and reconstruction projects, is the general to improve the site conditions through manual intervention, which can adapt to the biological characteristics of plants; the vegetation root has a fixed role, it can control soil erosion and soil evaporation in a certain extent, increase microbial activity and improve microclimate. Therefore, vegetation restoration has become an important means of reclamation of coal mine waste land and environmental protection measures [34].

The selection of plant species is very important in the initial stage of land reclamation and ecological restoration. Using the local crop varieties in the reclaimed area, planting such plants can quickly adapt to the growing environment of reclaimed land and can grow and reproduce well [35]. Based on the principles of restoration ecology, landscape ecology and vegetation community theory, the composition, structure and density of vegetation communities are designed to create suitable living space and avoid interspecific competition [36]. The simulation of natural vegetation structure of vegetation community structure, the implementation of mixed planting of deciduous trees, shrubs, using a small amount of soil with planting of evergreen trees in the soil ball transplant, herbaceous plants should be dipped in mud or soil mix seeding planting [37]. In addition, by distinguishing the advantages and disadvantages of the surface soil development in the mining area, we can cover a certain thickness of soil, sludge and fly ash on the surface of the abandoned land, which will greatly improve the soil environment and facilitate plant growth.

In Shendong mining area, the loess slope is relatively slow, thick, rich in plant species, and vegetation types in loess area are *Cleistogenes squarrosa*, *Lespedeza*, *Astragalus*, *Thyme*, *Alfalfa*, *Sweet clover*; vegetation to desert shrub of *Artemisia ordosica* and *Yang Chai* as the dominant species associated with rice, insect, P. tricuspidata sand 1 to 2 years of herbaceous plants. According to the climate and soil in Shendong mining area, rainfall and other natural conditions and adjacent ecological geographical area of the vegetation, such as Table 1 preferred plant species and vegetation restoration.

Table 1. Plant species of re-vegetation at Shendong Mining Region.

	Sand area	loess area
	Populus	*Populus*
Trees	*Pinus sylvestris var. mongolica*	*Robinia pseudoacacia*
	Pinus tabulaeformis carr	*Sophora japonica*
Shrubs	*Salix mongolica*	*Hippophae rhamnoides*
	Amorphafruticosa	*Caragana microphylla*
	Artemisia ordosica	*Amorphafruticosa*
	Hedysarum mongolicum	
	Astragalus adsurgens	*Melilotus officinalis*
Herbaceous	*Alfalfa*	*Crown Vetch*
		Alfalfa

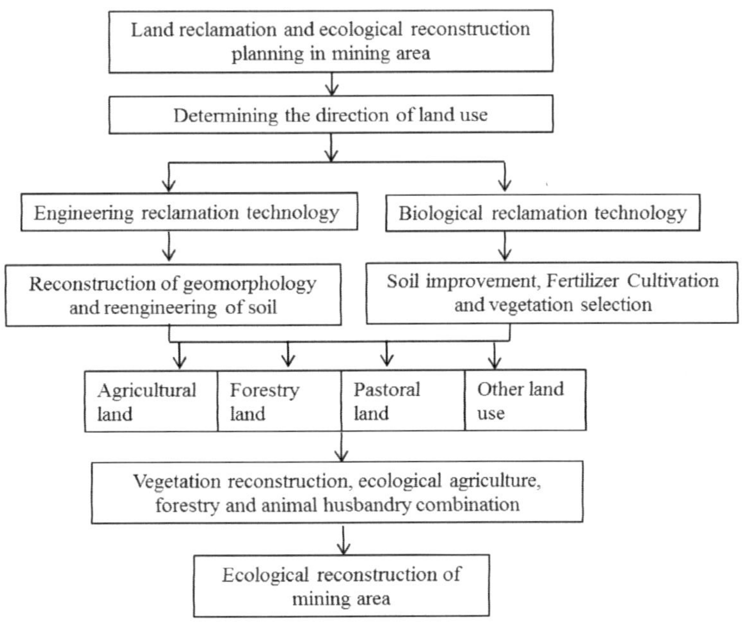

Figure 6. Ecological restoration system of mining wastelands.

In the ecological reconstruction in mining area, the main choice of Herbage vegetation are *Alfalfa* and *Astragalus adsurgens*, *Coronilla varia*, which are pioneer plant species in Shendong mining area.

Its drought resistance is higher, the growth rate is faster, and has the ability of nitrogen fixation to improve soil; For shrub species, *Salix*, *Caragana*, *Amorpha fruticosa*, *Sabina vulgaris* are the first choice. These shrubs have good adaptability, resistance and high survival rate; the main tree species are considering *Pinus tabulaeformis, Platycladus orientalis, Sophora japonica, Robinia pseudoacacia*, Xinjiang Yang, which are excellent local native species. Shendong mining area is water shortage area, cultivating suitable alien species is difficult, so the excellent local plant species are preferred. Overall, the goal of land reclamation and eco reconstruction is to take practical measures to make the completion of the sustainable utilization of land damaged due to mining. The ecological reconstruction system of land reclamation in mining area is like Figure 6.

5. Land regeneration Modes in Shendong mining area

5.1. Ecological agriculture model

Based on China's large population, less cultivated land per capita, insufficient reserve resources and prominent contradiction between people and land, most of the land damage is agricultural land, or even the vast majority of cultivated land. Therefore, the purpose of land reclamation is to take the priority of agricultural land. The combination of climate, light, water and other natural conditions, use comprehensive analysis and design of ecological reclamation mode, agriculture reclamation mode of fruit industry and comprehensive breeding mode.

5.1.1. Ecological agricultural park reclamation model. The mode of Eco Agriculture Park focuses on creating plant landscapes from the perspective of ecology and gardens, highlighting the landscapes of local rural plants, and rationally laying out production projects and facilities, making them an ornamental sight, aiming at developing courtyard economy and eco-tourism and leisure agriculture. The agricultural products are mainly planted with corn, potato, sunflower and other high quality varieties, such as mixed grain, melon and fruit, vegetables and so on, forming a standardized planting base.

Making full use of modern agricultural technology, building “four areas a garden” of modern agricultural park, that is the production of pig breeding and vegetable planting area, planting grasses and sheep farming area, grain planting area, processing and distribution area and Cultural Industrial Park, and making the construction of ecological agriculture park as the core, comprehensive national park, modern planting breeding, processing, science and technology, tourism and development, to build a modern agricultural circulation northwest paradigm.

5.1.2. Forestry and fruit farming reclamation model. The forestry and fruit farming reclamation model is ecological reclamation on abandoned land and damage and collapse area in Shendong mining area. On the choice of forestry tree species, we should consider not only the high economic benefits, but also the multi-functional benefits, such as the well-developed roots, the strong adaptability to drought and the good effect of water and sand fixation. Suitable economic forest planting in Shendong mining area mainly include Ginkgo, Salix, Caragana, Yang and Xinjiang both ecological forest and economic forest of peach, apricot and jujube [38].

In wasteland and mining subsidence in loess hilly gully region area, desert area, building reclamation mode of forestry and fruit farming can not only improve the ecological landscape in desert areas, but formed in apricot, peach, jujube forest based industrial chain. This is conducive to the preservation of soil moisture and preventing soil erosion, desertification, also can produce considerable economic benefit and good ecological benefits.

5.1.3. Comprehensive breeding model. In Shendong mining area where water is sufficient for a generation, according to the design standard of ecological integrated farming, aquaculture and reclamation with plant watering function also can use the level terrace of cultivated land or shallow flat crop planting, dryland farming reclamation will be before a single operation mode for the transformation of three-dimensional ecological agriculture mode of water reclamation the integration of the land.

Taking a dig deep shallow, flat land reclamation, in the vicinity of the Yellow River and its tributaries Linhe area, it can build ecological agriculture comprehensive breeding project, aquaculture ponds, pig farms, sheep farms and backyard farms and high standard greenhouses, planting potatoes, small grains and other crops quality. In this ecosystem, feed fermented manure pond and pond fertile, plant straw and grass, to achieve multi-level recycling plant breeding system, ecological and reasonable structure, complete functions, a higher proportion of the whole energy input-output culture system, has the superiority of ecological agriculture.

5.2. Mine park model

Taking land reclamation ecological park as a means to optimize the mining area ecological sustainable development as the goal, using the unique natural landscape and remains mining landscape as the background, we can built to reflect the Shendong mineral resources exploration and mining development history and has regional characteristics, Erdos northern prairie style, at the same time with the wetlands, desertification research study on the value of historical and cultural education function, and it also can be used for people viewing and leisure tour in scientific research experiment of ecological reclamation development model [39-40].

Land reclamation of ecological park model in ecological fragile mining area in Shenfu mining, is based on mining land reclamation, environmental governance and protection, taking the mining landscape as the axis, with green (agriculture, forestry, animal husbandry) landscape as the main body, implemented by the local cultural landscape, relying on the scientific connotation of the landscape. The mining environmental governance culture, typical ecological landscape, historical and cultural, recreational and other landscape are the organic combination of the series. A show of Shendong mining area special mineral resources exploration and development history, more important is through the land reclamation of this model to promote the transformation of social economy development in Shendong mining area, in order to promote the social, economic, Shendong mining area ecological harmony sustainable development.

5.3. Building land model

In view of the contradiction between natural conditions in Shendong mining area restrictions and increasingly people tense trend, in light of a large area of ground subsidence, press accounts, combined with the county industrial development planning and regional natural conditions, we can use filling reclamation mode in the steady subsidence area, to supplement the construction land. The use of coal gangue, fly ash and other solid waste production, living garbage filling reclamation as construction land reclamation is a more mature mode, which can dispose of a large number of coal gangue, fly ash and other solid waste production and living waste material, avoid the cause of soil, environment and water pollution. It can increase the construction land for industrial park planning and construction, mining, urban construction and new rural construction land.

Especially in the industrial Park, logistics park location relatively flat area such as Jinjie, Dian Ta, Yan Jia TA of the "Shenfu economic and Technological Development Zone", taking the subsidence filling reclamation process at the same time by solving packing, pumping dredging, in-situ soil, gangue backfill layer of dams, and combing the policy of land use overall planning requirements and the construction of new rural life and production infrastructure land, industrial park, wind power generation infrastructure land after land reclamation. At the same time, the park should make full use of coal gangue and fly ash as materials of building material recycling economy.

6. Discussion and conclusion

This paper uses Shendong mining area as the research object, making an understanding of the natural and social conditions in the project area through literature research, on-the-spot investigation and analysis of the damage situation of mining wasteland reclamation area. Large area mining in Shendong mining area resulted in evident ground fissures, increasing open-pit mining wasteland and soil and water pollution, and serious damage of ecological vegetation. Based on the current situation of land destruction in Shendong mining area, the ecological remediation technology is proposed to control the damaged land in the mining area. The main engineering measures included the restoration of landform reshaping, soil, vegetation and ecological reconstruction. To make full use of coal mining of coal gangue and fly ash as basic material of pit filling, cover the surface soil and subsurface soil; the physical properties improvement, pH improvement and improvement to improve the nutritional status can improve the soil texture and structure, enhance microbial activity, improve soil nutrient status; planting advantage cultivation in Shendong Mining Area such as *alfalfa*, *Artemisia*, *Salix*, *poplar* and other local superior crop varieties, can restore the soil, create artificial ecological community at the same time, improve the ecological environment of the abandoned opencast mining area.

Based on the ecological restoration of abandoned mine land in Shendong mining area, this paper explores the development mode of the future, including ecological agriculture mode, mine park mode and construction land use pattern. Ecological agriculture mode was dominated by agricultural land use, land reclamation ecological agriculture park by planting corn, potato, sunflower and other agricultural products, aiming at the development of courtyard economy and Ecological Sightseeing Leisure Agriculture; Forestry and fruit farming is growing with the reclamation of ecological forest and economic forest of peach, apricot and jujube tree, to improve the ecological environment of the reclamation area at the same time, to build the economy forestry industry chain; integrated culture to dig deep shallow water is relatively abundant in the region, the establishment of aquaculture, poultry raising, agricultural planting and other eco cycle industrial chain integration. Mine park mode making the sustainable development of Shendong mining area as the goal, the mining landscape as the axis, build ecological tourism industry both in Loess Plateau of Northern Shaanxi Province customs and

Erdos prairie style. The mode of building land is to supplement construction land resources, make rational use of land, and build land for land use planning based on filling reclamation. To sum up, from the mining land remediation, reclamation of land use development, economic development and protect the environment, it was to fully explore the sustainable development way for Shendong mining area of Shaanxi Province and the Inner Mongolia Autonomous Region.

Mining damage control, ecological restoration and renewal development mode is a multi-disciplinary comprehensive control project involving mining, geology, surveying, ecology, environment, land use and planning. This article gives the ecological restoration and regeneration of Shendong mining area reclamation mode from the macro aspects, but in some detail problems in ecological restoration, such as selecting appropriate exotic plant species, high technology and utilization of water resources in arid area, the mining land matrix modified heavy metals removal technology and so on, which will encounter in the implementation process. These are the difficulties in future research of ecological restoration and reconstruction of western areas.

References

[1] Hao B Y and Kang L X 2014 Mine land reclamation and eco-reconstruction in Shanxi Province: Mine land reclamation model *The Sci. World J.* **9** 1-9

[2] Lei K, Pan H Y and Lin C Y 2016 A landscape approach towards ecological restoration and sustainable development of mining areas *Ecol. Engi.* **90** 320-25

[3] Sun J 2003 Examing key issues in recovery of mine environments *Chin. Acade. Land Res. Econ.* **21**(10) 4-12

[4] Wang L and Zhang H S 2013, The Research progress of land reclamation in mining area in domestic and abroad. Res. Soil Water Conserv., **20**(1):294-30.

[5] Bian Z F 2000 A review of land reclamation in coal mining areas at home and abroad *China Land Sci.* **25**(14)6-11

[6] Gao L, Miao Z, Bai Z, Zhou X, Zhao J and Zhu J, 1998 A case study of ecological restoration at the Xiaoyi bauxite mine, Shanxi Province, China *Ecol. Engi.* **11** 221–29

[7] Cheng P. 2011 Applications of ecological principle on land reclamation *Liaoning Tech. Univ. (Nat. Sci.)*.**30** 408–11

[8] Li M S 2006 Ecological restoration of mine-land with particular reference to the metalliferous mine wasteland in China: a review of research and practice *Sci. Total Environ* **357** 38–53

[9] Zhou L B. 2007 Investigation and Practice on Mining Land Rehabilitation and Ecological Reconstruction in China *Nonferrous Metals* **59**(02) 90-94

[10] Bian Z F. 2005 Research on the recultivation and ecological reconstruction on coal mining area in China *Resour. Industries* 7(02)18-24

[11] Yang J Z, Chen S S, Wang Y, Gao D Y, Hou Z C and Qiang L I 2013 Green mining technology of Shendong Mining Area *Coal Sci. Tech.* **41**(09)34-39

[12] Meng J J. 2008 Ecological environment problems caused by coal producing and comprehensive control measures in Shendong mining area *Coal Geology Explor.* **36**(3)45-47

[13] Gao B and Qiang J T 2010 Application of cleaner production in coal mining in Shendong Mining Area *Energy Envir. Protect.* **42**(1)31-32

[14] Bian Z F, Xu J L and Lei S G. 2007 Discussion on mine ecological construction *J China Soc.*

32(1)13-19

[15] Li R F, Hu C Y and Wang Y. 2007 Study on the variability in fertility of Majiata reclaimed regions *J. Inner Mongolia Agri. Uni.* **10**(02) 242-48

[16] Yu M, Bi Y L and Zhang C Q 2013 Lasting improvement effects of arbuscular mycorrhizal fungi and Bradyrhizobium japonicum on rhizosphere soil environment in miming subsidence *Transa. Chin. Society Agri. Engi.* **29**(08)242-48

[17] Xie S S,Ma C and Tian S J. 2015 NPP Changes and Climate and Chinese Impact of Shendong Coalfield from 2000 to 2010 *J. Geomatics Sci. Tech.* **32**(01)47-51

[18] Fan L M and Yang H K. 2000 Present situation and cause of ground collapse in Shenfu mining area *Coal Tech. Shaanxi* **18**(1)8-9

[19] Fan L M. 1996 Research on ground fissures in coal mines. Study of Environmental geology (Part third) *Beijing Seismological Press*

[20] Chao M A, Tian S J, Zou Y F, Guo Z Z, Han R M and Xie S S 2016 Dynamic responses of the coalfield ecosystem to mining intensity, spatio-temporal variation, and climate change derived from AVHRR/NDVI in Shendong coalfield *China Envir. Sci.* **36**(09) 2749-56

[21] Hu Z Q. 1994 On the planning and design of mining subsidence land reclamation Mine Survering **21**(6)32-34

[22] Wei Y, Gu H B, Xue L. *et al*, 2012 Review of studies on reclamation and ecological restoration of abandoned land of mine *Sci. Soil Water Conser.***10**(2)107-14

[23] Lu C D, Li J H, Jin D S and Gao S H 2013 Current situation of land reclamation and ecosystem reconstruction in shanxi mining area and its development measures *J. Shanxi Agri. Sci.* **41**(2) 152-54

[24] Wang Y H and Chen X J. 2007 Ecological restoration technology for metal mine wasteland *Metal Mine* **21**(6)4-12

[25] Hu Z Q, Zhao Y L and Cheng L L 2004 Extension of goal and meaning of land reclamation in China *China Land Sci.* **18**(3) 3-8

[26] Wang T 2017 Study on reclamation and ecological reconstruction of coal mining land in Shanxi Province *Huabei Land Res.* **77**(2) 118-19

[27] Su G Q, He S J and Guo H C. 1998 Suitability assessment on the wasteland resources of China's Mining Areas *Progr. Geogra.* **17**(4) 39-45

[28] Zhong X B, Yu G M and Lu D. 2006 Carbon storage loss during land readjustment and optimization of ecological compensation *Chin. J. Ecol.* **25**(3) 303-08

[29] Peng J,Jiang Y J and Wu J S. 2005 Eco-enviroment effects of mining and related land reclamation technologies in China *Progr. Geogra.* **24**(2) 38-47

[30] Jitendra A and Subodh K M 2016 Assessment of soil propertities of different land uses generated due to surface coal mining activities in tropical Sal(Shorea robusta) forest, India *Catena* **140** 155-63

[31] Julia K and Thomas R 2012 Development of soil physical parameters in agricultural reclamation after brown coal mining within the first four years *Soil Tillage Res.* **125** 109-15

[32] Chen S C, Hu C Y and Gao Y. 2015 Preliminary study on soil cultivation technology on regional reclamation at the excavated coal pits with sandy soil-A case study of Huojitu and

Bulianta excavated coal pits of Shengdong Diggings *J. Inner Mongolia Agri. Uni. (Natural Science Edition* **36**(03) 61-69

[33] Li M S. 2006 Ecological restoration of mine-land with particular reference to the metalliferous mine wasteland in China: A review of research and practice *Sci. Total Envir.* **357** 38-53

[34] Singh J. et al. 2016 Changes in the physico-chemical properties of soil under shifting cultivation with special references to Karbi Anglong district of Assam *Ind. J. For* **26** 116-22

[35] Wang J Q. 2003 Study on the species selection and ecological monitoring of land reclamation in open pit dump *China Mining Magazine* **25**(6) 254-58

[36] Liu G and Hao M Z 2011 Preliminary study on ecological restoration technology of spraying mixed plants in abandoned mines *West-China Explor. Engi.* **15**(2) 153-57

[37] Li P B, Hu Z Q and Wu J *et al*, 2016 Study on Model of Vagetation Rehabilitati on Techniqueon Coal Waste Pile *J. Shandong Forest. Sci. Tech.* **164**(04) 13-15

[38] Liu Y,Hou S K and Yue H. 2017 Dynamic monitoring and trend analysis of vegetation change in Shendong mining area based on MODIS *Remote Sensing Land Resour.* **36**(04)113-15

[39] Neele L and Dagmar H. 2012 Valuing post-mining landscapes using an ecosystem services approach-An example from Gernmany *Ecolog. Indicators* **12** 567-74

[40] Ma H W, Wu D, Chen Y P, et al, 2012 Some thoughts on land reclamation and ecological reconstruction in the small and medium open coal mines in Ordos Area *Opencast Mining Tech.* **S2**(01) 89-93

IOP Conf. Series: Earth and Environmental Science **185** (2018) 012028 doi:10.1088/1755-1315/185/1/012028

Extraction and characterization of pectin from the peel powder of *Aloe barbadensis*

K Chen, L X Zhu[1], X F Zhan and S A Zhang

Zhongkai University of Agriculture and Engineering, Guangzhou 510225, China

E-mail: 14660662@qq.com

Abstract: In this study, pectin was extracted from aloe peel dried powder using acid extraction and alcohol precipitation method. The results showed that the amount of acid-insoluble ash, galacturonic acid, the loss on drying of pectin was 12.72%, 10.791% and 9.48%, respectively. There were several monosaccharides in the pectin, with the amount order of mannose > glucose > galactose > arabinose > rhamnose > fucose. This suggests that the method was suitable for the extraction of pectin from the aloe peel, but it still needs to be improved. Furthermore, the obtained aloe pectin had obvious ability of anti-oxidant, indicating that it might be a promising material for food and medicine. Our study will pave a new way for handling the aloe peel resource in the future.

1. Introduction

Aloe (*Aloe barbadensis*) is a plant which has been widely used as food or medicine. However, aloe peel has been discarded during the process of aloe into food or medicine, resulting in the waste of resource and pollution to the environment. How to handle the aloe peel has become an issue in the aloe culture industry. It has been shown that aloe pectin was widely used as thickener, emulsifier, gel, stabilizer, etc. In this study, the pectin from aloe peel was extracted by acid extraction and alcohol precipitation method [1]. The ingredients and the anti-oxidation activity of the extracted pectin have been elucidated.

2. Materials and methods

2.1. Materials and equipment

The peel powder of Aloe barbadensis was provided by an aloe plant company in Guangdong province, China. The main chemical reagents used in the experiments were as follows: hydrochloric acid, anhydrous ethanol, salicylic acid, ferrous sulfate, hydrogen peroxide, ammonium persulfate, sodium nitrate, trifluoroacetic acid, acetonitrile, phosphate. The equipment of the experiments included analytical balance, chromatograph, multi-function mill, ultraviolet spectrophotometer, acidity meter, centrifuge, rotary evaporator, dryer and pump.

2.2. The extraction and the ingredients of pectin from aloe peel

The pectin was extracted from aloe peel powder by acid extraction and alcohol precipitation method according to Chinese National Standard (GB 25533-2010). Thereafter, the main ingredients in the extracted were measured [2-4].

2.3. The anti-oxidant activity of the extracted pectin

The anti-oxidant activity of the extracted pectin was measured by two methods [5-8]. The clearance of hydroxyl radical (·OH) and the clearance of ABTS+ of pectin was tested by Fenton action method and ultra-violet method, respectively.

3. Results and analysis

3.1. The physical and chemical features of extracted pectin

As shown in Table 1, by the acid and alcohol precipitation extraction method [3], the amount of acid-insoluble ash, galacturonic acid, the loss on drying of pectin was 12.72%, 10.791% and 9.48%, respectively. The amount of acid-insoluble ash was higher and the amount of galacturonic acid was less than those proposed by the Chinese National Standard, indicating that the extraction method still needs to be optimized.

Table 1. Physical and chemical features of the pectin from aloe peel.

Indexes	Used method	GB 25533-2010
Acid-insoluble ash/%	12.72	≤1
Galacturonic acid/%	10.791	≥65
The loss on drying of pectin/%	9.48	≤12

3.2. The molecular weight distribution of extracted pectin

To test the purity of the obtained pectin, the molecular weight of the pectin was tested by HPLC. The result showed in Table 2 that there were several molecular weights with distinct values in the pectin, indicating that the pectin was not a pure chemical compound.

Table 2. Molecular weights distribution of aloe pectin.

Indexes	Aloe pectin		
Amount average of molecular weight/ Mn	232621	13111	2699
Heavy average of molecular weight /Mw	1.015e6	17102	3097
Average molecular weight/ Mz	3.153e6	21553	3415
Summit molecular weight/ Mp	452497	18556	3461

3.3. The monosaccharide composition of pectin

To further elucidate the compositions in pectin, the monosaccharides in the pectin were subjected to further analyzed. As shown in the Figure 1, 2 and Table 3, there were several monosaccharides in the pectin, with the amount order of mannose > glucose > galactose> arabinose > rhamnose > fucose.

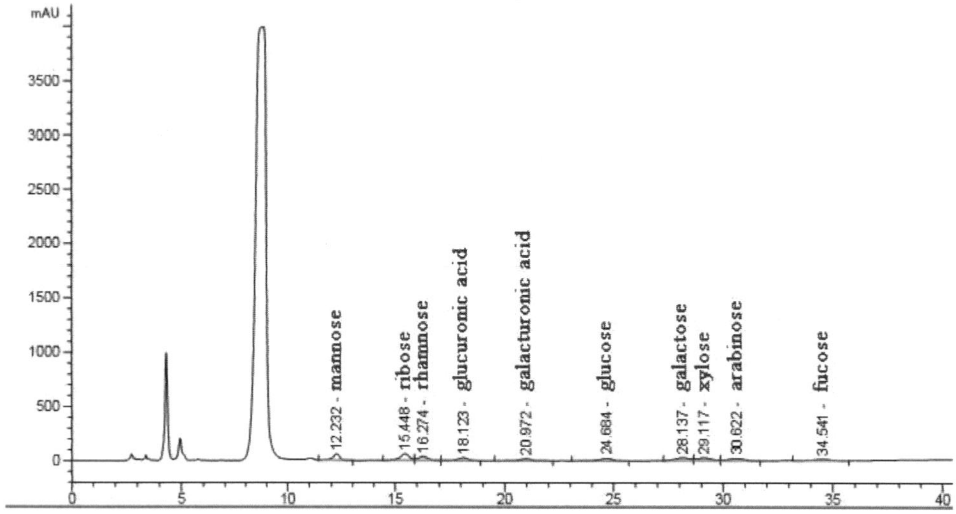

Figure 1. The HPLC chromatogram of internal standard monosaccharide.

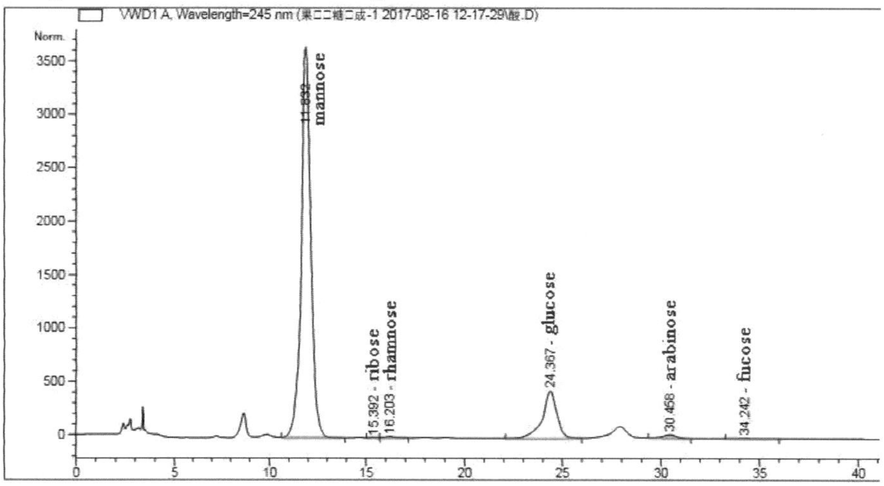

Figure 2. The HPLC chromatogram of the pectin.

Table 3. The monosaccharide composition of pectin.

contents（mg/Kg)	mannose	rhamnose	glucose	galactose	arabinose	fucose
Alone pectin	110453.40	414.00	21580.47	4368.87	1667.17	391.49

3.4. The anti-oxidant activity of the extracted pectin

The clearance of hydroxyl radical (OH) of pectin was shown in Figure 3, the results showed that the clearance ability of ·OH was pectin dose dependent, with 61.17% of ability of reducing ·OH when the pectin concentration was at 10 mg/mL.

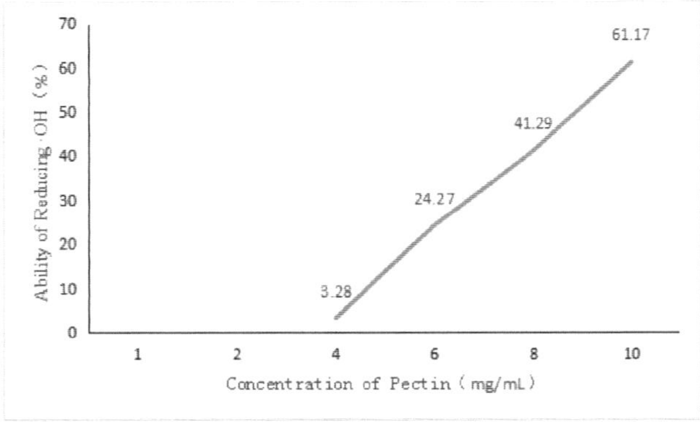

Figure 3. The clearance of hydroxyl radical of pectin.

Furthermore, the clearance of ABTS+ of pectin was shown in Figure 4, the results showed that the clearance ability of ABTS+ was also pectin dose dependent, with 19.59 % of ability of reducing ABTS+ when the pectin concentration was at 1mg/mL. All above results showed that the aloe pectin had obvious anti-oxidant ability.

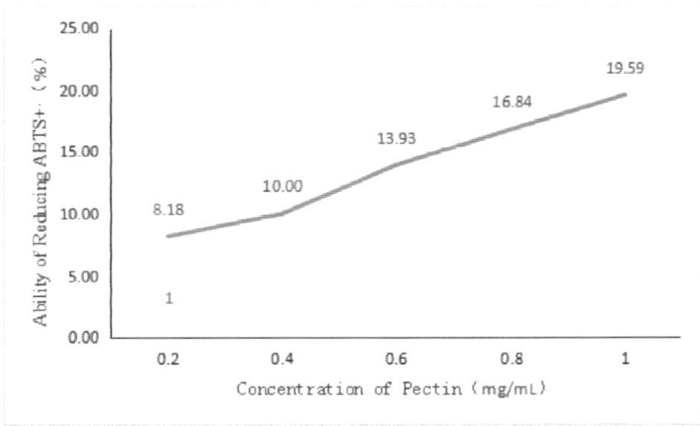

Figure 4. The clearance of $ABTS^{+}$ of pectin.

4. Discussion and conclusion

In this paper, the pectin from aloe peel was extracted by the acid extracting and alcohol precipitation method. However, this method still needs to be improved so as to meet the Chinese national standard. The main contents of the pectin were tested, and the results showed that there are several monosaccharides in the pectin. Furthermore, the aloe pectin had obvious ability of anti-oxidant, indicating that it might be a promising material for food and medicine. Our study will pave a new way for handling the aloe peel resource in the future.

References

[1] Lixue Z, Ke C and Xufeng Z. 2018 Optimized pectin extracting technology from aloe skin. CIGR. Antalya Turkey: 203-5

IOP Conf. Series: Earth and Environmental Science **185** (2018) 012028 doi:10.1088/1755-1315/185/1/012028

[2] Yeoh S, Shi J and Langrish T A G. 2008 Comparisons between different techniques for water-based extraction of pectin from orange peels. *Desalination*, **218(1-3)**:229-237

[3] Liu Z D, Wei G H, Guo Y C and Kennedy J F 2006 Image study of pectin extraction from orange skin assisted by microwave. *Carbohyd. Polym.*, **64(4)**:548-552

[4] Iglesias M and Lozano J E. 2004 Extraction and characterization of sunflower pectin. *J. Food Engin.*, **62(3)**:215-23

[5] Pagán J and Ibarz A. 1999 Extraction and rheological properties of pectin from fresh peach pomace. *J. Food Engin.*, **39(2)**:193-201

[6] Thakur B R, Singh R K and Handa A K 2009 Chemistry and uses of pectin - A review. *Critical Revi. Food Sci. Nutri.*, **37(1)**:47-73

[7] Kalapathy U and Proctor A. 2001 Effect of acid extraction and alcohol precipitation conditions on the yield and purity of soy hull pectin. *Food Chem.*, **73(4)**:393-6

[8] Wang S J, Fang C, Wu J, Wang Z, Liao X and Hu X, 2007 Optimization of pectin extraction assisted by microwave from apple pomace using response surface methodology. *J. Food Engi.*, **78(2)**:693-700

IOP Conf. Series: Earth and Environmental Science **185** (2018) 012029 doi:10.1088/1755-1315/185/1/012029

Cinchona cells performance in *in vitro* culture: quinine alkaloid production with application of different elicitors

D R Pratiwi[1], Sumaryono[2], P T Sari[3] and D Ratnadewi[4,5]

[1]Plant Biology Study Program, School of Post-Graduate Program. Bogor Agricultural University, Darmaga Campus, Bogor 16680, Indonesia

[2] Indonesian Research Institute for Biotechnology and Bioindustry, Jalan Taman Kencana No.1, Bogor 16128, Indonesia

[3] Biology Undergraduate Study Program, Bogor Agricultural University, Darmaga Campus, Bogor 16680, Indonesia

[4] Department of Biology, Faculty of Mathematics and Natural Sciences, Bogor Agricultural University, Darmaga Campus, Bogor 16680, Indonesia

E-mail: diahbiologi.ipb@gmail.com/dratnadewi@ipb.ac.id

Abstract. Quinine is a dominant alkaloid in *Cinchona ledgeriana*. It is widely used in anti-malaria, anti-cramping, and anti-arrhythmia medication. In addition, it is commercially demanded for its bitter taste in certain soft drinks. The main source of quinine is the bark of Cinchona plant, which consists of 4-7% of the alkaloid. Land use shifting from perennial trees to cash-crops has resulted in a shortage of the plant material for quinine extraction. Plant cell culture has proven its success in producing secondary metabolites, including alkaloids. In principle, plant cells synthesize secondary metabolites when they are exposed to stress conditions. Therefore, in cell culture of *C. ledgeriana*, two stress-inducing agents i.e. abscisic acid (ABA) and paclobutrazol (PBZ) in combination with mannitol or sorbitol were applied to increase quinine yield. ABA and PBZ depressed the cells growth. Sorbitol caused more stress to cells than mannitol. Its combination with ABA 3 mg L^{-1} or with PBZ 7 mg L^{-1} mixed three weeks after culture, produced the highest concentration of quinine followed by mannitol with ABA 3 mg L^{-1} or PBZ 7 mg L^{-1} combined at the third week after culture. Cultured cells showed three different shapes: round, oval, and elongated. The elongated shape pre-dominated in the mature cells culture, while round cells pre-dominated in young cells culture. Those mature cells were found to contain alkaloids.

1. Introduction

Cinchona ledgeriana is one of Cinchona species containing high amount of quinine, ranging from 4-13% [1]. It is still used as an antimalarial because the resistance of *Plasmodium falciparum* parasites to quinine is lower than in synthetic drugs [2], and it is considered to be safer for pregnant women suffering from malaria [3]. In addition, quinine is active as an antibacterial. Growth inhibition of Gram-positive bacteria like *Staphylococcus aureus* and *Streptococcus pyogenes* as well as Gram-negative bacteria like *Pseudomonas aeruginosa, Escherichia coli, Proteus mirabilis,* and *Salmonella typhii* has been achieved by quinine applications [4]. Quinine has also been utilized to treat leg or muscle cramps [5, 6] and used for anti-arrhythmic purposes. In the soft drink industry, quinine is employed as a tonic and bitter substance [7, 8].

The increasing demand for quinine, both in the food and drug industries, is still insufficiently satisfied through cinchona bark harvesting. Moreover, quinine is commonly extracted from the bark after the tree becomes at least 7 years old and the highest alkaloids content can be obtained after 20 years [1]. Many studies have been done with various plant cells by regulating alkaloid production through different manipulations to cell suspension cultures [9, 10, 11]. For quinine, the results have not yet been satisfying [9, 12, 13]. The application of elicitors as stress-inducing agents has been suggested to enhance secondary metabolites biosynthesis in plant cells [14, 15]. This research was aimed at revealing the effects of double elicitors on the growth and quinine content in cultured cells of *C. ledgeriana* while determining cell viability and cell shape. They would be further beneficial in cell selection in order to get cell lines of higher quinine production potential.

2. Materials and methods

2.1. Callus initiation

Primary callus was induced from fresh leaves of *Cinchona ledgeriana*. Young leaves were surface-sterilized with 0.4% benlate for 30 min, followed by 20% commercial NaClO for 20 min, then rinsed thrice with sterile distilled water. Their laminas were used as explants for callus production by placing them on semi-solid Woody Plant (WP) medium [16]. This medium was enriched with 15 μM picloram, 2 μM benzyladenine (BA), 1 μM phloroglucinol, 30 gL^{-1} sucrose (standard level), and was solidified with 3.5 gL^{-1} gelrite. The pH of medium was adjusted to 5.7. Explants were incubated at 25 °C in the dark for 8 weeks.

2.2. Cell suspension cultures

Fast-growing callus was transferred to baffled flasks containing WP liquid medium and homogenized for 2 weeks on a horizontal shaker, before being used for suspension culture. The liquid medium had the same formula as the callus initiation medium, except that BA was reduced to 0.5 μM [17]. Furthermore, about 0.5 g of cells were put into each Erlenmeyer flask containing 20 mL of liquid medium with various treatments. The medium composition was the same as for cell homogenization, with the addition of a growth inhibitor abscisic acid (ABA) or a growth retardant paclobutrazol (PBZ) combined with sucrose and mannitol or sorbitol. Each treatment consisted of 12 replications. The treatments were listed in Table 1.

Table 1. Treatments applied in the research and composition.

Treatment Code	Treatment composition				
	ABA (mgL^{-1})	PBZ (mgL^{-1})	Sucrose (gL^{-1})	Mannitol (gL^{-1})	Sorbitol (gL^{-1})
A1K	1		30		
A1M	1		20	5.3	
A1S	1		20		5.3
A3K	3		30		
A3M	3		20	5.3	
A3S	3		20		5.3
P7M		7	20	5.3	
P7-3M		7 (at 3rd week)	20	5.3	
P7-3S		7 (at 3rd week)	20		5.3
C (Control)			30		

The molarity of 5.3 gL^{-1} mannitol or sorbitol is equivalent to that of 10 gL^{-1} sucrose. The cell suspension cultures were maintained for 7 weeks on a horizontal shaker, at 80 rpm, 26 ± 1°C, under light intensity of 30 μmol photon/m^2/second for 14 hours per day.

Cell growth was measured once a week, until the time of cell harvest. Cell volume after sedimentation (CVS) was used to represent cell growth, according to [18]. Cell weight was measured only at the harvest time; each treatment was represented by 5 flasks. Cell viability was also measured on the seventh week-old culture by using 2, 3, 5-triphenyl tetrazolium chloride (TTC), according to [19] with slight modification. A total of 50-100 mg of fresh cells were put into a test tube and 3 mL of 0.6% TTC solution in distilled water were added. The cells were left overnight at 25 °C in the dark room. Then 7 mL of 96% alcohol were added into the tube, and incubated for 7 minutes in a water bath at 60 °C. The cells were centrifuged twice at 2000 rpm, for 6 minutes at a time. Absorbance was determined at 490 nm on the supernatant using a spectrophotometer. Percentage of cell viability was calculated by the formula:

$$\text{Cell viability} = \frac{\text{Absorbance at 490 nm}}{\text{mg fresh cells}} \text{ X } 100\%$$

2.3. Quinine analysis

Quinine analysis was performed using UFLC (Ultra-Fast Liquid Chromatography; Shimadzu Prominence 20 AD type). The analysis was conducted in 3 replications. Cells were analyzed in 7-week-old cultures.

An oven-dried sample (0.1 g) was extracted with 1 mL of phosphate buffer pH 7, then filtered using filter paper. This extract was processed through gel chromatography column F 254 (7 cm length; 1 cm diameter). The eluent used was a mixture of phosphate buffer pH 7 and acetonitrile (7:3). The third fraction which resulted in the first clean extract was used for quinine analysis. HPLC with ODS C-18 VP column was used with an eluent mixture of H_2O:acetonitrile: glacial acetic acid (81:18:1); column length 250 mm, temperature 30°C, and flow rate 1 mL/min. Quinine content was calculated by comparing the peak area of the sample with that presented by quinine standard. The chromatogram was scanned with a UV detector at 250 nm.

2.4. Histochemical test of alkaloids in cultured cells

The cells were harvested after 7 weeks in culture, then centrifuged at a low speed of 20 rpm for 5 minutes. The pellets were suspended with a few drops of distilled water. A small amount of cells were placed in an Eppendorf tube and some drops of Wagner reagent were applied to let the reaction go for 1 minute. The presence of alkaloid in cells was characterized by a brownish red color. As a negative control, the sample was immersed in 5% tartaric acid in 95% ethanol for 48 hours at room temperature, and then it was incubated with Wagner reagent for 1 minute [20].

2.5. Cell observation on size and shape

Cell observation was performed when the cultures aged three and seven weeks. The parameters observed were the length and width of cell to determine the shape category according to [21]. The number of round, oval, and elongated cells were counted from 5 image fields for each sample; triplicate observations were carried out. Measurement of length, width and diameter was performed using image analysis software Image Raster v 2.1.

A completely randomized design was used in this study. The collected data were analyzed in ANOVA; significant differences among the treatments were further proceeded to Duncan's Multiple Range Test (DMRT) at the 5% level of significance.

3. Results and discussion

3.1. Cell growth and viability in suspension culture

Callus initiated from leaf explants appeared on the second week. The callus was bright yellow and friable. The administration of elicitors showed its effect on cell growth. In the seventh week of culture, the highest cell growth (volume) was shown by A3K treatment, which was 11.51 mL, while the most depressed cell growth was shown by P7M, A1K, P7-3S and A3S (Table 2). Moderate growth rates were shown by cells in A1M, A3M, P7-3M, and Control. Cell growth reached its maximum in the sixth week, and simultaneously began to decrease in the seventh week (Figure 1).

Table 2. Cell volume and cell viability of *C ledgeriana* in suspension culture at the seventh week of culture.

Treatments	Cell Volume (mL)	Cell Viability (%)
A1K	5.27[a]	23.33[bcd]
A1M	7.64[cd]	24.84[cd]
A1S	7.09[bc]	21.38[bcd]
A3K	11.51[e]	26.09[cd]
A3M	8.06[d]	13.78[ab]
A3S	5.74[ab]	23.49[bcd]
P7M	5.20[a]	10.89[a]
P7-3M	9.35[d]	16.82[abc]
P7-3S	5.73[a]	31.25[de]
C	9.69[d]	40.85[e]

*The numbers followed by the same letter in the same column were not significantly different based on DMRT at the 5% level of significance.

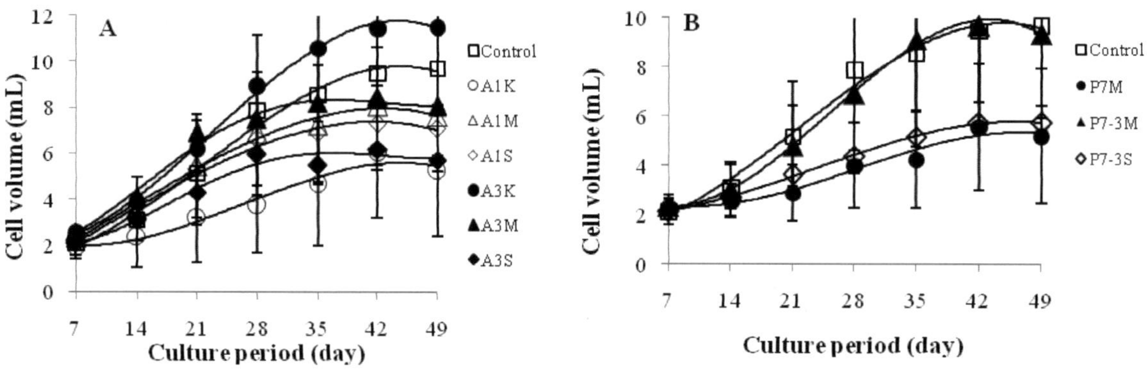

Figure 1. *Cinchona ledgeriana* cells growth in various elicitors treatment combinations.
Notes: (A) Combination of ABA and mannitol or sorbitol; (B) Combination of PBZ and mannitol or sorbitol.

Superior cell growth was observed in 3 mgL^{-1}ABA with standard concentration of sucrose (A3K). ABA at 3 mgL^{-1} promoted cell growth compared to 1 mgL^{-1}. PBZ exhibited more severe inhibition on cell growth, but when it was incorporated in the third week of culture, which was the beginning of the linear phase of cell growth, lighter suppression occurred whereas the inhibition process was more pronounced when PBZ was given earlier (Figure 1). When PBZ was added into the culture media in the fifth week of culture, it did not significantly reduce the cell growth [13]. Sorbitol was more disadvantageous than mannitol when it was combined with either ABA or PBZ. ABA is also known to play a complex role in plant growth, either as a promoter or inhibitor. The addition of ABA in cell culture helps cells to cope with water deficit and osmotic stress [22] that may be due to sugar content.

The use of alcohol sugar (mannitol or sorbitol) as a partial substitution of sucrose diminished energy supply to the cells since alcohol sugar is not easily metabolized, but it contributes in the medium's osmotic potential that reduced the cells capacity to absorb water from the media [23]. Consequently, cell growth in mannitol or sorbitol treatment combinations became lower. This explains why ABA increased the rate of cell growth on a medium without sugar substitution, increasing cell volume and cell weight compared with other treatments, including Control (C).

Without elicitor, Control (C) gave the highest cell viability (Table 2). The time of PBZ administration provided a different effect; PBZ added since the beginning of culture (P7M) caused lower cell viability compared with that when PBZ was added after three weeks (P7-3M and P7-3S). Cell volume represented the accumulating growth result during seven weeks of culture, while cell viability was the value of a one-time measurement at the seventh week. There seems to be inconsistence in several cases when the cell growth corresponded with cell viability. It might be due to the fact that six weeks of culture is the maximum for cells to grow, beyond which the cells rapidly decreased their metabolic activities. The decrease in metabolic activity may be caused by the cell age, and additionally the elicitors in cell cultures had created stressful conditions that had great impact on cell viability. The cells treated with elicitors showed a decrease in growth rate in the seventh week, while Control cells still showed a slight increase.

3.2. Quinine content

Various combinations of elicitors in cell culture were employed in order to obtain an optimal yield of quinine, higher than those previously gained. Quantitative analysis of quinine with HPLC showed that treatment with elicitors generally had a marked effect on increasing quinine content than without elicitors (C) (Table 3). The highest quinine content was obtained from A3S treatment (92.39 $\mu g g^{-1}$) followed by P7-3S (85.57 $\mu g g^{-1}$) and A3M (71.00 $\mu g g^{-1}$), but the total quinine produced in each culture flask showed different results. The A3K treatment with the most abundant cell biomass provided the highest total quinine of 21.87 µg per 20 mL of medium in a flask, followed by A3S (19.03 µg/20 mL).

Table 3. Quinine content of *C. ledgeriana* in seven-week-old cultures.

Treatments	Cell dry weight (g)	Quinine content ($\mu g g^{-1}$)	Total quinine/culture flask[1] (µg)
A1K	0.11[ab]	34.19[a]	6.17[a]
A1M	0.12[abc]	45.23[ab]	8.56[ab]
A1S	0.11[ab]	43.20[ab]	7.63[a]
A3K	0.21[d]	61.99[abc]	21.87[d]
A3M	0.12[abc]	71.00[bcd]	14.21[bc]
A3S	0.12[abc]	92.39[d]	19.03[cd]
P7M	0.07[a]	56.05[ab]	6.95[a]
P7-3M	0.14[bc]	63.82[abc]	14.22[bc]
P7-3S	0.08[a]	85.57[cd]	11.38[ab]
C	0.17[cd]	33.46[a]	9.27[ab]

[1]The culture flask contained 20 mL of media.

The numbers followed by the same letter in the same column were not significantly different based on DMRT at the 5% level of significance.

The use of ABA in low concentration of 1 mgL^{-1} and high concentration of 3 mgL^{-1} showed different patterns in quinine content. Treatment with 3 mgL^{-1} ABA induced quinine content more than 1 mgL^{-1}. P7M treatment had lower quinine content (56.06 $\mu g g^{-1}$), significantly different from P7-3S (85.57 $\mu g g^{-1}$), while the quinine content in P7-3M was in between. In reality, however, total quinine is more appreciable and useful; therefore, high content must be accompanied by high cell biomass. Indeed, PBZ induced high content of quinine but the biomass produced was very low due to its strong inhibition of the cell growth particularly if PBZ was provided from the beginning of the culture. P7-

3M treatment was promising since it produced sufficiently good cell biomass and quinine content which resulted in a moderate total amount of quinine (14.22 μg).

Overall, quinine content obtained in this study was smaller than those reported previously [13], which achieved the highest yield of quinine of 1206.70 $\mu g g^{-1}$, whereas in this study the highest yield was 92.39 $\mu g g^{-1}$. These results were obtained by different extraction methods and by the use of different clones of *C ledgeriana*. The plant materials used in this research were collected from an open plantation. In cell suspension cultures, the role of media composition is very important to support the yield of quinoline alkaloids, especially quinine. Some studies on *C ledgeriana's* quinine were carried out by using various combinations of culture media. As much as 0.11 μg quinine per g cell fresh weight was obtained by combining 0.5 mgL^{-1} 2,4-D and 0.1 mgL^{-1} IBA [9]; zeatin riboside and IAA were able to increase quinine to reach 2.2 $\mu g g^{-1}$ of cell fresh weight [24]; 1 $\mu g g^{-1}$quinine was obtained by cell suspension culture in a medium containing of 3.5 mgL^{-1} NAA and 0.1 mgL^{-1} BA [12].

Leaves of *Cinchona ledgeriana* from a one-year-old tree contained 90 $\mu g g^{-1}$ quinine [1]. Normally, quinoline alkaloids are extracted from tree bark of seven to twelve-year-old trees [25]. Samples from the Bukit Tunggul plantation, West Java, Indonesia have 3.9-6.1% quinine in *C ledgeriana* and 2.4-2.8% in *C succirubra*, extracted from 7-8 year-old trees (personal communication with Santosa TB, the Operational Director of PTPN VIII March 2018). Cell suspension culture of *C ledgeriana* possessed very low capacity in producing quinine. Several factors might be involved in the production process, such as the availability of precursors, important enzymes in the biosynthesis, and also maximal level of quinine required by individual cell to maintain its homeostasis. According to [26], quinine and quinidine are substantially more toxic to cultured cells than cinchonine and cinchonidine. In intact plants, the site of biosynthesis may be away from the site of its accumulation; a translocation is required to avoid toxic levels of a substance in an individual producing cell or tissue. Cultured cells have no such a system.

3.3. Cells size and shape

The cultured cinchona cells were characterized by three different shapes: round, oval, and elongated (Figure 2). These cells were single and clustered. Measurements in the third week showed cells with a round-shape predominated over oval- and elongated-shaped cells, while in the seventh week, cells with elongated-shape were found in larger number than oval- and round-shapes (Figure 3). Oval cells were present in moderate number either in the third week or in the seventh week. It is assumed that young cinchona cells were round-shaped, which progressively extended to oval then elongated as the age of the cell increased. *Peganum harmala* cells grown *in vitro* in the presence of 2 mgL^{-1} 2,4-D composed of round, oval and elongated cells at the exponential and early stationary phase in the cell growth curve, then elongated and large cells predominated in the exponential phase [27]. In cinchona cell cultures, the abundance of round and elongated cells was influenced by the treatment; ABA treatments which demonstrated growth promotion up to six weeks of culture provided round cells more than PBZ. It was thought that the round cells have the capacity to divide, not the long cells.

Certain cellular forms may indicate the presence of secondary metabolites, since the cell shape can change during their accumulation [23]. Previous research suggested that the process of cell expansion is a process of cell specialization to produce secondary metabolites [28]. The results obtained by [20] that quinine content in cells aged seven weeks were quantitatively much higher than in six-week-old cells in all treatments applied, indicated that elongated cinchona cells accumulate quinine alkaloid more abundantly than the round or young cells. The fact that strong elicitor combinations such as A3S, P7-3S and A3M enhanced quinine biosynthesis that led to higher concentration in quinine obtained, suggested that low growth rate of cells resulted in those treated media had more elongated cells. In *Peganum harmala* cultures where elongated cells predominated, they accumulated large amounts of β-carboline and serotonin alkaloids in their vacuoles [27]. Similarly, in strawberry cell suspension culture, elongated cells have higher anthocyanin level than round cells [29].

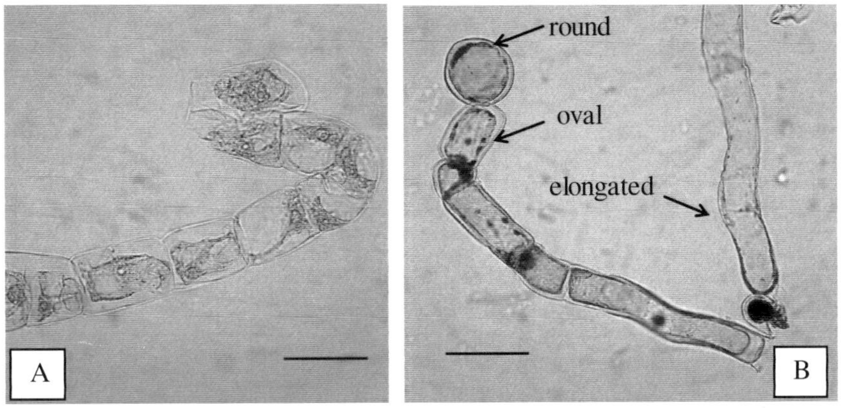

Figure 2. Cinchona cell shapes. Negative control (A), positive result (B) of histochemical test on various cell shapes accumulating alkaloids. Bar = 100 μm.

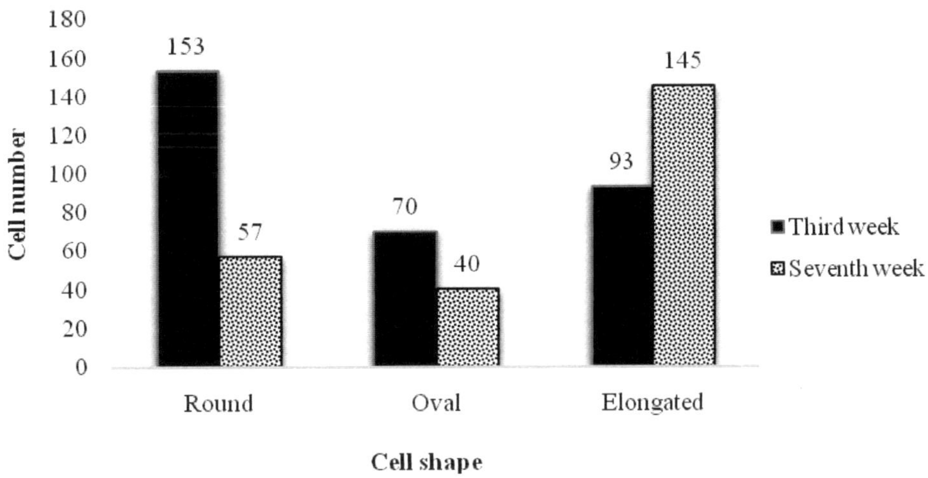

Figure 3. Distribution of cell shape variations in *C ledgeriana* suspension culture at the third and seventh week of culture.

4. Conclusion

Double elicitors in *C ledgeriana* cell culture inhibited cell growth but promoted higher concentration in quinine alkaloids. The best combination of elicitors for quinine content was 3 mgL^{-1} ABA with sorbitol as a partial substitution of sucrose. Sorbitol gave a higher abiotic stressing effect than mannitol. The large cells biomass and quinine content determined the total quinine yield; the highest was obtained from 3 mgL^{-1} ABA without sugar substitution. The characteristic of cinchona cell shape was round, oval, and elongated. All cell types were detected containing alkaloids, but cultures with predominant elongated cells contained quinine more intensively at the seventh week. In addition, round cells were widely found in ABA treatments, indicating active cell division that extrapolated through the cell growth rate.

Acknowledgement

We are grateful to the Indonesian Ministry of Research, Technology and Higher Education c.q. Directorate General of Research and Development Strengthening, for the research grant we received to carry out this work.

References

[1] Staba E J 1988 *Alkaloid production from Cinchona cell and organ systems Genetic Manipulation of Woody Plants* eds J W Hanover, D E Keathley, C M Wilson, G Kuny (New York: Plenum Pr) pp 313-328
[2] Zalis M G, Pang L, Silveira M S, Milhous W K and Wirth D F 1998 Characterization of *Plasmodium falciparum* isolated from the Amazon region of Brazil: evidence for quinine resistance *AJTMH* **58** 630
[3] Achan J, Talisuna A O, Erhart A, Yeka A, Tibenderana J K, Baliraine F N, Rosenthal P J and D'Alessandro U 2011 Quinine: an old anti-malarial drug in modern world: role in the treatment of malaria *Malaria J.* **10** 1
[4] Kharal S A, Hussain Q, Ali S and Fakhuruddin 2009 Quinine is bactericidal *JPMA* **59** 208
[5] Mandal A K, Abernathy T, Nelluri S N and Stitzel V 1995 Is quinine effective and safe in leg cramps? *J. Clin. Pharmacol.* **35** 588
[6] El-Tawil S, Al Musa T, Valli H, Lunn M P T, El-Tawil T and Weber M 2010 *Quinine for Muscle Cramps* (Hoboken (US): John Wiley & Sons)
[7] Anderson L A, Philipson J D and Robert M F S 1986 *Aspect of alkaloid production by plant cells cultures* (Cambrige: Cambrige University Pr)
[8] Sheldon R, Duff H and Koshman M L 1995 Antiarrhythmic activity of quinine in humans *Circulation* **92** 2944
[9] Robins R J, Payne J and Rhodes M J C 1986 Cell suspension cultures of *Cinchona ledgeriana*; I. growth and quinoline alkaloid production *Planta Med.* **52** 220
[10] El-Sayed M and Verpoorte R 2002 Effect of phytohormones on growth and alkaloid accumulation by a *Catharanthus roseus* cell suspension cultures fed with alkaloid precursors tryptamine and loganin *Plant Cell, Tiss. Organ Cult.* **68** 265
[11] Narasimhan S and Nair G M2004 Effect of auxins on berberine biosynthesis in cell suspension culture of *Coscinium fenestratum* (Gaertn.) Colebr-a critically endangered medicinal liana of Western Ghats*Indian J. Exp. Biol.* **42** 616-9
[12] Staba E J and Chung C T A 1987 Effects of age and growth regulators on growth and alkaloids production in *Cinchona ledgeriana* leaf-shoot organ cultures *Planta Med.* **53** 206
[13] Ratnadewi D and Sumaryono 2010 Quinoline alkaloids in suspension cultures of *Cinchona ledgeriana* treated with various substances *Hayati J. Biosci.* **17** 179
[14] Huang C, Qian Z G and Zhong J J 2013 Enhancement of ginsenoside biosynthesis in cell cultures of *Panax ginseng* by N,N'-dicyclohexylcarbodiimide elicitation *J. Biotechnol.* **165** 30
[15] Panwar G S and Guru S K2015 Stimulation of reserpine production in the whole plant culture of *Rauwolfia serpentina* L. by elicitors and precursor feeding*J. Biochem. Biotechnol.* **24** 49
[16] Llyod G and McCown B 1981 Commercially feasible micropropagation of mountain laurel *Kalmia latifolia* by use of shoot tip culture *Combined Proc. Int. Plant Propagators' Soc.* **30** 421
[17] Sumaryono and Riyadi I 2005 Pertumbuhan biak kalus dan suspensi sel tanaman kina (*Cinchona ledgeriana* Moens) *Menara Perkebun.* **73** 1
[18] Blom T J M, Kreis W, Van Iren F and Libbenga K R 1992 A non-invasive method for the routine estimation of fresh weight in batch suspension cultures *Plant Cell Rep.* **11** 146
[19] Benson E E 1994 *Cryopreservation Plant Cell Culture: A Practical Approach* 2th Ed eds R A Dixon, R A Gonzales (Oxford: IRL) chapter 7 pp 147-166
[20] Furr M and Mahlberg P G 1981 Histochemical analyses of laticifers and glandular trichomes in *Cannabis sativa J. Nat. Products*. **44** 153
[21] Vissenberg K, Feijo J A, Weisenseel M H and Verbelen J-P 2001 Ion fluxes, auxin and the induction of elongation growth in *Nicotiana tabacum* cells *J. Ex. Bot.* **52** 2161
[22] Seijo G 2000 Effects of preculture with sucrose and ABA on cell suspension water status and its relation with vitrification resistance *Revista Brasileira Fisiologia Vegetal* **12** 166

[23] Dumet D, Engelmann F, Chabrillange N, Dussert S and Duval Y 1994 Effects of various sugars and polyols on the tolerance to desiccation and freezing of oil palm polyembronic cultures *Seed Sci. Res.* **4** 307
[24] Rhodes M J C, Payne J and Robins R J 1986 Cell suspension cultures of *Cinchona ledgeriana*; II. the effect of a range of auxins and cytokinins on the production of quinoline alkaloids *Planta Med.* **52** 226
[25] Kurian A, Sankar M A and Peter K V 2007 *Medicinal Plants* (New Delhi: New Delhi Publ Agency)
[26] Walton N J, Parr A J, Robins R J and Rhodes M J 1987 Toxicity of quinoline alkaloids to cultured *Cinchona ledgeriana* cells *Plant Cell Rep.* **6** 118
[27] Khafagi I K 2007 Generation of alkaloid-containing idioblasts during cellular morphogenesis of *Peganum harmala* L cell suspension cultures *Am. J. Plant Physiol.* **2** 17
[28] Kim S W, Jung K H, Kwak S S and Liu J R 1994 Relationship between cell morphology and indole alkaloid production in suspension cultures of *Catharanthus roseus Plant Cell Rep.* **14** 23
[29] Forni C, Frattarelli A and Damiano C 1999 Different size, shape and growth behaviour of cells in suspension cultures of strawberry (*Fragaria* x *ananassa* Duch) *Plant Biosystem.* **133** 205

IOP Conf. Series: Earth and Environmental Science **185** (2018) 012030 doi:10.1088/1755-1315/185/1/012030

Significance of cultivation of new agricultural management subjects for the development goal of modern agriculture: a perspective review

L Chen[1]

Business School of the University of International Business and Economics, Beijing, China

E-mail: chenlian0208@126.com

Abstract: Along with the industrialization, urbanization and globalization deepening, China's agriculture faced with the development bottleneck of "who farming and how farming". Back to the founding of new China era, the traditional agricultural organizations are difficult to achieve agricultural modernization, such as farmers' mutual aid group, agricultural cooperatives, the people's Commune and family contract management. Therefore, it is imperative to cultivate and develop new agricultural management subjects for the goal of Rural Revitalization and agricultural modernization in China. A review combined with related literatures, the paper combs and refines the mechanism of different types of new agricultural management subjects to promote modern agriculture. This is of great significance for eliminating misunderstandings of new agricultural management subjects and allowing full play to their driving roles in China. The paper elaborates and proposes to vigorously promote the development of new agricultural management subjects such as professional households, family farms, farmers' professional cooperatives, farmer' shareholding cooperation and agricultural enterprises, which can promote the specialization, institutionalization, organization and marketization of agriculture, and effectively overcome the problems of low efficiency, high cost and weak effectiveness of traditional agricultural operation organization, so as to speed up the process of agricultural modernization in China.

1. Introduction

Since the twenty-first Century, with the rapid development of industrialization, urbanization, marketization and internationalization, the rural areas have undergone profound changes, and agricultural production is facing severe challenges. First, the transfer of rural labor force to cities and towns. The structural contradiction of rural labor is prominent. Agriculture is facing the problem of no successor. Second, the constraints of agricultural resources and environment are aggravated, and the contradiction between the rigid demand growth of agricultural products and the traditional agricultural development way, which is driven by resources and labor input, is unsustainable. Third, the market risk of agriculture and the pressure of international competition are increasing, as well as the increasing demand for food quality and safety of urban and rural residents, which poses a severe challenge to the traditional agricultural development mode, which is mainly characterized by small farmers. These challenges inhibit the future development of China's agricultural operation. How to further break through all bottlenecks restricting the operation of agriculture and effectively solve the problem of "who will cultivate land" and "how to cultivate land"?

The Report of the nineteenth National Congress of the Communist Party of China clearly put forward the "cultivation of new agricultural management subjects"; "Opinions of the CPC Central Committee and the State Council on implementing the strategy of Rural Revitalization" further proposed that "to give consideration to the cultivation of new agricultural management subjects and support small farmers", "pay attention to the driving effects of the new agricultural management subjects". To cultivate and develop new agricultural management subjects is imperative [1-2]. The basic viewpoints of the scholars are: on the basis of the household contract management, we should speed up the cultivation of the new agricultural management subjects, and focus on the development of professional households, family farms, farmers'' professional cooperative organizations and agricultural leading enterprises, and actively develop a variety of forms of moderate scale management, and build a combination of intensive, specialized, organized and socialized agricultural management system.

2. The professional households promote the efficiency agriculture by deepening the rural labor socialization division

Professional households were grown up under the background of Chinese government supporting the development of grain production and curbing land abandonment. Compared with the traditional mode of management of decentralized peasant households, the level of the scale, mechanization and intensification of agricultural production are increasing by professional households, which is significant in improving the production of agricultural specialization and promoting the increase of farmers' income and the progress of agricultural technology.

Firstly, the development of large professional households will help to reduce farmers' concurrent business behavior and improve the comprehensive benefits of agriculture. Previous research showed that, compared with the pure non-agricultural households and pure agricultural households, per capita income of households with combined occupations is lowest [3]. The long-term existence of farmers' concurrent business behavior is the main reason leading to the marginalization of agriculture and the low efficiency of agriculture. The data of Longwen District of Zhangzhou city in Fujian Province showed that the development of large agricultural households can increase agricultural output value, increase agricultural labor productivity and the commodity rate and the income level [4].

Secondly, the professional household has more favorable conditions for agricultural mechanization and technological progress. Under the support of the government, a large-scale grain-production household in Yueyang city not only realized mechanization of rice planting, but also provides mechanical operation services for other farmers around the areas [5]. Wang Xiaoping [6] investigated 520 households in Jingzhou city. The results showed that the technological progress of the professional households was significantly higher than that of the general farmers.

Thirdly, the development of professional households is more likely to evolve or generate other forms of agricultural organization. The professional households of the Guanyun County in Northern Jiangsu Province were observed, and some of the first wealthy professional households were invested in setting up a new type of joint stock cooperative private enterprise [7]. It was found that farmers' demand for professional co-operatives was stronger with the increase of the cultural level of the household and the enhancement of cooperation consciousness [8].

3. The family farms achieve agricultural Reducing cost and efficiency increasing by means of the institutional innovation in agricultural production organization

The institutesl change from family contract to family farm is a result of the potential profits induced by scale management, technological progress, and specialty division and so on. Zhu Xuexin [9] defines family farm as a modern agricultural organization form, which is characterized by modern technology, scale management and enterprise management. Compared with the traditional family contract system, family farm system has the superior advantage.

On the one hand, family farms not only inherit the advantages of family management, but also overcome the drawbacks of family management. The family farm, based on Farmers' families, can adapt well to nature of agriculture and also meet the requirements of high flexibility, consciousness and sense of responsibility for agricultural production. Family members have strong self-motivation, so as to ensure the enthusiasm of agricultural production. From retail farmers to family farms, scale of agricultural operation and capital investment are increasing continuously, and management and operation level is continuously improving. The family farm, characterized by the market-oriented orientation, the intervention of the modern production factors, the specialization of the division of labor and the mechanized scale economy, is free from the traditional household contract responsibility system to divide equally the land usage power, which is caused by the cultivated land broken bits, concurrent behavior and low efficiency in agriculture.

On the other hand, family farms can effectively reduce market transaction costs, and have obvious comparative institutional advantages. The research results of Wu Kaiqun [10] showed that family farms can use cooperatives to reduce the price of the agricultural means of production and the transaction costs of agricultural products market, and reduce the cost of financing by credit cooperatives. The effective labor incentive and constraint mechanism can reduce the transaction of the labor market. The improvement and implementation of China's national policies and policies will also help to reduce transaction costs in land market. It can be seen that the development of family farms will effectively overcome the problems of high cost, low efficiency in agriculture.

Moreover, family farms are more advantageous to ally with and cooperate with other agricultural organizations. Compared with the strength of capital, the family farm is relatively weak. With the help of technological innovation, the system innovation of the agricultural cooperative organization should be adopted to promote the development and expansion of family farms. For example, family farms and farmers' professional cooperatives can cooperate effectively through the integration, contract formation, and establishment of company farms [11].

4. The farmers' professional cooperatives promote the modernization of agriculture by improving farmers' organizational level

Before the implementation of the household contract responsibility system, the farmer mutual aid and cooperation organization was the main form of agricultural operation organization in China. However, no matter the farmer mutual aid group or the farmer cooperatives in 50s and 60s twentieth Century, they failed to make farmers rich, agricultural development and rural prosperity. The cause of the failure of the traditional cooperative economy lies not in the invalidity of the form of the cooperative organization, but in the cooperative system in the planned economy, which separates the labor and remuneration, the farmers and the market, and erased the enthusiasm of the farmers, and eventually leads to the loss of the agricultural cooperation. The fact is that after the reform and opening up, when a single peasant family faces unavoidable natural risks and market risks, farmers' demand for mutual aid and cooperation is more and more intense. In 2006, the government promulgated the law on farmers' professional cooperatives to comply with the needs of developing cooperative economy. Compared with the traditional cooperative organizations, the farmers' professional cooperatives will better build a bridge between the farmers and the market, stabilize the agricultural income and promote the development of agricultural industrialization.

For one thing, the development of professional cooperatives can effectively alleviate the contradiction between small production and big market. With the advancement of China's agricultural marketization process, facing the market environment of frequent fluctuation of agricultural products prices, small-scale farmers are in a weak position in the aspects of agricultural purchase and agricultural products sales, and the cost of market transaction leading to high, and difficulty to guarantee the stable agricultural income. This means that under the condition of lack of effective market intermediary, millions of households are not able to effectively connect with ever-changing market. As a mutual aid economic organization voluntarily participated by farmers and professional cooperatives have increased the income of farmers and reduced the expenditure of farmers by participating in centralized sales of agricultural products and the centralized purchase of agricultural means of production. With professional cooperatives, farmers have enhanced the ability of farmers to resist natural and market risks, and better achieve the effective docking with the market.

For another thing, professional cooperatives are the carriers of agricultural industrial organization suitable for the transfer of surplus labor. "Farmers' Professional Cooperative Law" clearly points out that "farmers' professional cooperatives, with their members as the main service object, provide the purchase of agricultural means of production, the sale, processing, transportation, storage of agricultural products, technology and information related to agricultural production and management". It can be seen that the professional cooperatives will turn the external transaction relationship into the internal transaction relationship in the various links of the agricultural chain, and share the risks with the farmers and reduce the transaction cost. The study of Lian [12] showed that the development of farmers' professional cooperatives can not only promote the development of the integrated agricultural industry, but also effectively absorb the employment of agricultural surplus labor.

5. The farmers' shareholding cooperation accelerates agricultural modernization through reforming property rights of collective assets

Rural collective economy is an important part of China's economic development. China has experienced profound lessons about how to manage rural collective assets and how to arrange the property rights of rural collective assets. The people's commune is a failed attempt to manage rural collective assets in China. The system arrangement of the people's commune is a form of public property right, that is, collective assets belong to the whole community, and "the Co-ownership is not and cannot be objectified on each member" [13]. "Since the common property rights are not exclusive within the community, the cost is shared by the members of the community, which leads to the congestion of the use of resources and reduces the efficiency of the long-term utilization of resources" [14]. Therefore, in order to develop the rural collective economy, it is necessary to change the form of public property rights, introduce the shareholding system into the cooperative system, and promote the reform of the collective ownership cooperative system in rural areas.

The farmers' joint stock cooperation quantifies the shares of collective assets through the agglomeration of land and capital, and changes the common property rights system of farmers to create a system innovation for farmers to enjoy the income of collective assets in accordance with their shared property rights system. This kind of property right system has reduced or cancelled collective shares, and has realized the "Pareto improvement" through folding shares to farmers to complete the individual property rights of them. Therefore, the development of farmers' shareholding cooperative system is not only conducive to safeguarding the material interests of farmers, but also of great significance for promoting rural urbanization and the development of collective economy.

In the first place, the farmer shareholding cooperative system effectively liberated the dependence of farmers on collective assets and promoted the transfer of rural labor force to urban areas. In the process of promoting industrialization and urbanization, it is inevitable for a large number of agricultural population to flow to cities and towns. However, many farmers do worry that when land is expropriated or transferred to the public, they will lose the right to enjoy the rights of collective assets, which will further obstruct the expansion of cities and towns. In order to eliminate these concerns, joint stock cooperation, by converting collective assets into stock and converting the ownership of asset into equity, and "transforms from physical form to value form", which guarantees the farmers' rights to the income of collective assets, and brings economic benefits by holding or transferring shares after being transferred to the public. This has formed a huge push for farmers to leave the countryside and effectively promoted the transfer of agricultural population to urban areas. For example, Nanhai District in Foshan City of Guangdong Province [15] and Jiangdong District in Ningbo City of Zhejiang Province [16] is the beneficiary area that implements farmer share cooperation system.

In the next place, the implementation of farmer share cooperation can better allocate all agricultural production factors and develop modern agriculture through advanced management. Zhang Yunhua [17] took Jiangsu Province as an example to introduce the cooperative farm formed by the village collective and the land share cooperative on the basis of the land share cooperation, and carry out the specialized contracting, which not only effectively uses the capital and organizational advantages of the village collective, but also makes full use of the land, the machinery and the labor force. As a result, the efficiency of agricultural production has been improved. Moreover, the innovative management system of "land share cooperation and production professional contract" adopted modern management methods such as professional contract, clear responsibility and performance appraisal to solve the outstanding supervision and incentive problems in agricultural production, and thus promoted the development of agricultural modernization.

6. The agricultural enterprises allocate agricultural resources in a market-oriented way to lead the modernization of agriculture

In ancient China or before the reform and opening up, there were few agricultural enterprises in China because enterprises evolved in the market competition, and they are the outcomes of market economy development. When the market subject allocates resources by the means of market with higher transaction costs, it will make the administrative arrangements of the bureaucratic organization and the internal organization of the organization to configure the resources. Thus, the market subject is inclined to replace the external market mechanism with the internal organization. In this way, the market subject will participate in the market transactions in the form of enterprise. Compared with the administrative plan of the co-operatives and the people's communes, the agricultural enterprises are engaged in intensive management in the way of clear property rights, independent management and equal contracts. Therefore, the organizational form of agricultural enterprises will optimize the allocation of capital, technology, labor and other production factors, and promote the farmers' specialization, scale and intensive production to improve the efficiency and competitiveness of agricultural production in an all-round way.

First of all, the successful experience of developing the large capitalistic companies and large enterprises to realize the agricultural scale benefit [18] can be the important reference for the development of agriculture in China. For example, in 1950s, American agriculture absorbed a large amount of social capital and established a vertical integrated agricultural company of production and marketing. It operated through internal integration, contract, and material service and so on. It had the characteristics of specialization, mechanization and socialization, which greatly improved the productivity of labor and the efficiency of social service. After World War II, the agricultural enterprises in Germany mainly participated in the process of agricultural industrialization through the way of cooperation with cooperatives. Since 1990s, France has made great efforts to develop companies or enterprises in the form of contract. By signing a contract for production and marketing with agricultural workers, companies or enterprises provide farm workers with seedlings or seeds, feed and mechanical equipment to preserve and develop the form of family labor. Despite the different forms, the advanced experience of developed countries in developing modern agriculture is inseparable from the successful operation of agricultural enterprises.

Secondly, according to the theory of Growth Pole, the development of agricultural leading enterprises can not only promote the growth of agricultural economy, but also benefit the other agricultural industrialization links because of the "Expansion Effect" [19]. Under the expansion effect, agricultural industrialization links with low development gradients will benefit from the spillover effects of agricultural leading enterprises. For example, Zhang Yongqiang [20],taking Heilongjiang Province as an example, found that the agricultural leading enterprises have a knowledge spillover effect on family farms. It is suggested that the "agricultural leading enterprises + family farm" or "leading enterprises + cooperatives + family farm" should be used to improve the market advantage of family farms.

Finally, by innovating with the operation mode of agricultural industrialization and avoiding the "Backwash Effect" of agricultural leading enterprises, we can earnestly safeguard the interests of farmers. The theory of Growth Pole enlightens us that the agricultural leading enterprises will increase their competitiveness and influence increasingly because of their continuous enhancement of self-ability and the continuous improvement of their investment environment, which compels the "low gradient" link to be in a relative disadvantage. In reality, the agricultural leading enterprises can take advantage of their own capital, technology, price monopoly and other advantages to buy agricultural products from the farmers at low prices, and then sell high value-added products at high prices after processing, and make the enterprises develop rapidly by partially depriving agricultural profits. Therefore, "leading enterprises + farmers" is not the best development mode, it is easy to generate "Backwash Effect", which will damage farmers' interests. In order to avoid the "Backwash Effect", we must innovate the organization mode of agricultural industrialization operation— "agricultural leading enterprise+ cooperative + peasant household" mode. By combining the retail groups to build up the cooperatives that represent their own interests, the co-operatives will negotiate with the enterprises and sign the contract, which can not only reduce the transaction costs of the enterprises and farmers, but also enhance the farmers' right of discourse and management, so as to better protect the farmers' interests.

References

[1] Chen X W, 2013,Building a new agricultural management system is imperative, *Seeks Truth*, **22**, 38-41

[2] Luo D and Chen J, 2013, It is imperative to build a new agricultural operation system, Seeks Truth, **14**, 29-31

[3] Duan Q L, 2002, Research on the economic type and division economy of rural families in China, *Economis*, **5**, 72-78

[4] Ji Y M and Chen Y G, 2007, Professional household should be the main force in building modern agriculture, *Chin. Rural Econ.*, **S1**,73-7

[5] Yang Y Q, 2012, Strive to achieve 50 thousand acres of rice planting mechanization--the heart of a large grain farmer, *Rural Work Commun.*, **1**, 22

[6] Wang X P, 2007, An empirical analysis of the technical progress of agricultural major households-Based on the survey of 520 agricultural professional households in Jingzhou City, *Sci. Tech. Prog. Poli.*, **7**, 47-50

[7] Shi Q Z, 1995, An investigation on the establishment of joint-stock cooperative private enterprises in rural areas, *Res. Econ. Manag.*, **5**, 63-4+37

[8] Zhu H G, Chen Z J, Weng Z L and Liu X C, 2008, Analysis of the factors affecting the demand for professional cooperatives in rice farming households, *Issues Agri. Econ.*, **12**, 71-8+112

[9] Zhu X X 2006, Family farm is a realistic choice for intensive agricultural management in South of Jiangsu, *Issues Agri. Econ.*, **12**,39-42+80

[10] Wu K Q, 2013, Optical fiber theory analysis, *Econ. Review*, **6**, 65-9

[11] Wang Y, 2014, Research on the cooperative relationship between family farms and farmers'

professional cooperatives, *China Rural Survey*, **2**, 39-48+93-94

[12] Lian G B and Li Y D, 2009, The agricultural industrialization approach of the transfer of agricultural surplus labor force: an analysis based on the perspective of farmers' professional co-operatives, *J Northwest Uni. (Philosophy and Social Sciences Edition)*, **5**, 66-9

[13] Kong J Y, 1995, An analysis of the cooperative economy and its system in rural areas, *Econ. Res. J.*, **3**,22-34

[14] Fu C., 2001, Research on the property right system of community-based rural joint stock cooperative system, Reform, **5**,100-9

[15] Jia C M and Ge Y, 2012, A study on the farmers' income increase effect on the agricultural land joint stock cooperative system--Based on the empirical analysis of the Foshan four cities(districts)in 1992 to 2009, *J. Nanjing Normal Uni. (Social Science Edition)*, **1**, 58-65

[16] Shi T T and Zhang R B, Joint-stock economic co-operatives promoted the urbanization of the Suburb--Taking the Jiangdong District of Ningbo as an example, *Zhejiang Social Sci.*, **7**,145-150+161

[17] Zhang Y H and Guo C, 2013, The case of Jiangsu's agricultural operation system innovation: land share cooperation and production professional contract, *Reform*, 2,151-8

[18] Li Q and Zhang B, 2011, Thinking about the localization of agricultural development mode driven by leading enterprises, *Guizhou Social Sci.*, **3**, 50-4

[19] Peng Y, He P C and Shao G R, 2005, Construction of the leading enterprise of agricultural industrialization--a view from the perspective of the theory of development polar theory, *J Zhejiang Uni. (Humanities and Social Sciences)*, **11**, 97-104

[20] Zhang Y Q, Cai Z and Zhang L, Research on the knowledge spillover effect of agricultural leading enterprises on family farm--Taking Heilongjiang Province as an example, *Issues Agri. Econ.*, **11**, 10-6+110

IOP Conf. Series: Earth and Environmental Science **185** (2018) 012031 doi:10.1088/1755-1315/185/1/012031

Poverty alleviation through rural e-commerce mode of "One Village, One Product and One Shop": a case from Suqian, Jiangsu Province

F Y Long[1,2,4] and C Wang[3]

[1] School of Economics, Shanghai University, 99 Shangda Road, 200444, China

[2] School of Business, Suqian College, 399 Huanghe Road, 223800, China

[3] School of Economics and Management, Nanjing Agricultural University, Weigang Road, 210095, China

Email: lfyfly916@163.com

Abstract: The rural E-commerce has become a hot spot in the current poverty alleviation in China. Suqian of Jiangsu Province took the lead in exploring the new path and has achieved good effects. Practice shows that the rural E-commerce mode "one village, one product and one shop" in Suqian has a favourable impact on increasing farmers' income. This paper firstly elaborates the theoretical mechanism of poverty alleviation through "one village, one product and one shop", then introduces a typical case: the "Party branch + E-commerce" poverty alleviation mode of Jidong community in Suqian, finally puts forward several enlightenments: Focus on "one product" to develop featured product brands; Strengthen "one shop" relying on multiple E-commerce platforms; Motivate "one region" by cultivating paradigms; Form "one chain" by integrating the policy resources. This paper not only can enrich the theoretical and case research of rural E-commerce poverty alleviation in China and abroad, it also can provide decision supports for government departments in targeted poverty reduction and make some suggestions for E-commerce enterprises to enhance operating efficiency.

1. Introduction

Poverty is a global problem, the elimination and alleviation of poverty is the common challenge to all countries including China. In order to achieve the ambitious goal of building a moderately prosperous society in all respects by 2020, the task of poverty alleviation remains severe during the "13th five-year plan"[1] period in China. In 2014, the state council's leading group of poverty alleviation and development formally incorporated E-commerce poverty alleviation" into poverty alleviation policy system, and regard it as one of the "top ten projects in targeted poverty alleviation"[2]. Since then, E-commerce poverty alleviation has become an important part of China's "targeted poverty alleviation" strategy.

[1] China has made its "first five-year plan "since 1953. The year of 2016 is the first year of its "13th five-year plan" (2016-2020).

[2] The state council's leading group of poverty alleviation and development :http://www.cpad.gov.cn/

Some Chinese scholars have studied this issue. The earlier research scholar Wang [1] made a definition of the role about E-commerce in poverty alleviation: integrating E-commerce into the system of poverty alleviation, acting on the targeted objects, innovating the ways of poverty alleviation, and improving the performance of poverty alleviation. Li *et al.* [2] proposed that the government should take the lead to set up the E-business poverty alleviation experiment areas; summarize and promote the successful experience; encourage experiment areas and enterprises to invest heavily to develop E-commerce and give preferential policies to the pilot areas. Meng and Zhao [3] introduced four main modes of poverty alleviation through rural E-commerce: farmers' professional cooperative dominating mode; leading enterprises dominating mode; government agencies dominating mode; and Professional operator dominating mode. They also introduced the characteristics of these modes respectively. Wang *et al.* [4] believed that rural E-commerce could not only change the local economy pattern and the social relation network, but also stimulate the enthusiasm of the floating population to return home to start a business, which can expand the employment channels and promote the precision poverty alleviation.

Several scholars abroad have studied the issue as well. Nadarajan and Ismail [5] put forward a comprehensive and applicable E-commerce framework that serves as an enabler to the development of rural agriculture sector in Cambodia, and he believed the E-commerce plays an important role in the economic and social growth by promoting rural livelihoods, food security and poverty reduction. Kwak and Jain [6] deemed E-commerce has a great potential to reduce poverty by improving the availability to education, health, government, financial and other services due to easy access to mobile devices.

2. The Origin and development of the e-commerce model "one village, one product and one shop"

"One village, one product and one shop" is a practical mode of rural E-commerce. Suyu District of Suqian is the earliest pilot area of "one village, one product and one shop". The so-called "one village, one product and one shop" means each village makes efforts to develop a special industry on the basis of local resources endowment, to build leading products, then to open a standard online shop to expand the product market[3]. Through the development of rural E-commerce, it will finally form the "Agriculture + Internet" new format, which broadens new channels for low-income farmers and economically weak villages to increase income, and explores a new path of E-commerce poverty alleviation.

In 2016, the per capita disposable income of rural residents in Suqian was 13,929 RMB, a 9.1% increase from last year, in which 70% comes from E-commerce. In this year, 146,000 out of 662,900 low-income populations are out of poverty smoothly. From January to July of 2017, the total amount of E-commerce transactions is 50 billion RMB, and various online platforms directly absorb more than 5,000 low-income populations. Till the November of 2017, the city's low-income farmers have set up 5,600 online shops, which directly and indirectly generate 52,000 labor forces. At present, the coverage rate of "one village, one product and one shop" in economically weak villages is up to 46%, directly and indirectly drive 65,000 people out of poverty. The experience is popularized by Jiangsu provincial government to the whole province[4].

3. The poverty alleviation mechanism of "one village, one product and one shop"

3.1. Innovating agricultural products trading modes and reducing transaction costs

Different from the traditional trade modes, E-commerce platforms link farmers, consumers and markets together, make the buyers and sellers complete the whole transactions online, which reduce the transaction procedures and costs. In addition, through the E-commerce platform, producers can

[3] The government network of Suyu District of Suqian: http://www.suyu.gov.cn/col/col3101/index.html

[4] The data comes from the statistics of Suqian municipal government.

rapidly and accurately acquire the market demand information. They can produce appropriate amount of agricultural products, which could avoid the storage and processing costs caused by excess products. This can greatly change the unmarketable situation of agricultural products.

3.2. Improving the agricultural practitioners' structure and releasing the invisible demographic dividend

The E-commerce mode "one village, one product and one shop" can realize the situation that every family can open an online shop, each of the young and the old has something to do. The leading enterprises, competent people and professional cooperatives engaged in E-commerce fields could set up special jobs for elderly people, the disabled and housewives to involve in the E-commerce work, which makes them become industrial workers with stable income. This makes jobless people find right positions, and further releases the rural invisible demographic dividend.

The E-commerce mode "one village, one product and one shop" also attracts a large number of young people and the floating populations return home to start their own business, college graduates, returning farmers and village officials have become the main force. In order to encourage these people to the countryside to start their own business, Suqian has introduced a system of starting business vouchers[5] for them, which has greatly improved the structure of agricultural workers.

3.3. Embedding in E-commerce industry chain and creating more job opportunities

Rural E-commerce benefits a large number of farmers through agricultural products processing industry chain, farmers could directly participate in agricultural planting, processing, packaging, sales and after-sales works. In addition, the development of rural E-commerce forms an integrated industry chain including product supply, platform sales, logistics, financial payment and service support. The whole chain can create more job opportunities and provides a large number of local jobs for poor groups.

3.4. Building featured agricultural products brand and enhancing the added value

Based on the local resource endowment, the featured agricultural product brand can be built to enhance the added value of agricultural products. In the practice of "one village, one product and one shop", each town majorly supports one featured industry with better endowment and builds the featured product brands. Making full use of the green agriculture resources, Suqian has constructed a batch of standardized farming and processing bases, organized professional talents for brand designing, package planning and promotion, finally created a batch of agricultural product brands.

4. A typical case: the "Party branch + E-commerce" poverty alleviation mode of Jidong community[6] in Daxing town, Suyu District

The "Party branch + E-commerce" is a typical mode dominated by primary-level village Party organization in poverty alleviation. Jidong community has developed this new mode based on the festival industry. At present, the festival industry represented by wedding candy boxes, glow sticks

[5] There are two kinds of business vouches: the first type is a special funding to encourage the initial behavior of young talents to start a business. The second type is the targeted funding in five stages of starting a business, including training vouchers, internship vouchers, loan discount coupons, agricultural insurance subsidy vouchers and publicity subsidy vouchers. Each kind of the vouchers is issued a maximum number of 1,000 copies per year.

[6] Jidong community is identified as "Taobao village" of China (2017) by Ali Research. According to the definition of Ali Research, "Taobao village" mainly includes three principles: the operation place is in the rural area, and the operation unit is administrative village; The annual turnover of E-commerce is over 10 million RMB; The number of active online shops in the village reaches more than 100, or accounts for more than 10% of the local households. When the number of villages meeting the standard is more than or equal to 3 in one town, the town is identified as "Taobao town".

IOP Conf. Series: Earth and Environmental Science **185** (2018) 012031 doi:10.1088/1755-1315/185/1/012031

and silk flowers begins to take shape, forming an integrated production-supply-marketing industry chain.

According to the preliminary statistics, festival industry has driven more than 600 people of Jidong community to get a job in the hometown and the average monthly salary is 2000 RMB. It has solved the employment problems of a large number of elderly people who stay at home, and helped 43 low-income families out of poverty, accounting for 57% of the low-income families in the village. In 2016, Jidong community has 38 online stores engaged in festival products sales, and the E-commerce turnover are 158 million RMB, of which 64.8 million RMB is from the wedding candy boxes, 53.2 million RMB is the glow sticks, and 40 million RMB is the silk flowers. The wedding candy boxes and glow sticks account for more than 90% of the market share among Taobao's similar products. Jidong community takes advantage of the great demand in online market and makes a lot of efforts to develop the festival industry[7].

First, it has a policy support. The branch has issued the policy of "four exemptions": the entrepreneur who is engaged in the festival industry could enjoy two years of rent free, free installation of broadband, free provision of storage and free E-commerce skills training. Until the end of July, 2017, Jidong community has invested a total of 2.1 million RMB supporting funds, and amount to 117 entrepreneurs have enjoyed the preferential policy of "four exemptions". At the same time, the Party branch keeps contact with the local banks, and takes the initiative to guarantee loans for the residents who has urgent fund needs. In 2016, Jidong community coordinated more than 2 million RMB bank loans for 14 entrepreneurs, and guaranteed 320,000 RMB poverty alleviation loans for 32 low-income rural households.

Second, it has standardized management. Facing the unordered competition, the Party branch takes the lead to establish the industry in accordance with the "uniform standard, unified production, unified price, unified logistics" four unified model. The industry association sets up the R&D center specialized in R&D and promotion for new products, which makes the festival industry always remain full of vitality.

Third, it provides the availability of talent training. The residents' committee successively organizes village cadres, head of the manufacturing enterprises and the shop owners totaling 110 people to investigate and learn from the E-commerce developed area Yiwu[8], Zhejiang Province. The committee also sets up E-business training center in order to provide E-commerce training regularly and impart online shops operating skills. Since 2016, it has trained a total of 140 people, helped 68 young people successfully open online shops and selected eight excellent online shop owners to receive senior E-commerce training in Zhejiang.

5. The enlightenments of E-commerce mode "one village, one product and one shop" in poverty alleviation of Suqian

5.1. Focus on "one product" to develop featured agricultural product brands

Rural areas should regard "one product" as the core of developing rural E-commerce. Firstly, we should select the advantageous local products as the main cultivation industry based on the natural resources and vigorously develop the rural specialty products. Suqian set up a special support fund 20 million RMB to promote the "1 + X" characteristic agriculture industry system construction (i.e., a leading industry + several characteristic industries), form a pattern of "one town has one industry, one village has one product ". In 2016, the number of the online specialty products reached more than 320 kinds.

Secondly, we should accelerate the development of standardized process, and guide farmers to carry out standardized production. We should acquire the certification of agricultural products to

[7] The data comes from the statistics of Suqian municipal government.

[8] Yiwu is the largest "Taobao village cluster" in China (2017). "Taobao village cluster" is defined as consisting of 10 or more "Taobao village", and the E-commerce transactions reach or exceed 100 million RMB.

improve the ability and level of agricultural processing. In 2015, Suqian took the lead to build and run the quality traceability platform. Suyu District Government also signed the product quality assurance agreement with nationally known E-business platforms (Tesco, Jingdong, and Dangdang) to ensure the local product quality.

Thirdly, we should encourage the professional talents to take part in brand designing, package planning and publicity for local products, to create a batch of agricultural products brands and raise the added value of agricultural products. At present, Suqian has created 12 agricultural products brands at and above the provincial level, including three national well-known trademarks.

5.2. Strengthening "one shop" relying on multiple E-commerce platforms

E-commerce platform is a key element for E-commerce activities. With the rapid development of E-commerce, platform competition has become the battleground. Firstly, we should settle the superior and special products into large-scale E-commerce platforms, like Jingdong Mall, Tmall and Suning relying on their competition advantage. Currently, most of the products sold by online stores can't enter the high threshold E-commerce platforms due to lack of brand and standard, which can only go to platforms with lower thresholds, such as Taobao. Jingdong Information Technology Park is located in Suyu, and Suqian utilized this resource advantages to accelerate the strategic cooperation with Jingdong Mall. The government constructs the rural E-commerce service sites in every towns and villages and promotes in-depth integration between the Internet and farmers.

Secondly, we should actively set up E-commerce associations to promote joint operation, and integrate all the resources, technologies and information to maximize their utilities. Besides, the associations could unify the production and processing standard to ensure a sustainable competitiveness.

Thirdly, we should innovate the online marketing modes. For example, we could explore the personalized customization services and establish the O2O experience stores to enhance consumers' experiences. We can also utilize the "travel + shopping" idea to develop the agricultural by-products, and form rural featured tourism products.

5.3. Motivating "one region" by cultivating paradigms

Firstly, the primary Party branch should popularize the "Party branch +E-commerce" mode and put more investment into infrastructure, equipment and human capital, initiate more agricultural projects and set up more online stores. Suqian has successively developed some advanced Party branches like Shunhe Street and Guangming village, forming fierce competition atmosphere between villages.

Secondly, college village official should play the demonstration role. Take them as the main body to form a powerful team who will undertake the duty of rural E-commerce promoters and remote education administrators. Support these village officials to take the lead to start their own business to set up online stores. Currently, 92% of college village officials in Suqian have their own online start-up shops.

Thirdly, the government department should set up incentive system for paradigms and introduce professional talents and special support funds for the outstanding townships and villages in order to form demonstration effects [7].

5.4. Forming "one chain" by integrating the policy resources

Firstly, the government should develop the supporting industry systems, such as the logistics, information technology, financial and after-sale services, especially the logistics system. The government should optimize the layout of logistics and expand the service scopes to build a comprehensive rural logistics system.

Secondly, they should establish a perfect supporting system of E-businessman operation. Suqian has founded the online training platforms to provide information supporting, strategies references and practical trainings for Internet entrepreneurs. With the help of the "big data" information of Jingdong,

Suqian could predict the trend of the market changes of agricultural products, and provide scientific guidance about agricultural product planting.

Thirdly, the government should provide support funds through financial discount, start-up subsidies and sales incentives. They could establish a cooperation bridge between Banks and enterprises, guarantee loans for rural E-commerce operators to solve practical problems .Suyu District of Suqian has set up 15 million RMB start-up support funds, in which 5 million RMB is specialized for "one village, one product and one shop".

6. Conclusion

Rural E-commerce is growing vigorously, which opens up new channels for poor groups to increase income, and becomes an important new path for poverty alleviation. However, we should be aware that its development is in early stage, the policy system is not perfect, and the market environment is not strong enough. In addition, the lack of E-commerce talents and farmers' low participation in E-commerce activities will reduce the radiating and driving effects on increasing income. Next, we need to concentrate on constructing the poverty alleviation system and optimizing the hard and soft environment, in order to make rural E-commerce be the "booster" of poverty alleviation.

In general, the theoretical and empirical researches on rural E-commerce poverty alleviation are relatively deficient, which give scholars more opportunities to explore in this field. This paper is also shallow, which lacks the support of empirical data and only demonstrates case analysis and its enlightening. It will provide the direction for my future research.

References

[1] Wang X D 2015 Four problems of E-commerce to poverty alleviate *Gansu Agri.* **13**:18-20
[2] Li Z M, Qi J L and Sun H Q 2015 Poverty alleviation exploration of rural e-commerce *Market Modernization* **02**:74-75
[3] Meng B C and Zhao X 2017 The targeted poverty alleviation mode of rural e-commerce *Enterprise Tech. Develop.* **01**:58-60
[4] Wang Y, Xie Y and Wang M.2017 Rural e-commerce relational network and place making under the background of targeted poverty reduction *World Regional Studies* **26(06)**:119-130
[5] Nadarajan SV and Ismail R 2011 E-Commerce framework to improve rural agriculture sector in Cambodia *International Conference on E-business,management and Economics* **25**:287-291
[6] Kwak DH and Jain H 2016 The role of web and e-commerce in poverty reduction: A framework based on ecological systems theory *Lecture Notes Busi. Inform. Process.* **258**:143-154
[7] Wang C and Long F Y 2017 A analysis of rural e-commerce mode "one village, one product and one shop" *Jiangsu Agri. Sci.* **45(04)**:293-295

IOP Conf. Series: Earth and Environmental Science **185** (2018) 012032 doi:10.1088/1755-1315/185/1/012032

Enumeration of ice nucleation active bacteria and severity of frost injury (*embun upas*) on potato in Wonosobo, Dieng Plateau

A Susilowati[1], L Y Oktiningtiyas and R Setyaningsih

Study Program of Bioscience, Graduate Program, University of Sebelas Maret, Jl. Ir. Sutami 36A Kentingan Jebres Surakarta 57126, Indonesia

E-mail: arisusilowati@staff.uns.ac.id

Abstract. The altitude of the Dieng plateau is between 1,500-2,500 meters above sea level (m ASL), with an average temperature of 14ºC. During the dry season, the temperature in this area becomes very cold that reaches -2ºC, therefore frost injury may occur in potato, which is often referred to as *embun upas* (frozen dew). Frost injury causes the production of potatoes in Dieng to decline. Frost injury in plants is characterized by spots symptom, which can be brown, yellow, or black on leaves surface with the chlorotic ring that surrounds them. The phenomenon of frost injury is thought to be related to the activity of ice formation by the Ice Nucleation Active (INA) bacteria and triggered by cold temperatures (-2°C). This study aimed to determine the number of INA bacteria and its correlation with frost injury spots on potato leaves and to discover the class of INA bacteria based on ice formation temperature. The sampling of the potato leaves used purposive sampling method. Leaves showed frost injury symptom with different severity were taken from a different altitude of ± 2,000 m ASL (Dieng Wetan village) and ± 2,500 m ASL (Sembungan village). Bacterial isolation was done by spread plate method on Nutrient Agar supplemented with 2.5% glycerol. The activity of ice nucleation was determined using tube nucleation method. Estimation of INA bacteria number was done by multiple-tube nucleation method 3.3.3. Correlation between the number of INA bacteria and percentages of frost injury spots was analyzed using SPSS bivariate. The results showed that the highest INA bacteria number i.e. >2.75 x 10^4 was found on scale 3 of leave spots while the lowest value of 7.25 x 10^3 on scale 1. Correlation between the populations of INA bacteria with the scale of frost injury spots showed 0.844. The correlation values ranging from 0.80 to 1.00 indicating a strong association between the INA bacterial number and the scale of frost injury spots of potato leaves. Based on the ice forming temperatures, INA bacteria were classified into three classes, i.e., class A that can freeze water at temperatures of -2 to -4°C, class B -5 to -7°C and freezing point of class C below -8°C.

1. Introduction

Dieng Plateau is located in Central of Java Province and has three regencies, which are Banjarnegara, and Wonosobo and Temanggung. The altitude of the Dieng plateau is between 1,500-2,500 meters above the sea level (m ASL). *Embun upas* is the locally Indonesian name of ice nucleation diseases on potato leaves on Dieng plateau. *Embun upas* generally appears in the Dieng plateau on July or August of each year. During that time, the temperature is at the critical point reaching zero and even minus causing the appearance of *embun upas*. The lowest temperature that occurs at a critical point can reach -8°C. The emergence of *embun upas* results in frost injury on the potato that ultimately leads to the death of the potato and the risk of crop harvesting failure [1]. *Embun upas* becomes of the limiting factors in potato production in the Dieng Plateau, causing loss to the potato farmers reaching hundreds million rupiah, after more than 35 ha of potato plantation xperienced frost damage in 1994 [2]. The research location of Dieng Plateau (Central Java) and potato plantation in Sembungan village can be seen in Figure 1A and 1B.

Figure 1A. Location of Dieng plateau which includes Banjarnegara, Wonosobo and Temanggung (Central Java).

Figure 1B. Sampling Area: Sembungan Village (7°14'24.94"S-109°55'0.97"T) is dominated by potato plantations in terraces.

The frost injury on potato has not been well understood by most farmers and that the numbers of INA bacteria are responsible for its formation. INA bacteria have been considered as the main causative factor of frost injuries on potato because of their ice nucleation property that can initiate theformation of ice crystalin plant tissue. Bacteria that cause frost injury are called INA bacteria [3]. The INA bacteria are bacteria that live on the leaves surface (phyllosphere) and able to catalyze the formation of frost injury on the leaves surface.

Reports of frost injury mostly come from subtropical countries, like in Japan, which occurred on tea plants [4], cabbage, broccoli and mulberry [5] and in the United States, which occurred on corns' plants [6], oranges, wheat, and tomatoes [7]. Hirano and Christen later proved that frost injury that occurs on some species of plants are caused by the activity of bacteria in the leaves surface and is harmful because it causes plants death [8]. Research on frost injury bacteria from the tropics has not been widely reported or published. The data of INA bacteria studied on potato in Wonosobo, Dieng Plateau will provide information about the novelty of this bacterium in the tropics.

Further research concerning bacteria in the tropics needs to be conducted, considering its negative impact on agriculture due to the *embun* phenomenon caused by ice nucleation activity. Basic knowledge revealed about INA bacteria on potato plant can later be used to arrange strategies to overcome and prevent the occurrence of damage due to frost injury on potato plants in Dieng plateau. In this study, we covered about the estimation of INA bacteria number in potato leaves, determination of disease severity scale caused by INA bacteria, and the identification of INA class based on its freezing temperature.

2. Material and methods

2.1. Estimation of INA bacterial number

Estimation of INA bacteria number can be done using multiple tube nucleation method. Samples of potato leaves weighed, 2 grams were homogenized in 50 mL phosphate buffers in 0.1% protease-peptone. Taken as much as 1 mL and inserted into a test tube containing 9 mL sterile phosphate buffers. Afterwards, 3 series of dilutions i.e., 10^{-1}, 10^{-2}, 10^{-3} were performed into a tube containing 9 mL of phosphate buffers. In each series of dilutions, duplicate replications were performed so that the dilution of the 3-tube series for each dilution series was obtained. The bacterial suspension as the result of dilution was introduced into the circulating alcohol bath at the temperature of -10ºC for 10 minutes. umber of bacteria per gram of fresh leaves weight was estimated based on the Most Probable Number (MPN) method. The number of total INA bacteria was determined by counting the number of tubes giving positive frozen reaction and then comparing the pattern of positive results (the number of tubes showing frozen at each dilution) with standard MPN statistical tables [9-10].

2.2. Disease severity of frost injury on potato

The sampling of the potato leaves used purposive sampling method. Leaves showed frost injury symptom with different severity were taken from a different altitude of ± 2,000 m ASL (Dieng Wetan village) and ± 2,500 m ASL (Sembungan village). Disease severity was expressed in the spotting scale (Table 1).

The association of INA bacterial population with disease severity was analyzed using SPSS with correlation bivariate test.

Table 1. Diseases severity of frost injury on potato leaves expressed in the spotting scale.

Spot Scale	Description
0	No symptoms on the leaves
1	Little spots, not very real, about 10% of the leaves area infected

2	Little spots spread on the leaves and easily observed, but do not cause any apparent damage, 11-30% of the leaves area infected
3	Symptoms spread on the leaves area, damage and chlorosis are limited, 31-50% of the leaves area infected
4	Widespread symptoms occur, leaves necrosis, 51-71% of the leaves area infected
5	Spotting almost or spreading throughout the leaves area, causing necrosis to death, 71% or of the leaves area infected

2.3. Isolation of INA bacteria
For each leave, a sample of 2 grams was cut with a size of 0.5 cm^2 and the sample was put into a 250 mL Erlenmeyer flask containing 50 mL of 0.1 M phosphate buffer pH 7 and 0.1 M protease peptone. Erlenmeyer flask was shaken on a rotary shaker at a speed of 150 rpm for 2 hours. One mL of the samples were taken and put into a test tube containing 9 mL of sterile distilled water. Three series of dilution i.e., 10^{-1}, $10^{-2,}$ 10^{-3} were performed, and from each dilution 100 μL of the sample was taken then plated on NA medium containing 2.5% glycerol (NAG) using spread technique into the petri dish. Glycerol-containing media is a common medium used to isolate ice-forming bacteria because it can optimize the growth of cultures [11]. Glycerol as a major carbon source is capable of enhancing nucleation activity, according to [12] studies suggesting that some bacteria with weak or lost nucleation activity are able to restore their activity after growth in nutrients agar added glycerol at 22ºC. Each dilution series was spread on two Petri dishes. After the surface of the media dried sufficiently was then wrapped in paper and incubated at 18-24ºC with the reversed position for 2 days [13].

2.4. Collection of pure isolates
Colonies of bacteria that have grown in NA media + 2.5% glycerol that had different morphologies were transferred into an agar slant by a scratch method using inoculating needles in an aseptically Laminar Air Flow Cabinet. Single bacterial cells formed a single colony was then grown for 4-6 days in the incubator at a temperature of 22ºC, then stored in the refrigerator at 5ºC [14].

2.5. Activity Test for Ice nuclei formation
The activity of ice nucleus formation was conducted using multiple nucleation tube tests. Four to six days old bacterial colonies on an agar slant medium were transferred using inoculation needle and re-suspended in 400 μl sterile phosphate buffers and tested for the ice nucleation activity at temperature -4°C, -7°C and -10°C in a circulating alcohol bath for 10 minutes. Positive test results indicated that phosphate buffers when dropped by bacterial suspension freeze within 30 seconds at a temperature of -4°C or less, it was considered that the colony contains active ice nuclei or contains INA bacteria [15-16].

3. Results
Based on the positive results of Most Probable Number (MPN) with three series of dilutions at a different altitude of Dieng Wetan Village (± 2000 m ASL) and Sembungan Village (± 2500 m ASL) respectively, obtained a freezing tube showing the number of INA bacteria (Figure 2).

IOP Conf. Series: Earth and Environmental Science **185** (2018) 012032 doi:10.1088/1755-1315/185/1/012032

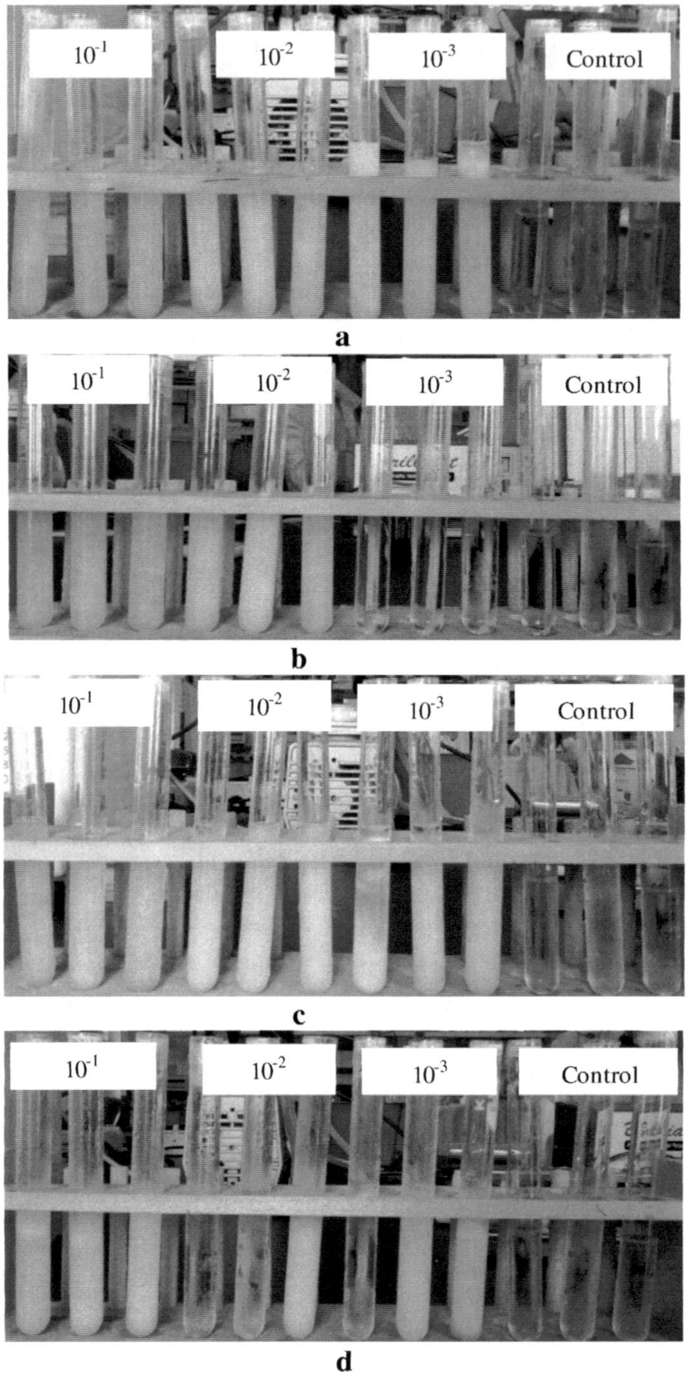

Figure 2. Positive frozen tube on MPN test. Dieng Wetan station I a). Tube combination 3-3-3; Dieng Wetan station II b) Tube combination 3-3-2; Sembungan stasiun I c) tube combination 3-3-3; Sembungan stasiun II d) tube combination 3-1-2.

MPN values in replication one and two at each at sampling area of Dieng Wetan and Sembungan (Table 2 and 3). MPN value indicated the highest INA bacterial population in potato plant in Dieng plateau with tube combination 3.3.3 i.e >2.75 x 10^4. Average number of bacteria based on MPN on the spotting scale of 3, 2 and 1 (Tabel 4). The average number of bacteria on a scale of 3 is > 2.75 x 10^4, the number of bacteria on scale 2 is >1.67 x 10^4, scale 1 in Dieng Wetan > 1.52 x 10^4, and scala 1 in Sembungan 7.25 x 10^3.

Table 2. Estimation of bacterial number based on MPN table, replication 1.

Sampling Station	Scale ofSpots	Replication 1			
		10^{-1}	10^{-2}	10^{-3}	Estimation of Bacteria number MPN/g leaves
Dieng Wetan Station I	3	3	3	3	>2.75 x 10^4
Dieng Wetan Station II	1	3	3	0	6 x 10^3
Sembungan Station I	2	3	3	3	>2.75 x 10^4
Sembungan Station II	1	3	1	2	3 x 10^3

Table 3. Estimation of bacterial number based on MPN table, replication 2.

Sampling Station	Scale of Spots	Replication 2			
		10^{-1}	10^{-2}	10^{-3}	Estimation of Bacteria number MPN/g leaves
Dieng Wetan Station I	3	3	3	3	>2.75 x 10^4
Dieng Wetan Station II	1	3	3	3	>2.75 x 10^4
Sembungan Station I	2	3	1	2	3 x 10^3
Sembungan Station II	1	3	3	1	1.15 x 10^4

Table 4. The average of the estimated number of INA bacteria based on the table MPN, Thomas formula series 3.3.3.

Sampling Station	Spotted Scale	Estimation of Bacteria number MPN/g leaves
Dieng Wetan Stasiun I	3	>2.75 x 10^4
Dieng Wetan Stasiun II	1	>1.67 x 10^4
Sembungan Stasiun I	2	>1.52 x 10^4
Sembungan Stasiun II	1	7.25 x 10^3

Potato leaf samples from different altitude have the varying severity of disease based on spotting scale scaling, i.e. 3 and 1 scale at ± 2,000 m ASL, 2 and 1 at an altitude of ± 2,500 m ASL (Figure 3)

Figure 3. Spot scales on the leaves of potato plants based on the scoring, a) scale 3: 31-50% infected leaf area; b) scale 2: 11-30% infected leaf area; c) scale 2: 11-30% infected leaf area; d) scale 1: 10% infected leaf area.

Based on the correlation analysis showed the correlation value of 0.844 and significance 0.156 (Table 5).

Table 5. Analysis of correlation of bacterial estimation with spot scale using SPSS Bivariate Correlations.

		Bacterial Population	Spot Scale
Bacterial Population	Pearson Correlation	1	.844 (*)
	Sig. (1-tailed)		.156
	N	4	4
Spot Scale	Pearson Correlation	.844 (*)	1
	Sig. (1-tailed)	.156	
	N	4	4

The bacteria isolated from potato leaves were 84 pure cultures, but based on the nucleation activity test only 20 isolates were clotted. From the results of ice nucleation test in this study, We found three classes of INA bacteria based on freezing temperature, i.e. class A that freezes at temperature -4°C, class B at temperature -7°C and class C at -10°C (Table 6).

Table 6. Classification of class INA based on a temperature difference of ice nucleation activity from potato plant leaves.

Bacterial isolate code	Number of bacterial isolate	Freezing Temperature	Class of INA
SM2-a1a, SM2-b1a	2	-4	A
DW1-a1a, SM1-b2b, SM2-b2a, SM2-c1a	4	-7	B
DW1-a2b, DW1-b1b, DW2-b1a, SM1-c1c, SM1-c1d, SM1-c2b, SM2-c1b, SM2-a1c, SM2-a2c, SM2-b1d, SM2-c2e, SM2-c2c, SM2-c2a, SM2-a2c	14	-10	C
Total number of INA bacteria	20		

Note: DW1 = Dieng Wetan station 1, DW2 = Dieng Wetan station 2, SM2 = Sembungan station 1, SM2 = Sembungan station 2

4. Discussion

4.1. *Estimation on the number of ina bacteria dieng plateau potato plant leaves*

The calculation of INA bacteria was done using Multiple Tube Nucleation (MPN) method, which is a method used to estimate the amount of ice from bacterial suspension and bacterial number on plant [13]. A test was conducted using the tubes to determine the number of ice nucleation based on the number of frozen tubes at each dilution. This method assumes that one of the frozen tubes contains at least one ice nucleus [17-18]. The optimum temperature for the estimation test was -5°C for 10 minutes in a circulating alcohol bath, which was sensitive enough to detect the formation of ice nuclei by INA bacteria [19-21].

Based on the calculation of MPN value, the highest INA bacteria population in potato plant in Dieng plateau from eight samples of >2.75 x 10^4 were found on the scale 3 of leaf spots while the lowest value of 7.25 x 10^3 on scale 1. Lindow states that the population of INA bacteria can reach 10^2-10^7 cell/g plant tissue [3]. In this study, the population of INA bacteria from potato plants ranged from 10^3 to 10^4. Gross *et al*, has estimated that the population of INA bacteria in fruit plants ranges from 10^2 to 10^4 [22]. In addition to the availability of bacterial number in the sample, determination of the magnitude of the INA bacteria number was also affected by the dilution series at the time of estimation. According to the result of positive tubes MPN 3-3-3 that corresponding with an uncertain amount of the bacteria (more than 1100 MPN/g), it seems to need analysis more than 10^{-4} to get a more precise estimate number. Regrettably, we did not do further dilution in this study.

4.2. *Correlation of INA bacterial population with disease severity of potato plant frost injury*

Analysis of disease severity was done by calculating the correlation between the scale of the leaf spot and the bacterial population obtained from the estimation of the bacterial population using the scale or disease severity scoring method [13]. Based on the scoring method on potato plant samples with spot scales varied on the four stations.

Based on scoring towards the severity of potato disease, the highest spot scale was scale 3 with 31-50% spot width, while the lowest scale was scale 1 with 10% spot width. The correlation results between INA bacteria number and spot scale on potato plant was analyzed using SPSS with Bivariate Correlations test. The Pearson Bivariate Correlations test assumes that if the r-value of correlation is > 0.05 then there is a correlation between the two variables, and if the sig (1-tailed) <0.05 then both variables have significant correlation [23]. The results of this study showed that the value of r = 0.844 means there was a correlation between the two variables. Sugiyono states that if the correlation value ranges from 0.80 to 1.00 then there is a strong relationship between the two variables [24]. For the significance value of 0.156>0.05 means the value of significance is lower than 95%, the value is about 84.4%.

4.3. *Classification of INA bacteria class*

The nucleation activity can be discovered when the bacterial suspension in the microtube freezes after its insertion into the circulation alcohol bath. The test was performed on three different temperatures namely -4°C, -7°C and -10°C for 10 minutes. The difference in temperature was due to differences in the activity of each bacterial suspension in accordance with its class group [25]. The results of the nucleation activity test by INA bacteria were obtained from several classes of INA bacteria based on the freezing temperature.

Based on the test of nucleation activity, not all isolates were found positive for INA bacteria. There were several classes of INA bacteria from the existing isolates, 2 isolates were classified into class A with freezing temperature of -4°C, 4 isolates were classified as class B with freezing temperature -7 °C and 14 isolates were classified as class C with freezing temperature -10 °C. Four bacteria isolates were found at an altitude of ± 2000 m ASL while 16 bacteria isolates were obtained at an altitude of ± 2500 m ASL. INA bacteria were classed based on freezing temperature as class A which can freeze water at the temperature of -2 to -4°C. While the freezing point of class B is -5°C to -7°C and below -8°C for class C [26]. It is also supported by Stephanie report of isolated INA bacteria from rainwater which was classified as class B INA bacteria that freeze at -7°C [27].

5. Conclusion

Highest of INA bacteria population in potato leaves i.e. 2.7×10^4 was found on scale 3 of leave spots while the lowest value of 7.25×10^3 on scale 1. There is strong correlation of INA bacteria population and the scale of frost injury spots even the significance is lower than 95%. Based on the ice forming temperatures, INA bacteria are classified into three classes, i.e., Class A can freeze at temperatures of -2°C to -4°C, class B -5°C to -7°C and class C below -8°C.

Reference

[1] Turasih 2016 *Adaptation strategy to climate change on farmer's household in Dieng Plateau* (Bogor: Bogor Agricultural Institute)
[2] Arwiyanto T 1996 Isolation of ice core bacteria on potatoes *J. Perlindungan Tanaman Indonesia* **2**:12-4
[3] Lindow S E 1993. Novel method for identifying bacterial mutant with reduced epiphytic fitness *J. Appl. Environ Microbiol* **36** 831-8
[4] Goto K, Inaba T and Goto M 1988 Factors affecting frost damage of vegetable crops by artificial: spraying of ice nucleation-active bacteria in a freezing chamber *J. Proc Assoc PI Protec Shikoku* **23** 47-55
[5] Goto M, Gotto T and Inaba T 1989 Identification of ice nucleation sites measured in situ low radiation inactivation analysis *J. Proc Natl Phytopath Soc Japan* **55** 330-5
[6] Lindow S E, Amy D C, and Upper C 1978. Erwinia herbicola: A bacterial ice nucleus active in increasing frost injury to corn *J. Phytopathology* **68** 523-7
[7] Lindow S E 1983 b The role of bacterial ice nucleation in frost injury to plants *J. Annu Rev Phytathol* **21** 363-84
[8] Christner B C, Cai R, Morris C E, McCarter K S, Foreman C M, Skidmore M L, Montross S N and Sands D C 2008 Geographic, seasonal, and precipitation chemistry influence on the abundance and activity of biological of ice nucleators in rain and snow *J. Proc Natl Acad Sci USA* **58** 1334-8
[9] Fardias S 1993 *Analysis of Food Microbiology* (Jakarta: PT. Raja Grafindo)
[10] Blodgett R 2006. Appendix 2, Most Probable Number from Serial Dilution. BAM (Bcteriological Analytical Manual), Chapter 4. FDA (Food and Drug Administration), California.
[11] Lindow S.E. 1990. Bacterial ice nucleation in frost injury to plants *J. Ann. Rev. Phytopathol.* **21** 363-84
[12] Mahdieh R, Hasanzadeh N, Khodaygan P and Riahi- Madvar A 2018 Ice nucleation active bacteria from pistachio in Kerman Province, *Iran J. Plant Pathol.* **100** 51-58
[13] Manandhar H K, Ram D T, Sarala S, Sharada,J., Shrinkhala M, Suk B G, Sajal S, Epsha P, Anju P, Balkrishna J, Gyanu Manandhar, Devendra G, Devra I J and Bhuwon R S 2016 *A field guide for identification and scoring methods of diseases in the mountain crops of Nepal* (Nepal : Biodiversity International) 77-81
[14] Lindow S E, Arny D C and Upper C D 1978 Distribution of ice nucleation-active bacteria on plants in nature *J. Appl Environ Microbiol* **36** 831-6
[15] Nejad P, Ramstedt M, Granhall U, Roos S, Mclvor I 2006 Biochemical characterization and identification of Ice-Nucleation-Active (INA) Willow pathogens by means of BIOLOG®MicroPlate, INA gene primers and PCR-based 16S rRNA-gene analyses *J. Plant Diseases and Protection* **113** 97-106

[16] Cazorla F M, Olalla L, Toresl J A, Perez-Garcia A, Codina J C and de Vicente A 1995 A method for estimation of population densities of Ice-Nucleating Active Pseudomonas syringae in buds and leaves of mango *J. Applied Bacteriology* **79** 341-6

[17] Muriel J, Attard E, Sancelme M, Deguillaume L, Guilbaud C, Morris C E, Amato P and Delort A M 2013 Ice nucleation activity of bacteria isolated from cloud water *J. Atmospheric Envir.* **70** 392-400

[18] Govindarajan A G and Steven E L 1988 Size of bacterial ice-nucleation sites measured in situ by radiation inactivation analysis *J. Biochemistry* **85** 1334-8

[19] Montesions E and Viraldell P 1991 Relationships among population levels of Pseudomonas syringae, amount of the nuclei, and incidence of blast dormant flower in Commercial Pear Orchards in Catalunya, *Spain J. Phytopathology* **81** 113-9

[20] Baertlein D A, Lindow S E, Panapoulos N J M, Lee S P, Min-drinos M N and Chen T H H 1992 Expression of bacterial ice nucleation gene in plants *J. Plant Physiology* **100** 1730-6

[21] Hirano S S, L Stuart B and Christen D U 1985 Ice nucleation temperature of individual leaves in relation to population sizes of ice nucleation active bacteria and frost injury *J. Plant Physiol* **77** 259-65

[22] Gross D C, Cody Y S, Proebsting E L, Radamaker J K and Spotts R A 1983 Distribution, population dynamics, and characteristics of ice nucleation-active bacteria in Deciduous Fruit Tree Orchards *J. Applied Envir.Microb.* **46** 1370-9

[23] Singgih S 2016 SPSS Complete Guide Version 23 (Jakarta: PT Elex Media Komputindo) Chapter 14 335-43

[24] Sugiyono 2007 Qualitative-quantitative research methods and R & D " (Bandung: Alfabeta)

[25] Keift T L and Ruscceti T 1990 Characterization of biological ice nuclei from A Lichenes *J. Bacteriol* **172** 3519-23

[26] Yankofsky S A, Levin Z, Bertold T and Sandlermand N 1981 Some basic characteristics of bacterial freezing nuclei *J. Applied Meteorol.* **20** 1013-9

[27] Stephanie and Waturangi D E 2011 Distribution of ice nucleation-active (INA) bacteria from rain-water and air *J. Biosciences* **18** 108-12

IOP Conf. Series: Earth and Environmental Science **185** (2018) 012033 doi:10.1088/1755-1315/185/1/012033

The rating mode of eutrophication in Dalinuoer Lake

W G Hao[1,4], H Liu[1], Z G Li[1], R Jiao[1], H L Liu[1], J Shen[2] and Z G Zhang[3]

[1] Ministry of Water Resources of Pastoral Water Institute of Science, Hohhot, 010020, China

[2] Inner Mongolia Soil and Water Conservation Workstation, Hohhot, 010010,China

[3] Chifeng Hydrological Survey Bureau, Chifeng, 024000, China

E-mail: hao_weigang@163.com

Abstract. The Logistic Curve eutrophication level comprehensive evaluation model was used to generate model samples and verification samples based on the evaluation grade criteria and uniform random numbers and was solved using an accelerated genetic algorithm. The water quality monitoring indicators in July, August, September and October of 2017 were selected to evaluate the water environment quality of Darinol Lake and determine the eutrophication level. The results showed that the degree of eutrophication in Darinol Lake is close to five extremely eutrophic state, and the eutrophication grades in July, August, September and October showed a slowly decreasing trend.

1. Introduction

Dalinuoer Lake is the second largest freshwater lake in Inner Mongolia. It is known as "China Swan Lake". It is the largest migration route for migratory birds in northern China and one of the most important migratory birds distribution center in Northeast Asia.

Dalinuoer Lake is located in the western part of the Keshiketengqi, Chifeng City, Inner Mongolia, east longitude 116 ° 29 '~ 116 ° 45', north latitude 43 ° 13 '~ 43 ° 23', is a typical northern drought area with a lacus area of about 228km^2, water storage of about 1.6 billion cubic meters, the maximum depth of about 13m, the average depth of about 6.7m, its geographical location shown in Figure 1. Dalinuoer Lake has four supply rivers: the Gonggeer River, Shali River, Haolai rivers and Liangzi River, respectively, all inland rivers. The annual river supply is about 60.9 million cubic meters. Precipitation recharge of 0.714 cubic meters, groundwater recharge of 1152 cubic meters, higher water quality alkalinity, as a closed half brackish water [1-2]. The lacus is rich in Crucian carp and Archangel fish, is the largest fishery base of the Keshiketengqi. There are 160 species in 16 orders and 36 orders in Nature Reserve of Dalinor Lake [3]. Among them, there are 8 species of national protected birds such as red-crowned cranes, white

Figure 1. Dalinore Lake location picture.

storks, white-naped cranes and black storks, 18 species of secondary protection birds such as gray crane, Xie Yi crane, big swan.

Dalinuoer Lake belongs to a closed plateau lacus, the flow of water into the lacus has decreased over the years, the evaporation is on the rise, the water is not exported, the lacus is shrinking year by year, combined with the intensification of human activities, the pollutants into the lacus cannot be discharged. These questions make the lacus salinity gradually increase and the water environment continues to deteriorate. Mastering Darunor Lake Eutrophication is of great significance for the sustainable utilization of the aquatic resources, the protection of wetland ecological functions and the comprehensive management in Dalinuoer Lake.

2. Establishment of eutrophication rating model

Comprehensive evaluation of water eutrophication level, using some water quality indicators and the establishment of mathematical models to comprehensively evaluate water eutrophication level, provide a scientific basis for water pollution prevention and protection. At present, the models for water environment quality assessment mainly include Sheperd similarity evaluation model [4], fuzzy neural network evaluation model [5], projection pursuit evaluation model [6] and Logistic Curve evaluation model [7]. Since evaluation results of some models have many discrete water quality ratings, they are semi-quantitative and their resolution is too coarse. Even though they belong to the same water body, their corresponding water quality indicators often differ significantly, and Logistic Curve evaluation model can overcome the above shortcomings and achieved good results in the application [7-8]. Therefore, Logistic Curve evaluation model is used to evaluate the eutrophication level of Dalinuoer Lake.

2.1. Evaluation standard

Selection of evaluation criteria of water quality, evaluation level is divided into extremely poor nutrition, poor nutrition, nutrition, rich nutrition, extremely rich nutrition. See Table 1 for details.

Table 1. Water quality evaluation standards of lacus.

Water quality indicators	**Water quality level**				
	1 extremely poor nutrition	**2 poor nutrition**	**3 nutrition**	**4 rich nutrition**	**5 extremely rich nutrition**
transparency (/m)	>37	12	2.4	0.55	0.17
Oxygen consumption (mg/L)	<0.09	0.36	1.8	7.1	27.1
Total nitrogen (mg/L)	<0.02	0.06	0.31	1.2	4.6
Total phosphorus (μg/L)	<1	4	23	110	660

2.2. Evaluation factors

Water quality data use the average monitoring results at various monitoring points in Daloreau Lake in July, August, September and October in 2017. Evaluation factors include: transparency, oxygen consumption, total nitrogen and total phosphorus.

2.3. Establishment of Logistic Curve eutrophication rating model (L-C model)

2.3:1. L-C model.

$$z = \frac{N}{1+e^{c+\sum_{i=1}^{n} c_i x_i}} \tag{1}$$

In the formula: z is calculated value of eutrophication level model
N is the highest level of water quality
x_1, x_2, Λ, x_n are the indicators of eutrophication evaluation is the number of indicators

n is the number of indicators
$c_0, c_1, c_2, \Lambda, c_n$ are the model parameters

2.3.2. Generation of modeling samples. Using uniform random number between [0, 1] and evaluation criteria of eutrophication level applying formula (2) [9] to produce random 25 modeling samples. Table 2 shows the L-C model modeling samples.

Table 2. Randomly generated L-C model modeling samples.

Serial number	Trans-parency	Oxygen consumption	Total nitrogen	Total phosphorus	Standard grade	Calculatione grade
1	46.62	0.051	0.015	0.559	1	1.013
2	89.617	0.034	0.009	0.106	1	0.26
3	45.867	0.061	0.01	0.978	1	1.032
4	55.149	0.051	0.015	0.646	1	0.791
5	99.746	0.056	0.01	0.078	1	0.185
6	23.709	0.133	0.042	2.821	2	1.867
7	16.26	0.176	0.02	1.426	2	2.15
8	19.699	0.098	0.035	1.947	2	2.022
9	14.496	0.195	0.04	2.79	2	2.261
10	18.619	0.356	0.049	3.862	2	2.104
11	3.252	1.479	0.145	9.171	3	2.965
12	8.979	1.308	0.29	16.739	3	2.939
13	7.487	0.583	0.305	4.379	3	2.939
14	10.476	0.879	0.149	18.998	3	2.688
15	5.647	1.611	0.159	22.001	3	2.948
16	1.038	5.341	0.606	93.002	4	4
17	1.715	5.229	0.684	104.886	4	4.082
18	0.655	3.409	0.794	52.709	4	4
19	1.961	5.678	0.799	64.253	4	4.049
20	1.413	4.582	1.004	49.799	4	4.16
21	0.263	8.747	3.879	359.852	5	4.989
22	0.267	17.778	3.764	222.285	5	4.982
23	0.222	21.613	2.287	401.613	5	4.962
24	0.213	22.129	2.319	605.647	5	4.984
25	0.512	11.198	1.626	267.201	5	4.831

Note: The upper level of the selected transparency level is 100.

$$X(i,j)=y(i,j)\left(a(i,j)\right)-b(i,j)+a(i,j) \tag{2}$$

In the formula: $X(i,j)$ is the jth level sample value of the ith indicator
$y(i,j)$ is a uniform random number between [0, 1] of the jth level sample of the

ith indicator generated
a(i,j) is the upper bound of the jth grade of the ith index
b(i,j) is the lower bound of the jth grade of the ith index

2.3.3. Solve the model. Currently, there are many methods for solving the LC model, such as the successive search method [10], the four-point average method [11], the orthogonal design method [12], the Macquarite method, the enumeration optimization method [13] M (1,1) model method [7] and other conventional methods, the above methods in the parameter optimization are computationally heavy, complex operation, but compared with other methods, the genetic algorithm in the literature [7], [14] to solve LC model have achieved good results, so this study selects the genetic algorithm for parameter optimization.

Model parameter estimation use accelerated genetic algorithm, the establishment of the following objective function:

$$\min f(c_0, c_1, c_2, \Lambda, c_n) = \sum_{i=1}^{m} |z_1 - z_1'| \tag{3}$$

In the formula: z_1, z_2, Λ, z_m are the calculated value of m modeling samples respectively
$z_1', z_2', \Lambda, z_m'$ are rank value of m modeling samples respectively

In order to eliminate the influence of the dimension and to facilitate the calculation, modeling and validation samples normalization processing performed - equation (4) [9]:

$$X'(i,j) = \frac{1}{M(i)} X(i,j) \tag{4}$$

In the formula: $X'(i, j)$ is the normalized sample value of the jth level of the i–th indicator
M(i) is the maximum value of the ith indicator standard grade

Using accelerated genetic algorithm to solve equation (3) [9], the results of the model parameters and accelerated genetic algorithm related parameters are listed in Table 3.

Table 3. AGA solves model parameter results.

Acceleration frequency	26		The best objective function value		3.542
Optimize variables	c_0	c_1	c_2	c_3	c_4
Initial interval	[-10, 10]	[-10, 10]	[-10, 10]	[-10, 10]	[-10, 10]
Parameter result	-0.263	-2.729	-0.66	3.545	-4.896

In this way, an eutrophication rating model of L-C is established - equation (5) [9]:

$$z = \frac{5}{1+e^{-0.263+3.545x_1-0.66x_2-4.896x_3-2.729x_4}} \tag{5}$$

In the formula: x_1, x_2, x_3, x_4 are the normalization of transparency, oxygen consumption, total nitrogen, and total phosphorus respectively.

Table 4. Randomly generated L-C model validation samples.

Serial number	Total phosphorus	Oxygen consumption	transparency	Standard grade
1	0.658	0.059	60.594	1
2	3.225	0.349	24.079	2
3	20.269	1.681	2.606	3
4	106.776	5.898	0.993	4
5	281.502	22.646	0.317	5

2.3.4. Model verification. As with the modeling samples,using uniform random number between [0,1] and evaluation criteria of eutrophication level applying formula (1) [9] to produce random 5 validated samples. Table 4 gives the validation samples, eutrophication levels and calculated levels of L-C models. As can be seen from Table 2 and Table 4, the LC model determines the level of calculation of the sample and depicts the effect of the number of differences of eutrophication indicators on the determination of eutrophication level with finer resolution. Compared with the standard grade, calculating the grade is reasonable .Therefore, the established L-C model can be used to evaluate the water quality of Dalinore Lake.

3. Evaluation results

According to the analysis of the individual indicators of transparency, oxygen consumption, total nitrogen and total phosphorus in July, August, September and October in 2017 in Darylor Lake, in addition to the total nitrogen in October between grade IV and grade V water quality standards in 'The Surface Water Environmental Quality Standards' (GB3838-2002), in other time periods, total nitrogen, total phosphorus and oxygen consumption exceeded the grade V water quality standard in 'The Surface Water Environmental Quality Standards' (GB3838-2002). The evaluation indexes are normalized and put into the eutrophication level evaluation model (formula 5). The evaluation results showed that although the eutrophication level of Darinor Lake decreased slowly from July to October, the overall eutrophication level is close to levels five extremely rich nutrition. The curve of eutrophication level of Dalinor Lake in 2017 is shown in Figure 2.

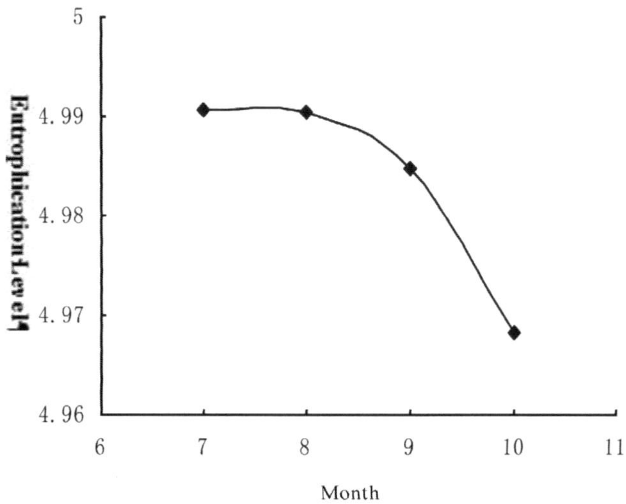

Figure 2. Eutrophication level changes in 2017, Darinor Lake.

4. Conclusion

Using evaluation model of Logistic Curve (LC) eutrophication, rating criterion and uniform random numbers to generate modeling and verification samples and real code accelerated genetic algorithm to solve evaluation model, finally, evaluate the eutrophication level of Dalinuoer Lake. The results show that Dalinuoer Lake eutrophication levels close to levels five extremely rich nutrition, eutrophication levels showed a gradual downward trend in July, August, September and October.

Dalinuoer Lake belongs to a closed plateau lacus, the main source of supply includes the four rivers into the lacus and atmospheric precipitation, the drainage of lacus has only evaporation. In recent years, the evaporation in the lacus area has been increased, the flow into the lacus has been reduced, pollutants entering the lacus have been increasing and accumulating, and human activities have been exacerbated. Total nitrogen, total phosphorus and salt content in lakes have been continuously increasing. Water quality is changing from freshwater → brackish water → salt water → salt lake. In the future, we should step up the monitoring of water quality in Dalinuoer Lake and take active management measures.

Acknowledgement

This study is supported by Water Conservancy Research Foundation of Inner Mongolia Autonomous Region of China (NSK2016-S5), Basic research business fee of China water resources and Hydropower Research Institute (MK2016J28、MK2018J02), National Natural Science Foundation of China (51009098), Supported by Natural Science Foundation of Inner Mongolia Autonomous Region of China (2017MS0516).

References

[1] Qin Y, Li X H. 2003. The magical Keshiketengqi. Yuanfang Press.
[2] Han F, Li X H and Gao L Y. 2007. The dynamic chrarcteristics of Dalinuoer wetland in Inner Mogolia based on remote sensing. *J. Inner Mongolia Agric. Univ.* **28**(1): 74-78
[3] Chen H Y, Yang G S and Xing L L. 2007. Study on the fauna and ecological distribution of birds in Dalinuoer nature reserve. *J. Inner Mongolia. Uni.* **38**(1): 68-74
[4] Jin J L, Ding J and Fu Q. 2002. An interpolation evaluation model for regional water resources sustainable utilization system. *J. Nat. Resour.* **17**(5):611-15
[5] Hu M X and Guo D Z. 1998. The method of fuzzy neural network of evaluation on lake eutrophication. *Res. Environ. Sci.* **11**(4):40-42
[6] Jin J L, Wei Y M and Ding J. 2001. Projection pursuit model for comprehensive evaluation of water quality. *Acta. Sci. Circumst.* **21**(4):431-434
[7] Jin J L, Yang X H and Jin B M. 2000. A new model of water quality evaluation. *China Environ. Monitor.* **16**(4):42-47
[8] Jin J L and Ding J. 2002.Water resources systems engineering. Sichuan Science and Technology Press.
[9] Li C Y, Liu T Q and Gao R Z. 2004. Study and synthetic assessment for the season-year change of the eutrophication main-control factors in Wuliangsuhai Lake. *Hydrology*. **24**(3):14-17.
[10] Zhou S H. 1992. The estimation of the parameters in logistic equation. *J. App. Stats Mgmt* . **11**(5):32-35
[11] Wang Z Z and Lin K X. 1987. Two-paired points method for estimating k value of logistic equation. *Acta. Ecol. Sin.* **7**(3):193-197
[12] Sima X S. 1994. The estimation of pear curve parameter. Prediction.
[13] Wan C X and Liang Z Y. 1983. A fitting method of the logistic curve. *Acta. Ecol. Sin.* **3**(3): 288-296
[14] Jin J L, Yang X H and Liu Y M. 1997. Applicaion of genetic algorithm for estimating logistic curve parameters. *Agric. Systems Sci. Comprehensive. Res.* **13**(3):186-190

IOP Conf. Series: Earth and Environmental Science **185** (2018) 012034 doi:10.1088/1755-1315/185/1/012034

Research progress of active compounds and pharmacological effects in *Akebia trifoliata* (Thunb) koidz stems

Y C Liu [1,2,3], H M Wang[1,2,4] and X H Zeng[1,2,4]

[1] Laboratory of Chinese Herbal Pharmacology, Affiliated Dongfeng Hospital, Hubei University of Medicine, Hubei 442008, P.R. China;

[2] Hubei Key Laboratory of Wudang Local Chinese Medicine Research, School of pharmacy, Hubei University of Medicine, Hubei, 442000, P.R. China;

[3] Wudang Animal Pharmacy Co., Ltd. Hubei, 442100, P.R. China;

E-mail: meirwang@126.com and zengken@126.com

Abstract. *Akebia trifoliata* (Thunb) koidz contains a variety of active ingredients, exerting the promising pharmacological actions. Its stems and fruits are legally recorded as herbal medicines in the Chinese Pharmacopeia. However, different parts of *Akebia trifoliate* stems have different effects. In this study, the literature in the last decade was retrieved to summarize the pure compounds and pharmacological effects of *Akebia trifoliata* stems. It showed that *Akebia trifoliata* stems mainly contain a variety of oleanolic acids, ivy saponin-type triterpenoid saponins, which have antitumor, anti-inflammatory, diuretic and antibacterial actions.

1. Introduction

Formerly known as Tongcao, Mutong was first published in "ShenNong's Classic" of Materia Medica, which has been widely used in China and other Asian countries for nearly two thousand years [1]. *Akebia trifoliata* (Thunb) koidz, as a genus Akebia plant of Laridizabalaceae, is one of the basic plants of the Chinese Pharmacopoeia in the 2015 edition of the Chinese Pharmacopoeia. *Akebia trifoliata* contains various active ingredients and has good pharmacology activities. The dried *Akebia trifoliata* stem has the effects of promoting diuresis and relieving stranguria, clearing away the heart-fire, and promoting lactation, and is used to treat stranguria, edema, dysuria, reddish urine, tongue sore, amenorrhea and lactation, and damp-heat [2], etc. However, the literature reports of the last decade mainly focus on anti-tumor, antidepressant, anti-inflammatory and antimicrobial effects. This article focuses on the chemical compositions and pharmacological effects of the stem of *Akebia trifoliata* in order to better study and use it.

2. Natural organic compounds in *Akebia trifoliata* stems

The stem of Akebia trifoliate contains various kinds of Akeboside and its aglycon, such as Hederagenin ($C_{30}H_{48}O_4$) or Oleanolic acid($C_{30}H_{48}O_3$). The stem [3-4] also contains phenylpropanoid chemical components, phenol alcohol and phenolic glycoside chemical components, caffeoyl quinic acid chemical components, and lignan chemical components. At present, pharmacologists have isolated 40 chemical constituents [3-7] (Table 1) from the cane of Akebia trifoliata, including lignans, phenolic acids and phenol glycosides, and various triterpenoids which accounts for the vast majority .

Table 1. Natural Oganic compounds in Akebia trifoliata stems

No.	Component	Extractant	Category
1	Oleanolic-acid-3-O-α-L-rhamnopyranosyl-(1→2)-α-L-arabinopyranoside	n-BuOH	triterpene saponins
2	Oleanolic-acid-3-O-α-L-rhamnopyranosyl-(1→4)-β-D-glucopyranosul-(1→2)-α-L-arabinopyranosyl	n-BuOH	triterpene saponins
3	Hederagenin-3-O-α-L-rhamnopyranosyl-(1→4)-β-D-glucopyranosyl-(1→2)-α-L-arabinopyranosul	n-BuOH	triterpene saponins
4	O-α-L-rhamnopyranosyl-(1→2)-α-L-arabinopyranosyl-Oleanolic-acid-28-O-α-L-rhamnopyranosyl-(1→4)-β-D-glucopyranosyl-(1→6)-β-D-glucopyranosyl	n-BuOH	triterpene saponins
5	3-O-β-D-glucopyranosyl-(1→2)-α-L-rhamnopyranosyl-(1→4)-α-L-arabinopyranosyl-Hederagenin-28-O-α-L-rhamnopyranosyl-(1→4)-β-D-glucopyranosyl-(1→6)-β-D-glucopyranosyl	n-BuOH	triterpene saponins
6	3-O-α-L-rhamnopyranosyl-(1→2)-α-L-arabinopyranosyl-Hederagenin-28-O-α-L-rhamnopyranosyl-(1→4)-β-D-glucopyranosyl	n-BuOH	triterpene saponins
7	3-O-β-D-glucopyranosyl-(1→2)-α-L-arabinopyranosyl-Oleanolic-acid-28-O-α-L-rhamnopyranosyl-(1→4)-β-D-glucopyranosyl-(1→6)-β-D-glucopyranosyl	n-BuOH	triterpene saponins
8	3-O-α-L-rhamnopyranosyl-(1→4)-α-L-arabinopyranosyl-Oleanolic-acid-28-O-α-L-rhamnopyranosyl-(1→4)-β-D-glucopyranosyl-(1→6)-β-D-glucopyranosyl	n-BuOH	triterpene saponins
9	3-O-α-L-rhamnopyranosyl-(1→6)-β-D-glucopyranosyl-(1-2)-α-L-arabinopyranosyl-Hederagenin-28-O-α-L-rhamnopyranosyl-(1→4)-β-D-glucopyranosyl-(1→6)-β-D-glucopyranosyl	n-BuOH	triterpene saponins
10	2α,23-dihydroxy-3β-sulfoxylean-12-en-28-oic acid-O-α-L-rhamnopyranosyl-(1→4)-O-β-D-glucopyranosyl-(1→6)-β-D-glucopyranosyl	n-BuOH	triterpene saponins
11	2α,23-dihydroxy-3β-sulfoxylean-12-en-28-oic acid-O-β-D-glucopyranosyl-(1→6)-β-D-glucopyranosyl	n-BuOH	triterpene saponins
12	3β-[(O-β-D-glucuronopyranosyl-(1→3)-O-[α-L-rhamnopyranosyl-(1→2)]-α-L-arabinopyranosyl)oxy] olean-12-en-28-oic acid	n-BuOH	triterpene saponins
13	(4-hydroxy-3,5-dimethoxybenzenemethanol)	water	phenol
14	[erythro-1-phenyl-(4'-hydroxy-3'-methoxy)-2-phenyl-(4''-hydroxy-3''-methoxy)-1,3-propanediol]	water	phenol
15	[threo-1-phenyl-(4'-hydroxy-3'-methoxy)-2-phenyl-(4''-hydroxy-3''-methoxy)-1,3-propanediol]	water	phenol
16	[(7S,8S)-1-(4-hydroxy-3,5-dimethoxyphenyl)-1,2,3-propanetriol]	water	phenol
17	[2-(4-hydroxy-3-methoxyphenyl)-ethanol-1-O-β-D-glucopyranoside]	water	phenol
18	[(7S,8S)-1-(4-hydroxy-3,5-dimethoxyphenyl)-1,2,3-propanetriol-2-O-β-D-glucopyranoside]	water	phenol
19	neochlorogenic acid methyl	EtOAc	caffeoylquinic acid
20	cryptochlorogenic acid methylester	EtOAc	caffeoylquinic acid
21	chlorogenic acid methyl	EtOAc	caffeoylquinic acid
22	methyl 3, 5-di-O-caffeoyl quinate	EtOAc	caffeoylquinic acid
23	methyl 3, 4-di-O-caffeoyl quinate	EtOAc	caffeoylquinic acid
24	methyl 4, 5-di-O-caffeoyl quinate	EtOAc	caffeoylquinic acid
25	Calceolarioside B	EtOAc	phenylpropanoid

IOP Conf. Series: Earth and Environmental Science **185** (2018) 012034 doi:10.1088/1755-1315/185/1/012034

26	Osmanthuside E	EtOAc	phenylpropanoid
27	dunalianoside C	EtOAc	phenylpropanoid
28	dunalianoside D	EtOAc	phenylpropanoid
29	2-(4-hydroxyphenyl)ethyl-(6-O-feruloyl)-β-D-		
30	Glucopyranoside	EtOAc	phenylpropanoid
31	Aesculetin	EtOAc	phenylpropanoid
32	caffeic acid	EtOAc	phenylpropanoid
33	Oleanolic acid	EtOAc	triterpene saponins
34	6,7,10-trihydroxy-8-octadecenoic acid	EtOAc	Organic acid
35	tetratriacontane	EtOAc	Other
36	2,6-dimethoxy-1,4-benzoquinone	EtOAc	Other
37	β-sitosterol	EtOAc	steroid
38	Vanillic acid	EtOAc	Organic acid
39	Aketrilignoside A	water	lignan glycosides
40	Aketrilignoside B	water	lignan glycosides

3. Pharmacological effects

3.1. Anti-tumor effect

The oleanolic acid aglycone (or ivy sapogenin) and its derivatives [8-14], the main constituen of *Akebia trifoliata*, have liver protection and anti-tumor effects. Among them, the anti-tumor mechanism of oleanolic acid mainly regulates signaling pathway, arrests cellcycle, induces apoptosis, inhibits tumor angiogenesis, and resists invasion [15]. In addition, some researchers studied the inhibitory effects of ivy aglycone on olon cancer, gastric cancer, head and neck cancer, prostate cancer, breast cancer, and lung cancer [16-20]; its anti-tumor mechanisms regulate signal pathway, promote apoptosis, and resist invasion and migration. *Aesculetin* isolated from the ethanol extract of stems has inhibit ryeffects on human hepatoma cells HepG2, A549 lung cancer cells, human T ymphoblastic leukemia cells and human gastric cancer cells *in vitro* [21]. The possible mechanism of aescin induces apoptosis of tumor cells: promotes the release of ytochrome C from mitochondria, activates Caspase-9 to induce apoptosis, activatescaspase-3 as a substrate of Caspase-3 to induce apoptosis, and regulates Bcl-2 /Bax gene expression causes programmed cell death; ADP-ribose polymerase (ADP-ribose) inactivates shear-induced cell apoptosis [22] (Figure 1).

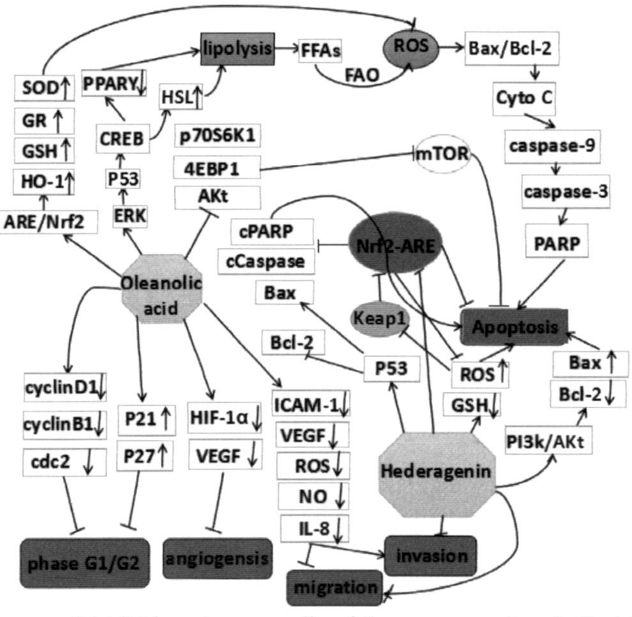

Figure 1. The signaling pathway of inhibiting tumor cells of the compounds of *Akebia trifoliata* stem.

3.2. Antidepressant effect
A preliminary study was conducted on the antidepressant activity of ivy glycosides using a corticosterone-induced PC12 cell injury model and a classic behavioral despair animal model (ie, mouse suspension and forced swimming) to detect hippocampal SOD. Activity and malondialdehyde (MDA) content, the results of the ivy saponin can improve the survival rate of PC12 cells damaged by corticosterone. The mouse experimental results showed that after an acute single dose, compared with the control group, the low dose (10 mg/kg) of ivy saponin significantly shortened the mouse tail suspension time. After 2 weeks of continuous gavage administration, ivy sapogenin reduced the immobility time of the mice without affecting the ability of the mice to move autonomously in the tail-suspended and forced-swimming experiments compared to the control group. Saponin has a certain anti-depressant effect, ivy aglycone can increase SOD activity in the hippocampus, and significantly reduce the MDA content. Its mechanism may be related to regulating the body's oxidative - antioxidative enzyme activity, reduce the peroxidation damage [23].

3.3. Anti-inflammatory effect
Inhibition of inducible nitric oxide synthase (iNOS) by phenolic alcohols and phenolic glycoside compounds was performed by measuring lipopolysaccharide (LPS)-induced NO production in RAW264.7 macrophage. Among them, nitric oxide is a strong signal molecule that reacts with superoxide anion to form eroxynitrite anion, which is a strong oxidant and can cause lipid peroxidation, protein tyrosine nitration, thiol oxidation Etc., mediate the occurrence of many diseases. As a result, it was found that phenolic alcohols and phenolic glycoside compounds (1-6) of *Akebia trifoliata* could inhibit the production of NO by RAW264.7 cells induced by LPS and inhibit the occurrence of inflammation [24].

3.4. Antimicrobial effect
The aqueous solution of the *Akebia trifoliata* was used as a continuous volumetric dilution of the nutrient broth in the test tube (the total volume per test tube was 1 mL), and the final concentration of the drug for inhibiting the bacteria was 200, 100, 50, 25, and 12.5 mg/mL. Add a certain amount of test bacteria dilution (10 μL of bacteria solution per test tube) to each test tube, and incubate at 37°C for 18 h. Compare with the negative control tube, then transfer to a solid medium to observe the presence or absence of bacterial growth, determination of the minimum inhibitory concentration (MIC) and the minimum bactericidal concentration (MBC). The results showed that the A. trifoliate water extract had significant effects on Streptococcus pneumoniae and Shigella, and had a certain antibacterial effect on Escherichia coli and Staphylococcus aureus [25].

4. Summary
The literature of the chemical constituents and pharmacological effects of *Akebia trifoliata* stems in the last decade indicates that the main active ingredients (see Figure 2 to Figure 6) found include triterpenoids, phenols and phenol glycosides, caffeoyl quinic acids, phenylpropanoids, and lignans, a total of 40 species. There are four main types of triterpenoid compounds, namely oleanane, oligostanane, lubutol, and wusuan. While the site modification and modification of oleanolic acid is mainly focused on 3-hydroxy and 28-carboxyl groups, wherein the esterification or amidation of 3-hydroxy glycosylation and the 28-carboxyl can effectively improve the anti-tumor activity of oleanolic acid [26]. Phenylpropanoid glycosides are usually based on β-D-glucopyranose as the central sugar nucleus, substituted phenylethyl as its aglycone, and substituted cinnamoyl as ester bond with sugar core. Additional sugars can also be linked to the C6 and C3 positions of the central sugar nucleus, affecting antitumor activity through structural modifications [11]. The pharmacological effects of the main active ingredients of *Akebia trifoliata* include anti-tumor, anti-inflammatory, anti-depression, and diuretic effects. However, some caffeoylquinic acids have not been found to have pharmacological effects at present, but their derivatives and analogues have been found to have

IOP Conf. Series: Earth and Environmental Science **185** (2018) 012034 doi:10.1088/1755-1315/185/1/012034

antioxidative activity, anti-inflammatory activity, and anti-microbial effect, etc., pending further researches [27].

In summary, the stem of *Akebia trifoliata* has high utilization rate and medicinal value, and has broad prospects for the development and application of medicine, and is worthy of in-depth study.

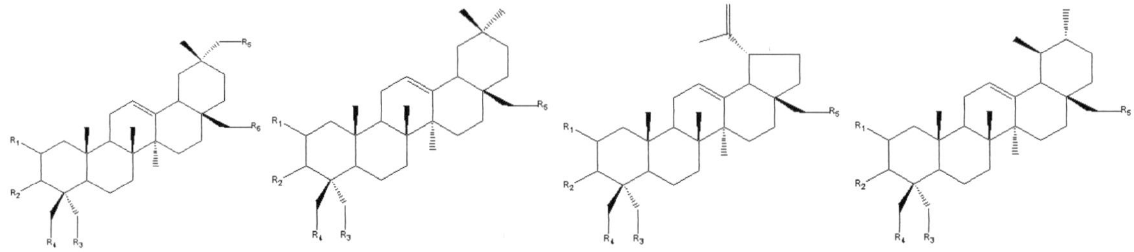

Figure 2. Polycyclic hydrocarbons' general formulas.

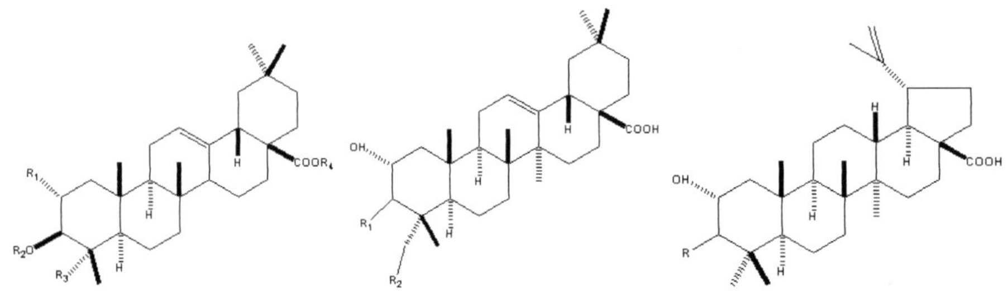

Figure 3. Triterpene saponins' general formulas.

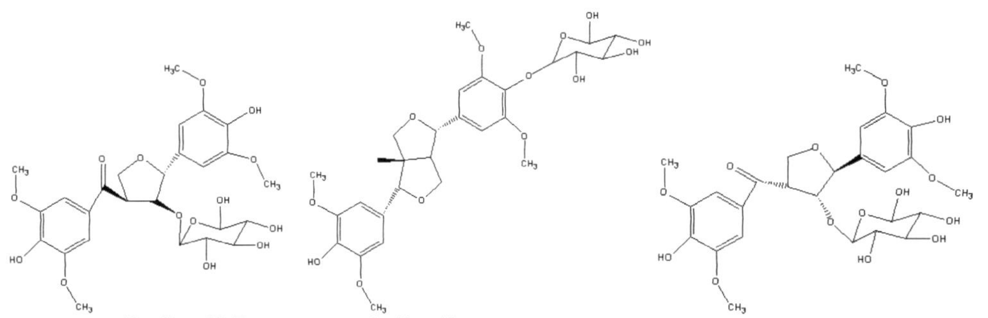

Aketrilignoside B aketrilignoside A **7R,8R,8'S-aketrilignoside B**

Figure 4. Lignans' structural formulas.

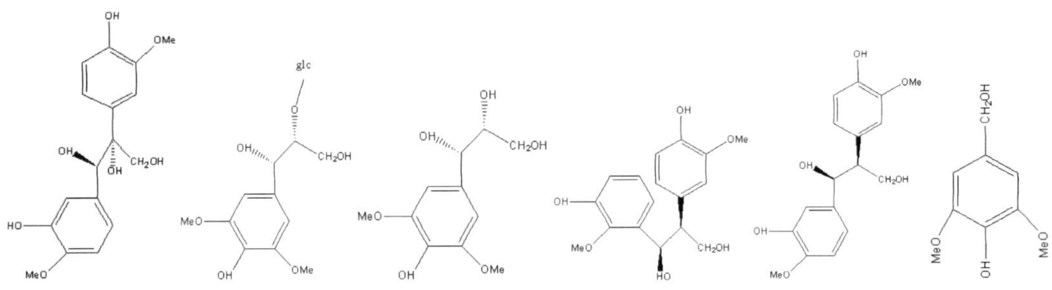

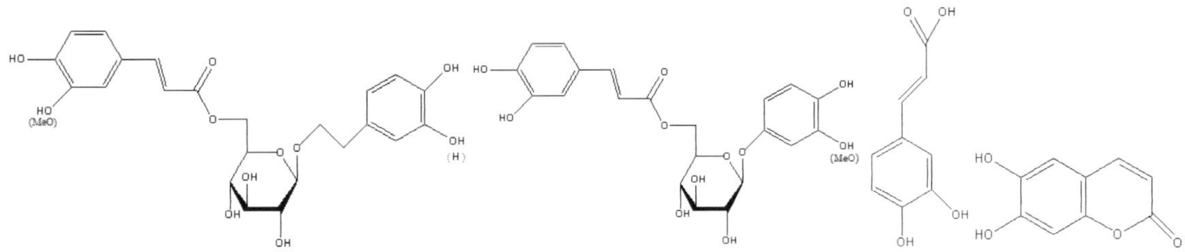

Figure 5. Phenols' and Phenylpropanoids' structural formulas.

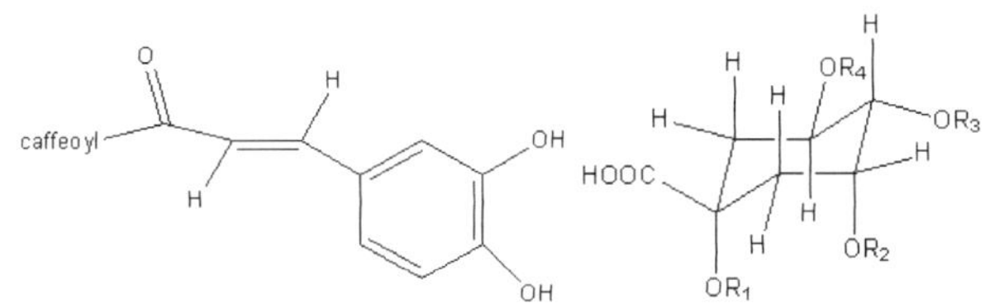

Figure 6. Caffeoyl quinic acids' general formulas.

Acknowledgments

We gratefully acknowledge financial support of this work by the Open Project of Hubei Key Laboratory of Wudang Local Chinese Medicine Research (Hubei University of Medicine, Grant No. WDCM009), the Hubei Province health and family planning scientific research project (Grant No. WJ2015Z113), the scientific research project of Educational Commission of Hubei Province of China (Grant No. B2018111), the Shiyan Municipal Science and Technology Bureau Science and technology project (Grant Nos. 18K79, 18Y01), and the scientific research projects of Hubei University of Medicine (Grant Nos. 2017QDJZR09, 2014QDJZR10, 2014CXZ01).

References

[1] Miao X Y. 2011 Shen Nong Herbal Classics. Beijing: China Medical Technology Press. 08. *First edition*.

[2] National Pharmacopoeia Commission. 2015 Pharmacopoeia of the People's Republic of China *Section 1*. Beijing: China Medical Science and Technology Press. 07.

[3] Guan S H, Xia J M, Lu Z Q, Chen G T, Jiang B H, Liu X and Guo D A. 2008 Structure elucidation and NMR spectral assignments of three new lignan glycosides from Akebia trifoliata. *Magn. Reson. Chem.* **46**(2): 186-90.

[4] Guan S G, Yu W B and Guan S H. 2010 Study on the chemical constituents of phenolic alcohol and phenolic glycosides in Akebia trifoliata lignin. *Lishizhen Med. Mater. Medica Res.* **21**(4): 905-6.

[5] Wang J, Xu Q L, Zhou Z Y and Tan J W. 2014 Caffeoylquinic acid derivatives from stems of Akebia trifoliata. *J. Chin. Med. Mater.* **37**(7): 1190-3.

[6] Wang J, Zhou Z Y, Xu Q L and Tan J W. 2014 Phenylpropanoids from the stems of Akebia trifoliata. *J. Trop. Subtrop. Bot.* (5): 511-5.

[7] Guo L X, Ma Y M and Qiao W. 2017 Chemical constituents from Akebia trifoliate and their anti-oxidant activities. *Chin. Tradit. Pat. Med.* **39**(2): 338-42.

[8] Wang X, Bai H, Zhang X, Liu J, Cao P, Liao N, Zhang W, Wang Z and Hai C. 2013 Inhibitory effect of oleanolic acid on hepatocellular carcinoma via ERK-p53-mediated cell cycle arrest and mitochondrial-dependent apoptosis. *Carcinogen.* **34**(6): 1323-30.

[9] Liu J, Wang X, Liu R, Liu Y, Zhang T, Fu H and Hai C. 2014 Oleanolic acid co-administration alleviates ethanol-induced hepatic injury via Nrf-2 and ethanol-metabolizing modulating in rats. *Chem. Biol. Interact.* **221**: 88-98.

[10] Zhou R, Zhang Z, Zhao L, Jia C, Xu S, Mai Q, Lu M, Huang M, Wang L and Wang X, *et al.*

2011 Inhibition of mTOR signaling by oleanolic acid contributes to its anti-tumor activity in osteosarcoma cells. *J. Orthop. Res.* 29(6): 846-52.

[11] Chu R, Zhao X, Griffin C, Staub R E, Shoemaker M, Climent J, Leitman D, Cohen I, Shtivelman E and Fong S. 2010 Selective concomitant inhibition of mTORC1 and mTORC2 activity in estrogen receptor negative breast cancer cells by BN107 and oleanolic acid. *Int. J. Cancer* **127**(5): 1209-19.

[12] Pratheeshkumar P and Kuttan G. 2011 Oleanolic acid induces apoptosis by modulating p53, Bax, Bcl-2 and caspase-3 gene expression and regulates the activation of transcription factors and cytokine profile in B16F. *J. Environ. Pathol Toxicol Oncol.* **30**(1): 21-31.

[13] Lin C C, Huang C Y, Mong M C, Chan C Y and Yin M C. 2011 Antiangiogenic potential of three triterpenic acids in human liver cancer cells. *J. Agric. Food Chem.* **59**(2): 755-62.

[14] Wang Y, Lu S W and Sun S. 2014 Studies on molecular mechanism of Oleanolic acid on HCC cells. *J. Harbin Univ. Commer. (Nat. Sci. Edit.)* (6): 645-8.

[15] Huang X X and Yang Y. 2014 Anti-tumor mechanisms of oleanolic acid. *J. Int. Oncol.* **41**(2): 96-9.

[16] Liu B X Z., Wang R P. and Zou W. 2013 The effect of hederagenin on the proliferation, adhesion, invasion and migration of human colon cancer cells LoVo. *J. Nanjing Univ. Tradit. Chin. Med.* **29**(1): 44-7.

[17] Liu B X Z, Wang R P, Zou W and Zhu J Y. 2013 Effect of hederagenin on biological behaviors including proliferation, adhesion, invasion and migration capabilities of human gastric cancer cell line. *Chin. J. Exp. Formulary* **19**(4): 212-5.

[18] Li X X, Yang Z L, Yang X L and Zhang C F. 2016 Effect of hederagenin on the growth of human breast cancer mcf-7 and lung cancer A549 xenogeafts in nude mice. *Strait Pharmaceut. J.* **28**(5): 16-8.

[19] Zhao Z X, Zhao Z M and Zhao C G. 2017. Effects of hederagenin on proliferation,migration and invasion of prostate cancer DU145 cell. *Chin. med. pharmac. clin.* **33**(1): 37-41.

[20] Kim E H, Baek S, Shin D, Lee J and Roh J L. 2017 Hederagenin Induces Apoptosis in Cisplatin-Resistant Head and Neck Cancer Cells by Inhibiting the Nrf2-ARE Antioxidant Pathway. *Oxid. Med. Cell Longev.* **2017**: 5498908.

[21] Wei Z M and Wang J M. 2012 Effects of aesculetin on the proliferation of HepG2 in vitro and investigation of its mechanism. *Shandong Medic. J.* **52**(39): 24-6.

[22] Wang J and Wei Z M. 2012 Mechanism of human hepatoma cell SMMC-7721 apoptosis induced by aesculetin. *Chin. Trad. Pat. Med.* **34**(11): 2059-63.

[23] Liang B F, Cheng Y F, Xue T, Guo H J, Yuan X and Xu J P. 2013 Pharmacodynamical evaluation of hederagenin as an antidepressant. *Mil. Medic. Sci.* **37**(4): 286-90.

[24] Guan S G, Yu W B, Pan W J and Xiu Z R. 2015 Study on the bioactivity of phenol and phenol glycosides in Akebia trifoliate. *Chin. J. Ethnomed. Ethnopharma.* (23): 12-3.

[25] Bai M R, Zhang B, Liu X Q, Gao Y Y and Ni L. 2008 A comparative study on pharmacodynamic action of Akebia Trifoliata(Thunb.) Koidz and Akebia Quinata(Thunb) decne and effect on drug metabol Izing-enzymes. *Chin. Arch. Trad. Chin. Med.* **26**(4): 732-5.

[26] Wang J, Zheng C Y, Wang Y, Liu L, Zhu C Y, Guo J Y, Shi D B, Nie X Q and Zhang L. 2017 Progress in structure modifications, antineoplastic activities and structure-activity relationships of oleanolic acid. *World Notes Antibiot.* **38**(3): 118-24.

[27] Liu Y T, Hou X J, Xie Y, Lu J and Lin R C. 2012 Overview of research on chemical constituents and pharmacological activities of Akebia Dence. *J. Jiangxi Univ. Trad. Chin. Med.* **24**(4): 87-93.

IOP Conf. Series: Earth and Environmental Science **185** (2018) 012035 doi:10.1088/1755-1315/185/1/012035

Golf course turf quality and maintenance management in Jinwan Golf Club of Zhuhai City, China

Q Wang[1] and L Fei[1, 2, 3]

[1] Zhuhai College of Jilin University, Zhuhai city, China

[2] Shanghai Jiaotong University, Shanghai city, China

E-mail: 14270088@qq.com

Abstract. Appearance quality such as color, density, coverage, texture, mowing height and uniformity of turf in the Zhuhai Jinwan golf course were studied in the present investigation. In addition, the functional quality like elasticity, ball speed on the green and recovery ability were also studied. After evaluating the results, a maintenance protocol for the quality of the turfs of tees, fairways and greens of Zhuhai Jinwan golf course has been suggested. The study indicated a medium level of turf in respect to color, density, coverage, texture and mowing height and rather long recovery period. Tee had a composite score of 25 points and the evaluation level achieved was III. The coverage, density, texture, and mowing height of the fairway were evaluated as well, and the uniformity, color, and recovery cycle were generally evaluated. The comprehensive score was 25 points and the evaluation level has been given a score of level III. The uniformity, coverage, density and texture of green were basically good, and had a short recovery period. The score of mowing height was poor and ball speed was low. Green's composite score was 27 points with an evaluation level III. The overall turf quality of Zhuhai Jinwan golf course has been recognized to stay at an intermediate level.

1. Introduction

Each functional area of golf course is composed of turf, which is formed by surface that covered by one to three different grass species. Turf is the foundation and has been considered as a rigid index to evaluate the quality of golf course. Golf course with ideal turf quality is competitive in the market. The quality of turf affects the operation and existence of the golf course which will directly affect the playing mood of players. Proposing a sound management schedule for the turf components of the golf course is very vital.

The study of Zheng *et al.* points out that a comprehensive system of turf quality assessment should include appearance quality evaluation, use quality evaluation and ecological quality evaluation [1]. Liu *et al.* propose an equation to explain the comprehensive evaluation value and weight assignment of the golf course turf quality. The equation is $v=0.25d+0.20t+0.10c+0.15u+0.30s$, in which d for density, t for texture, C for color, U for uniformity, s for smoothness [2]. As the quality evaluation system of the green has not been established in China, Wang *et al.* refer to the "Golf Course Quality evaluation standard" developed by the New Zealand Sports Turf Research Institute (NZSTI) to evaluate the quality of the golf course in Beijing Klauver base and make suggestions for improvement [3]. Yang *et al.* demonstrated that golf course green with function as the basis, based on the characteristics of the green, 6 index factors were determined, such as the green speed, hardness, disease area ratio, mowing height, smoothness and appearance quality, and an objective and simple evaluation method for the quality of the green turf was established through the experiment [4]. The method is used to evaluate the green of Qinling International Golf Club. The results show that the system can evaluate the real situation of the quality of green turf with accuracy, objectivity and simplicity. The conservation improvement measures based on the measured data are effective and have higher use value. The study on turf quality evaluation and maintenance management of Changsha Green Bamboo Lake Golf Course by Liu provides a program reference for this study [5]. Duan's research points out that the course must conduct regular professional inspection and analysis, find out the problem in time to maintain customer satisfaction and train the staff to promote the improvement of turf maintenance management [6]. Zhang mentioned that high quality turf cannot be separated from efficient turf machinery management, professional turf machinery storehouse management and mechanical maintenance management is also an effective way to control the increase of turf maintenance cost in terms of long-term benefits [7]. Dr. James B Beard comprehensively analyzes golf course management skills and provides recommendations in his book [8].

The schedule must contain the results of assessment of growth condition for the turf. Zhuhai Jinwan golf course has been considered to be an attractive course but it needed an immediate assessment on the turf growth condition and quality. The present research has therefore been undertaken to assess the current quality grade of turfs and to suggest the management schedule. This would help the management team of Zhuhai Jinwan golf course to implement the management procedure showing the further requirements of turf stuff, turf equipments, etc.

2. Research methods and contents

2.1. Experiment object

The experiment was carried out in Zhuhai Jinwan golf course located in Jinwan district Zhuhai city. The climate of the area is subtropical monsoon having high annual temperature, humidity and with abundant rainfall. The annual relative humidity of the area is 79.75% and the total rainfall is 2009.3 mm. The study area has basically no winter, on an average seven months are summer (i.e., April to October) and 5 months are spring and autumn (i.e., November to March). The tee, fairway and green of the golf course are composed of Seashore Paspalum (*Paspalum vaginatum* Sw., Fam.: Poaceae). It is a warm season turf grass species

2.2. Experiment indexes and methods

Similar to standard golf course, Zhuhai Jinwan golf course has 27 holes throughout the tees, fairways and greens. A total of randomly selected 6 holes from tees, fairways and greens bearing numbers 2, 8, 10, 19, 23 and 26 were selected for the present investigation.

The experiment was started on 20 October 2017 and took 4 days to complete. Among all the experimental indices, the experiment period of the recovery was extended until the selected turf of a quadrat returned completely returned to the initial states of the experiment by itself. All measurements were taken during the morning, which is most players' tee time.

An iron framed quadrat of 10 cm^2 area was used to collect data from the tees, fairways and greens. The quadrat was tossed randomly to each area of the golf course and data on color, density, coverage, texture, mowing height, uniformity, elasticity, ball speed on the green and recovery ability of the turf were recorded. Five replicate measurements were carried out in each of the 6 holes of the golf course. Of the 9 measured parameters, the uniformity, color and coverage were graded after visual inspection. The density index's method was used, whereby a needle was stabbed into the turf and the leaves of the turf grass that touched the needle were counted (n=10). To grade the grass leaves width, texture index was used. Mowing height index was measured by using a ruler (n=6). The bounding height was measured following the elasticity index assessment method, which was carried out by dropping a golf ball freely from a height of 1.3 m on the turf and the bounding height was recorded (n=3). A Speedmeter was used to measure the ball speed index (n=3). Assessing the recovery ability was point out the pitches in the tees and fairways and ball marks on the green, then recorded week number that pitches and ball marks recovered same state compare with the surrounding turf, got 6 pitches and 3 ball marks at each hole. Assessment of recovery ability in the pitches of tees and fairways and the ball marks of greens was done by selecting having same size and degree of injury present. All pitches and ball marks were then monitored for weeks until the pitch and ball mark recovered to the same state as compared with the surrounding turf.

2.3. Turf quality appraise standard

In the process of evaluation of turf quality, it is difficult to quantify the importance indexes of turf color and uniformity in comparison, judgment, evaluation and decision making. In order to simplify and synthesize multi-research results, this experiment has been combined with customary 5-points grading evaluation standard to build golf course turf quality grading standard (Table 1). After rating by table 1, the Jinwan Golf Course turf quality grade was determined according to the rating of table 2.

Table 1. Grading standards of turf quality of golf courses.

Indices	Standards (points)				
	5	4	3	2	1
Color	Blackish green	Dark green	Green	Light green	Olivine
Density (leaves)	≥4	3	2	1	0
Texture (mm)	≥1.5	1.6-2.0	2.1-2.5	2.6-3.0	>3
Coverage (%)	>98	98-95	04-90	89-85	<85
Green mowing height (mm)	<3.0	3.0-3.5	3.6-4.0	4.1-4.5	>4.5
Fairway mowing height (mm)	<10	10-15	16-20	21-25	>25
Tee mowing height (mm)	<9	9-12	13-16	16-20	>20
Uniformity	High	Above	Normal	Lower	Lowly
Ball speed on the green (m)	≥3.04	2.74-3.03	2.44-2.73	2.13-2.43	<2.13
Recovery ability (week)	<1	1-2	2-3	3-4	>4

Table 2. Turf grades of golf courses.

Grade	Score	
	Tee and Fairway	Green
I (great)	32-35	36-40
II (good)	28-31	31-35
III (general)	24-27	26-30
IV (poor)	19-23	20-25
V (very poor)	≥18	≥19

3. Results and discussion

3.1. Appraise results of the turf quality on Tees

The first shot played on the tee in a hole, which is usually considered as an important one. Tee's uniformity, color and texture will affect players' visual impact and golfing mood as it is the first part of the golf course that golf players reach, making tee become quite important.

As can be seen from Table 3, the uniformity of tees was good and scored 4 points. The colour is dark green and scored 4 points. The average coverage on the tee of Jinwan golf course was around 95%, and a score of 4 can be assigned and so as to density. The average mowing height on tee of Jinwan golf course was 13.23 mm, and had an assigned score 4 points. According to the measurement, the average leaf width of the turf grass was 1.93 mm. This ensured better texture and a point of 4 was scored accordingly. Since the experimental period covered mid-to-late October, the growth season of *P. vaginatum* passed. The turf species was about to enter the dormancy period, therefore, average 3 weeks of recovery were needed. The score 2 points was got. The ball rebounding height on the tee was 6.37 cm.

Table 3. Appraise results of turf quality on tees of Jinwan golf course.

Indices	Experimental hole numbers						Average	Score
	2	8	10	19	23	26		
Uniformity	Above	High	Popularly	Above	Above	Above	Above	4
Colour	Green	Dark green	Green	Dark green	Dark green	Dark green	Dark green	4
Coverage (%)	93	99	92	94	95	98	95.1	4
Density (leaves)	4(±0.80)	3(±0.40)	3(±0.97)	4(±1.12)	3(±0.51)	2(±0.24)	3	4
Texture (mm)	1.77 (±0.05)	2.01 (±0.07)	1.86 (±0.05)	1.89 (±0.06)	2.06 (±0.09)	1.96 (±0.05)	1.93	4
Mowing height (mm)	15.6 (±0.05)	14.1 (±0.05)	12.2 (±0.05)	12.2 (±0.04)	13.2 (±0.06)	12.1 (±0.05)	13.23	3
Rebounding height (cm)	7.87 (±0.59)	7.3 (±0.39)	5.77 (±0.29)	5.43 (±0.31)	6.33 (±0.43)	5.53 (±0.28)	6.37	/
Recovery ability (week)	3	4	3	3	4	3	3.3	2

3.2. Appraise results of the turf quality on fairways

Usually, the fairway is the largest consistent part of any golf course. Therefore, the quality of fairway greatly influences the selection and use of the golf course by the consumers.

Table 4. Appraise results of turf quality on fairways of Jinwan golf course.

Indices	Experimental hole numbers						Average	Score
	2	8	10	19	23	26		
Uniformity	Normal	Above	Normal	Normal	Above	Normal	Normal	3
Colour	Olivine	Dark green	Olivine	Green	Above	Light green	Green	3
Coverage (%)	93	95	92	97	98	96	95.1	4
Density (leaves)	2(±0.68)	3(±0.49)	3(±0.32)	4(±1.29)	2(±0.93)	4(±1.29)	3	4
Texture (mm)	1.74 (±0.07)	1.8 (±0.07)	1.91 (±0.07)	1.83 (±0.08)	2.14 (±0.09)	1.79 (±0.06)	1.87	4
Mowing height (mm)	13 (±0.05)	12.6 (±0.04)	11.7 (±0.05)	11.1 (±0.05)	12.7 (±0.05)	11.5 (±0.06)	12.1	4
Rebounding height (cm)	11.43 (±0.37)	9.43 (±0.61)	13.6 (±0.42)	11.4 (±0.33)	18.97 (±0.78)	14.97 (±0.93)	13.3	/
Recovery ability (week)	2	3	4	3	3	4	3	3

Table 4 shows that the uniformity of the fairways was normal and individual fairway even appeared bare land so the score was 3 points. The color of the fairways turf by visually measured was green, the score was 3 points. The overall density of the fairways turf was above with 4 points of score. The turf grass on fairways was a little bit long and the mowing height was 12.1 mm that's why it got 4 points. Visual inspection showed a better coverage of the turf which acquired 4 points. The texture of turf grass at the fairways was above and stayed in the middle and upper level. Since the experimental period did not cover the growth season which usually requires 3 weeks to recover. The average ball rebounding height of the fairway was 13.3 cm.

3.3. Appraise results of the turf quality on greens

The green is the most important part of golf course. The putting on the greens is usually considered the key to winning a hole. It can be seen from Table 5 that Jinwan golf course's greens were highly in uniformity, no uneven color, uneven density and uneven grass length issues were observed. So, it scored uniformity point 5. The colour of the green's grass was green and got 3 points score, this was happened because of low mowing height. The coverage of the greens was high to 99%, higher than normal golf courses, which usually demand 98% coverage. The measured mowing height was 5.7 mm which was significantly different from the 3.2 mm which is recommended by the turf department. So, the score was 1 point according to the grading standard. High level golf competitions require the green speed faster than 3.35 m, the green speed of Jinwan golf course was 2.45 m, which is little lower than the recommended and so received only 3 points. The green recovery period was short and can be restored in about 1 week. In this case, the score was 4 points. The rebounding height of the tee and fairway was recorded as 6.67 and 13.3 cm, respectively. When these values are compared with the value of green (4.02 cm), the latter is found low. This kind of situation is more beneficial to the professional player because they can keep a control on ball. Due to the lacking of other literature information on the rebound height data of different golf courses, it became impossible to discuss and interpret the information obtained in the present research.

Table 5. Appraise results of turf quality on green of Jinwan golf course.

Indexes	Experimental hole numbers						Average	Score
	2	8	10	19	23	26		
Uniformity	High	Above	Above	High	High	High	High	5
Colour	Green	Green	Green	Green	Olivine	Dark green	Green	3
Coverage (%)	98	100	98	98	100	100	99	5
Density (leaves)	2(±0.51)	2(±0.55)	2(±0.60)	3(±0.29)	2(±0.37)	2(±0.58)	2	3
Texture (mm)	1.75 (±0.06)	1.85 (±0.06)	1.77 (±0.06)	1.67 (±0.05)	1.68 (±0.05)	1.64 (±0.05)	1.73	4
Mowing height (mm)	5.8 (±0.03)	5.9 (±0.02)	5.8 (±0.02)	5.5 (±0.02)	5.3 (±0.02)	5.6 (±0.01)	5.7	1
Rebounding height (mm)	3.3 (±0.32)	3.8 (±0.24)	4.53 (±0.26)	3.6 (±0.20)	4.53 (±0.25)	4.33 (±0.23)	4.02	/
Ball speed on the green (m)	2.36 (±0.03)	2.71 (±0.03)	2.46 (±0.04)	2.49 (±0.06)	2.36 (±0.03)	2.31 (±0.05)	2.45	3
Recovery ability (week)	1	1	1	1	1	1	1	4

In summary, the tees of Jinwan golf course are graded with all moderate degrees, uniformity, colour, coverage, density and texture. It has a good appraisement in mowing height, but the recovery period is too long. The comprehensive score of tees was 25 with an evaluation grade III level. Fairways got good appraisement in coverage, mowing height, density and texture, but got medium appraisement in uniformity, colour, and recovery ability. The overall score was 25 points having evaluation grade III level. And green's uniformity, coverage, density, texture and recovery ability were basically good, however, poor appraisement in mowing height, ball speed on the green a little low, comprehensive score of greens was 28, level III.

The overall quality of Jinwan golf course is at a medium level.

4. Investigation on maintenance and management of golf course

4.1. Irrigation

Moisture content is the key to high quality golf turf, so everyday irrigation is an essential task for the turf departmet of Jinwan golf course. When turf is dry, its surface will harden, color will turn to olivine, trampling footprints becomes more visibly and ball's rebounding height will be higher. Especially, when the green turf is at low mowing height every day, the roots system will become shallower and the water absorption and holding capacity will decrease. Automatic sprinkler system is utilized in Jinwan Gold Course, through which the tee and fairway are irrigated when it is dark and the green is irrigated before business time. The practice is continued once in a day for 15-20 mins, if not rainy. At the commencement of hot weather, supplemented artificial irrigation is carried out in the golf course as turf lacks of water.

4.2. Mowing

The turf maintenance department of the Jinwan golf course depends upon the seasonal climate. Last year, the growth of the grasses was set in an annual turf mowing-height plan (Table 6). However, the actual green mowing-height was measured 5.7 mm in October, which is far higher than the annual plan (Table 6). In the meantime, measured ball speed on the green was 2.45 m and the mowing-height has severely affected the speed of the green. It happened because of the mowing technique of the green keeper was not qualified and the mower blades were older which caused ball speed slower than expected.

Table 6. Jinwan golf course annual turf mowing-height schedule.

Area	Mowing-height (mm)	Frequency		Time
		May-Oct.	Nov.-Apr.	
Tee	May-Oct: 10-11; Nov.-Apr.:13	3-4 times/week	1 time/week	5:30-6:00
Fairway	May-Oct: 10-11; Nov.-Apr.:13	3-4 times/week	1 time/week	
Green	May-Oct:2.1; Nov.-Apr.:2.5	1 time/day	3-4 times/week	

4.3. Topdressing and rolling

Theoretically, there is no seasonal limit for topdressing and rolling. In order to ensure green grass's benign growth and faster ball speed, the frequency of topdressing should be directly proportional to the growth rate of green grass. The topdressing and the rolling frequency should be average i.e., once in a week and the thickness of topdressing should be maintained 2-3 mm each time. However, in Jinwan golf course, the topdressing of greens were carried out once in a moth except the competition seasons. The thickness of the topdressing has been recommended as 1 mm on topdresser's scale, but in reality it did not fulfill the scale in the Jinwan golf course. This is another reason that the ball speed is not qualified. The authority used to carry out topdressing tees 2-3 times in a year and fairways once in a year. The same practice has also been carried out in case of the maintenance of the greens. As a result, the actual thickness of topdressing is less than 2 mm.

4.4. Coring

Coring not only alleviates the degree of soil compaction, but also cuts the stolons to stimulate the growth of the new stolons to the turf grass. This practice promotes the formation of a dense turf, which is suitable for the spring when the grass turns to green. At the same time, coring increases the permeability of the soil and the ability to seep water, allowing the root system to absorb nutrients faster and better, and further promotes the growth of greens. In addition, coring can control the turf thatch to avoid excessive moisture in the thatch layer which might cause rotting in the roots of the turf. Rotting can affect the growth of the new root system of the turf and eventually may lead to infectious diseases. Table 3 indicates the coring performance of the golf turf in the Jinwan golf course.

Table 7. Jinwan golf course turf coring survey.

Area	Core diameter (mm)	Core depth (mm)	Core type	Frequency
Tee	5	8	Hollow	2 times/year (spring/winter)
Fairway	10	8	Hollow	2 times/year (spring/winter)
Green	4	8	Hollow	2 times/year (spring/winter)
			Solid	Only when turf water permeability deteriorates

5. Conclusion

Through the appraise study of the turf quality of Jinwan golf course and comprehensive analysis of the work of the turf maintenance department, it revealed that the turf quality of Jinwan golf course was at intermediate level. The limiting factors of scoring are the uniformity, mowing height and recovery ability of the tee and fairway turf and mowing height, ball speed of the green.

Jinwan golf course should appraise turf quality regularly, and the maintenance schedule should be adjusted in time according to the results. It is also necessary to establish professional turf maintenance equipment, regular inspection and a strict follow up maintenance schedule. Inspecting and repairing all the turf related equipment on the golf course regularly and timely. Otherwise, upgrading will be required when equipment failed to meet the needs of turf maintenance standards. To pay attention to the details of the implementation of turf maintenance measures, such as whether, the thickness of the topdressing is made up to the standard, and whether the mowing height calibration on mower is consistent with the actual mowing height. What's more, the time of the green irrigation should be fixed as far as possible, avoid the excessive numbers of people playing golf in morning to delayed irrigation time to noon and led to grass burns. The particle size of sands of topdressing should be chosen in accordance with the green's construction and quality.

References

[1] Zheng H J, Hua L and Gao Z G, 2003, A study on indices system and assessment of turf quality. *J. Capital Normal Uni. (Natural Science Edition)*, **24(1)**, 78-82.
[2] Liu J D, Chen Q Q and Jiao N Z, 1999, Approach of assessment mthod of turf guality. *J. Inner Mongulia Agri. Anim.Husband. College Edition*. **20(2)**, 44-8.
[3] Wang J C, Lu T, Liu C Q and Wang Y H, 2008, Evaluation and improvement of the green quality. *Pratacult Sci.*, **25(8),** 126-9.
[4] Yang C, Liu Z H, Gao C N, Suo T T, Bai W Z and Zhen Q, 2016, The simple evaluation method establishment of golf green turf quality. *Pratacult Sci..*, **33(2)**, 192-200.

[5] Liu Z W, 2013, Study on assessment of Changsha golg course turf quality and maintaing management. Hunan, Hunan agriculture university., 1-37.
[6] Duan Q W, 2015, Discussion on the maintenance management of golf course turf. *Crop Res.*, **29(8)**, 924-6.
[7] Zhang Z S, 2011, Mechanical management of th golf course turf. *Golf Punk*, 102-4.
[8] James B Beard, 2001, Turf Management for Golf Courses, Wiley; 2 edition.

IOP Conf. Series: Earth and Environmental Science **185** (2018) 012036 doi:10.1088/1755-1315/185/1/012036

The Australian digital farmer: challenges and opportunities

G Cho[1]

Faculty of Science & Technology, University of Canberra, Canberra, Australia

E-mail: George.cho@canberra.edu.au

Abstract. Any agricultural farm enterprise is a complex undertaking. Physical, economic, social and technological factors demand equal attention if a farm enterprise is to succeed. To be successful a farmer needs to ensure the optimal use of farm inputs to achieve high productivity. Technological developments have dragged farmers into the digital age and have significantly altered the management of such inputs. This paper considers the disruption brought about by digital farming. Major technical, legal, policy and social challenges and opportunities are outlined in the context of digital farming in Australia.

1. Introduction

Any agricultural farm enterprise is a complex interaction of inputs and outputs. Some of these are within the control of the farmer such as farm inputs whereas environmental factors such as the weather, climate, soils and water availability are sometimes beyond the control of the farmer. The digital disruption promises unclear impacts but potential gains. The adoption of digital technology in the farm is best described as uneven [1]. The potential for the Australian economy through modeling methods show positive gains. The estimated Australian farm investments in digital technology is less than $5 million compared to global investments of more than $4 billion in 2015 [2]. Digital agricultural technology is in infancy in Australia compared to the US, Denmark and Israel [3].

As with the Industrial Revolution Industry 4.0, the agricultural sector too has evolved from the Agricultural Revolution 1.0 characterized by labor intensity, Agriculture 2.0 of high productivity gains from the Green Revolution and then to Agriculture 3.0 of precision farming deploying guidance and sensing systems, telematics and data management. Agriculture 4.0 features digital smart farming with physical products enhanced by additional non-physical services and the emergence of an agricultural eco-system with a subset of this revolution extending to autonomous unmanned decision systems of robotics and artificial intelligence.

In 2015-16 there were 85,681 agricultural businesses in Australia [4]. Both the area and the number of businesses using precision agriculture techniques increased during 2015-16. Controlled traffic farming continues to be the most prevalent of these practices and is undertaken by 6,600 businesses (a 14% increase on 2014-15) over an area of 6.7 million ha – a 28% increase from 2014-15 estimates. The total agricultural land is estimated at 391.86 million ha [4].

This paper describes the challenges and opportunities facing the Australian digital farmer. The gains from digital farming are given together with an assessment of adoption and maturity levels and considerations of 'big data' issues. Then technical barriers facing the digital farmer are outlined to suggest some of the reasons for the uneven and slow adoption of technology in the farm. The legal and institutional factors such as ownership of farm data, privacy, and transformational barriers describe unique challenges. The next section addresses policy matters in the new digital economy. The discussion is followed by addressing the 'next gen' digital agriculture in Australia and suggests some recommendations to enable digital agriculture to grasp the opportunities. Access to the marketplace through higher productivity, leadership and value propositions are a part of the mix in cross-industry collaboration within a strategic framework.

Any digital farm leverages latent value from data gathered from smart tools and equipment on farms. Uses of the data include phenomics, farm use software engineering, data analytics, precision agriculture and farm management systems. Phenomic tools offer non-destructive and high throughput methods measuring and analysing phenomes and genomic traits to quantify crop performance in response to the environment. Software engineering help map and analyze land use and crop performance on farms. Applied data analytics contribute to quality assurance and longitudinal datasets. However, unresolved challenges include the use of 'big data' on how to store the information and managing the raw data as intelligence.

The use of digital tools to manage Australian agriculture has continued to increase with ongoing reduction in chemical use, careful use of fertilizers and flexible approaches to grazing management to reduce erosion and increase productivity [5]. Economic modeling has shown that the implementation of digital agriculture in all Australian production sectors could lift gross value of agriculture by $20.3 billion, an increase of 25% from 2014-15 levels [3].

Definitions

Digital farming is means the connected knowledge-based farm production system that employs intelligent network and data management tools that enables the automation of sustainable processes in agriculture [6]. It is synonymous with smart agriculture and digital agriculture.

Precision farming uses global positioning systems (GPS) to enable the guidance of farm vehicles that are used for seeding, fertilizing and harvesting operations. This may involve the expansive use of telematics and data management not only in broadacre farming but also managing exact requirements for individual heads of animal stock, individual trees or plants to thrive optimally.

Decision agriculture is the end-point or action resulting from the use of knowledge and/or information that may be derived from digital farming.

Skinner's evaluation of Australian digital agriculture maturity levels from precision to decision agriculture models has shown that there is a low maturity level across all categories of digital agriculture [7-8]. Skinner suggests an actionable cross-industry digitization plan. Indeed value-adding to farm data may be achieved by the pooling of individual farm data as 'small data' [7].

The immaturity of digital agriculture in Australia is the result of interrelated factors including leadership, trust and legal barriers, information technology infrastructure and the availability of appropriate data. To this list may be added the lack of digital literacy in data analytics and a lack of an understanding of how to garner decision-support systems and tools [3].

In 2016 the Australian government's Rural R&D project on Accelerating Precision Agriculture to Decision Agriculture (P2D) Heath evaluated the current and desired state of digital agriculture in Australia and to make policy recommendations [9]. The evaluation specifically sought to estimate the impact of unconstrained digital agriculture on the Australian economy. 'Unconstrained' here is taken to mean a situation where any current limiting factors have been overcome and the full potential of digital technology realized by all Australian farmers.

The P2D study estimated that unconstrained digital agriculture would add a 25% boost to Australian agriculture with a $20.3 billion increase from 2014-15 gross value products for all sectors. Labor savings from automation was estimated at $7.4 billion, genetic gains from the use of objective data $2.9 billion, and savings from the careful monitoring of inputs to needs $2.3 billion and profits from the enhancement to market access and biosecurity $1.0 billion.

The assessment is that digital technology has the potential to transform Australian agriculture and profitability for farm businesses. Productivity gains of 10-15% in cropping systems have been separately estimated with half gained through input efficiencies and the other half through increases in output [1] .Economically there appears that there is much to be gained from digital agriculture.

2. The digital farmer: challenges and opportunities

To be a fully-fledged digital farmer there needs to be a conscious decision to embrace technology in every aspect of the farm enterprise. It would require a personal re-education program to know and understand aspects of the technical and institutional environment as well as an investment strategy to get the best possible equipment available for the farm. Such technical and institutional requirements take place within a social milieu where the sharing of information and knowledge is bounded by one's network and social circles. These are challenges and barriers that any digital farmer must overcome to grasp the opportunities.

It is not uncommon to find that the older generations of farmers often rely on net-generation millennials, for help to solve simple IT problems. But on some farms the net-savvy children may be at boarding school or university. In the absence of such help farmers may have to turn to their 'significant other' or to their neighbors. On the farm, external connectivity to the internet may not be available all the time. Even if the internet connectivity were available, the requisite speeds may be insufficient at critical times of the day or the speeds might be unreliable over different times of the week. These technologies come at a price, given that the nature of the market for digital agriculture products is one where there may be more price and service-takers than there are price or service makers [10]. The suggestion is that there are multifaceted problems that digital farmers must grapple with over and above the primary activity of farming.

2.1. Digital agriculture and information technology

Over two-thirds of Australian farmers rely on the mobile network to access the internet. The 3G or 4G (third or fourth generation wireless mobile telecommunications technology) towers located throughout the farming areas in Australia either suffer from unstable connectivity or the inability to get or stay connected [11]. Even when connected the user might experience fluctuating speeds for effective through-put when uploading and downloading data. External data connectivity is a significant challenge for farmers seeking to adopt and to deploy many technology-based innovations on the farm. Previously, telecommunication service providers had a universal service obligation (USO) that obliges providers to ensure that telephony services and prescribed carriage services are reasonably accessible to all people in Australia on an equitable basis (http://www.acma.gov.au).

On the issue of connectivity, a study of the implications of digital agriculture concluded that Australian farm businesses are operating with broadband connections that are slower, more expensive to acquire and less reliable than those available in other countries that compete for similar agricultural export markets.[1]

Suggested 'work-arounds' to resolve the external connectivity requirements include *Farm Mobile* that stores collected data in memory and then deploy a passive uplink connection (PUC) when an internet connection becomes available [12]. But even with this solution, the lack of mobile phone reception or access to the internet to transfer data remains. The challenge is to find better ways to link to the internet to integrate farm information

Another suggestion is to develop an on-farm connection using forms of 'radio networks' to supplement current standard WIFI networks. But there is still the need to connect to the internet to view the live feed and access to cloud-based analytics [10].

2.2. Digital agriculture and legal issues

Digital agriculture is poised to positively change the industry by enabling informed farm management decisions. While there is the availability of a large cache of data and information together with software to facilitate the analysis of the data, there is however the need to pay attention to emergent legal issues. The control, ownership and maintenance of the privacy of the data provide challenges as well as opportunities. Seemingly innocuous, questions of ownership of farm data, the lack of confidence and certainty in the quality of the data raise several interrelated legal questions.

Generally, the legal framework governing Australian agricultural data are intellectual property, privacy and statutory obligations. Under intellectual property law, confidentiality, trade secrets and data ownership are most relevant to agricultural data. Privacy, with reference to the Australian *Privacy Act* 1988 (Cth), refers to the handling of personal information about individuals is as relevant as the obligatory submission of farm data to the Australian Bureau of Statistics for statistical purposes [1].

It may be argued that data generated on the farm belongs to the farmer. However, this begs the question of who is the farmer – the owner of the property on which the farm enterprise takes place, the contract harvester or the manufacturer of the machinery that has on-board real-time sensors collecting production data?

Unmanned aerial vehicles (UAV) or drones have been used to remotely sense farms to capture data on cropping patterns, heat and water stress, fertilizing regimes and other useful information. Farmers may commission contractors to perform such functions which may also include an analysis of the data or they may perform the analysis themselves. In such situations the farmers who commission such work 'owns' the collected data.

While an argument that the data collected by drones over any farm may be considered personal information given the close connection of the farm as private property, a counter-argument may be run to suggest that the provisions of the *Privacy Act* 1988 (Cth) may not apply to individuals who operate drones in their private capacity. The Act applies to the handling of personal information about individuals by governments and to those organizations to which the Act applies.

There is uncertainty in the law in the imagery captured by satellites and drones. A view has been put that UAVs fall within the definition of an optical surveillance device of the *Commonwealth Surveillance Devices Act* 2014 and if so existing Australian surveillance laws could apply [13].

Ownership rights of farm data involving parties other than the owner of the property may be managed and controlled by agreement either by contracts of service or contracts for service. The agreements should spell out the ownership of the data explicitly by specifying rights over the data generated, and the permitted uses of that data by the contractor.

Another legal facet to the ownership of farm digital data is that of responsibility and legal liability. A key element in the food production chain is that digital data may demonstrate compliance with legislative obligations. Given society's heightened expectations for food safety and production methods, additional elements of transparency and traceability will become readily available to consumers, manufacturers and regulators alike.

In Australia most farm operations are owned and controlled either by individuals, families or corporations who also manage most aspects of the supply and production chain [14]. Digital technologies adopted in these operations also rely heavily on basic contractual and licensing agreements that describe how data can be collected on the farm, how it is controlled, shared and accessed by third parties.

In the US and in New Zealand the regulatory environment to protect farmer's ownership rights over digital and other data have been in the form of industry Codes of Practice [1]. Such codes spell out the limits for the use of the farm data and third-party usage. For example, the American Farm Bureau's *Privacy and Data Principles 2014* [15] and the New Zealand Dairy Industry's *Farm Data Code of Practice 2016* [16] are designed to encourage best practices in the ownership, control and access of agricultural data.

There is a need to re-think the collection of farm data, data management practices and improving transparency. Datasets that need manipulation into useful forms create barriers that make the datasets accessible only to a small set of experts. Such datasets need to be FAIR – findable, accessible, interoperable and reusable [12]. To prevent technology galloping away from practice and regulation human interactions with the technology should be better understood rather than to regulate the technology itself [17].

2.3. Digital agriculture and policy matters

For policy makers the challenge is to identify the assistance to digital farmers to achieve their potential and in so doing also provide some security for the national economy. [1 Executive Summary] New skills training and development will be necessary when introducing current technologies to enterprises. It implies accepting the high throughput costs and the greater reliance on smarter farm machinery.

The digital farm in the 21st C will be one where there is a heavy reliance on GPS to provide live information on where herds of cattle are dispersed and the direction of their movements on the farm. Electronic collars not only provide data of the stock's location but also other identification information of each head of cattle. Robotic meat processing may replace the need for human intervention and manual methods. Selected examples of farm technology in practice are summarized in [2, 18].

What are the policy requirements that may cement the place of digital agriculture within the national economy? Cross-industry collaboration has been suggested – agriculture, information technology, farm machinery – in concert with government and strategic developments and its implementation, that is situated within a dedicated, consistent and unified regulatory framework [19]. There is also a need for a data management code of practice so that both digital farmers and those supplying the digital equipment are at one in terms of data use, data ownership and data security. To achieve the higher standards of practice there will inevitably be the need for certification and accreditation [20]. The adoption of such practices will bring about monetary benefits as well as confidence and other social benefits to digital farmers. Two other policy recommendations arising from the Precision Agriculture to Decision Agriculture (P2D) project include investments in telecommunication infrastructure and the adoption of new investment models, again involving private and public entities [3].

2.4. Digital agriculture and social aspects

Digital agricultural technology is a skills-based industrial model accompanied by digital business and production methods. The model significantly challenges traditional farmers who may require upskilling and the adoption of a mindset of constant change and digital literacy. Education and capacity building in digital literacy is sorely needed to fill the identified capability gaps.

In Australian farms in 2016, the average age of farm providers (farmers) was 56 years and the average time the provider was involved in farming was 35 years. Such demographics suggest an 'ageing' of farmers in Australia which may mask the reinvigoration of Australian agriculture [4].

A survey has shown that there are more than 4,000 jobs per year in agriculture according to advertising statistics [21]. However, the number of graduates supplied by Australian universities has declined significantly by more than 20% to the number needed to satisfy the job market. Digital farming could become far more interesting, rewarding and challenging to the new generation of farmers. Digital farmers as the future custodians of the Australian landscape will assume responsibility for the management of native vegetation, landscape sustainability and biodiversity and contributing to addressing global issues of carbon emissions and the carbon economy. Automated farm vehicles using GPS could be fully autonomous, as remote sensed and UAV data guide farm management decisions in precision agriculture. At the same time, farmers would need increased personal skills and capacities together with expert advice in IT and other aspects of the farm economy from others.

Australian farming in general may yet become an attractive investment and career choice for many young new graduates from universities. Every Australian farmer now relies on an average of six local paid professionals and consultants to advise them on diverse aspects of their business from crop and agronomic advice to IT, grain trading and financial planning rather than a farmer trying to do everything themselves as in the past [22].

The data suggests that a dearth of graduates taking up agriculture as a career [21]. The challenge will intensify if there are no concerted efforts by all, including governments. One observer has said that farming should be a career choice and not a sentence [22].

3. Future of digital agriculture in Australia

The next generation of digital agriculture in Australia may see a growing maturity with leadership demonstrated both by the private and public sectors. The value proposition is the confidence in further investments and a greater adoption of digital, innovative practices in all aspects of farming. A high degree of trust and confidence will follow the lifting of legal and other institutional constraints presently restraining digital agriculture. These developments are premised on a level of sophistication in digital literacy, data analysis and decision support systems [3].

The P2D project made thirteen recommendations regarding digital agricultural policy, strategies, leadership, digital literacy and industry enablers [3]. The recommendations enable digital agriculture to be more competitive in the international marketplace. The projected future in Australia is an agricultural landscape that will be characterized by one which is digitally integrated at all stages of production from the use of genomics and genetic modification of crops and livestock to transport logistics at the farm, at the farm gate and through to the abattoir and market place.

Modern farming problems may require smart technologies and solutions that rely more on science and technology rather than on farming know-how that has been passed on from one generation to another. The know-how may in the future be digital and data driven that rely on sensors, information systems and computer decision support systems.

4. Summary and conclusion

The farmer in Australia is at a crossroads today with a choice of either taking the well-trodden path or the path provided by the digital revolution. The agricultural revolution 4.0 brought about by digital technology promises potential gains but with unclear impacts. The agricultural disruption in the farm has introduced new tools that produce value laden data for objective decisions and management systems. The data also are fed into biological studies of productivity, software engineering applications, data analytics and precision agriculture. Economic modeling has shown the lifting of the gross value of agriculture with an increase of 25% from the 2014-15 levels.

A critique of digital agriculture is that the use of technology on the farm may sometimes be isolated from knowledge of agriculture and an understanding of on-farm realities and business problems. Technology might compound the problems and challenges instead of contributing to solutions and doing the actual work of the farm.

There is a need for a deeper understanding of big data and its impacts on digital agriculture and more importantly the manner it has changed farming societies. These societal impacts loom large as the application of big data analytics has fundamentally changed farming practices and farmer interactions.[1] A critical question is: What happens when the predictive analysis starts to match local knowledge? Do we have 'singularity' to borrow a phrase from artificial intelligence? The answer, it seems, is that there is a greater need to re-double efforts to undertake both the objective science-based research and social research in the farm to understand what is happening and to explore trends and drivers.

References

[1] Australian Farm Institute (AFI) 2016 The implications of digital agriculture and big data for Australian agriculture, Research Report at http://www.crdc.com.au/publications/implications-digital-agriculture-and-big-data-australian-agriculture.

[2] Locke S 2016 Farmer interest in digital agriculture technology experiencing 'extraordinary' growth this year, *ABC Rural ABC News*, 8 June 2016.

[3] Trindall J, Rainbow R and Leonard E 2018 Enabling digital agriculture in Australia, *Farm Policy J.*, **15**(1), 1 – 6.

[4] Australian Bureau of Statistics (ABS) 2016 *Agricultural Commodities 2015-2016,* Cat. No. 7121.0; & *Land Management and Farming in Australia 2015-16,* Cat. No. 4627.0.

[5] Australia, Department of Environment 2016 *Australia State of the Environment Overview 2016,* Canberra, DOE.

[6] CEMA European Agricultural Machinery 2017 Digital farming: What does it really mean? 13 February, at http://www.cema-agri.org.

[7] Skinner, A 2018 Big data maturity in Australian agricultural industries, *Farm Policy J.*, **15**(1)

61 – 9.

[8] Skinner A, Wood G, Leonard E and Stollery T 2017 Advancing precision agriculture to decision agriculture: a bid data reference architecture for digital agriculture in Australia, Data to Decisions CRC and Cotton Research and Development Corporation.

[9] Heath R 2018 An analysis of the potential of digital agriculture for the Australian economy, *Farm Policy J.*, **15**(1) 9 – 12.

[10] Lamb D 2018 Connecting Australian farmers: gateways, gatekeepers and a forgotten solution, *Farm Policy J.*, **15**(1) 51 – 9.

[11] Lamb D 2017 Accelerating precision agriculture to decision agriculture: a review of on-farm telecommunications challenges and opportunities in supporting a digital agriculture future for Australia, University of New England and Cotton Research and Development Corporation and at https://www.crdc.com.au/sites/default/files/P2D%20telecommunications%20-%20UNE%20Final%20Report.pdf.

[12] Darnell R, Robertson M, Brown J, Moore A, Barry S, Bramley R, Grundy M and George A 2018 The current and future state of Australian agricultural data, *Farm Policy Journal,* Vol. 15(1) pp. 41 – 49.

[13] House of Representatives (HR) Standing Committee on Social Policy and Legal Affairs (2014) *Eyes in the sky: Inquiry into drones and regulation of air safety and privacy*, available at http://www.aph.gov.au/Parliamentary_Business/Committees/House/Social_Policy_and_Legal_Affairs/Drones/Report.

[14] Wiseman L, Sanderson J and Robb L 2018 Rethinking Ag data ownership, *Farm Policy J.*, **15**(1) 71 – 7.

[15] American Farm Bureau Federation 2014 *American Farm Bureau Federation's Privacy and Security Principles for Farm Data* available at http://www.fb.org/issues/technology/data-privacy/privacy-and-security-principles-for-farm-data.

[16] Dairy New Zealand 2014 *New Zealand Farm Data Code of Practice* available at http://www.farmdatacode.org.nz/wp-content/uploads/2014/06/farm-data-code-of-practice-final.pdf.

[17] Moses LB 2007 Recurring dilemmas; the law's race to keep up with technological change, *UNSW Law Research Paper* no. 2007-21. See https://www.researchgate.net/publication/228183058_Recurring_Dilemmas_The_Law's_Race_to_Keep_Up_With_Technological_Change.

[18] Wallace A 2017 Digital agriculture takes centre stage, *Spatial Source* at https://www.spatialsource.com.au/locate-isde/digital-agriculture-takes-centre-stage.

[19] Wiseman L & Sanderson J 2017 The legal dimensions of digital agriculture in Australia: An examination of the current and future state of data rules dealing with ownership, access, privacy and trust, Griffith University, USC Australia and Cotton Research and Development Corporation.

[20] Zhang A, Baker I, Jakku E and Llewellyn R 2017 Accelerating precision agriculture to decision agriculture: the needs and drivers for the present and future of digital agriculture in Australia. A cross-industries producer survey for the Rural R&D for Profit Precision to Decision (P2D) project, CSIRO and Cotton Research and Development Corporation.

[21] Pratley J 2012 *Professional agriculture – A case of supply and demand*, Australian Farm Institute Occasional Paper no. 12.01, February.

[22] Neales S 2018 'A new crop is on the rise. Farming is now, believe it or not, almost trendy', *The Weekend Australian,* May 19-20, 19.

Fluctuation characteristics and influence factors of rice price in Guangdong

J F Liang[1,2], Z Wan[1, 2,3], W Fang[1,2] and L Ma[1,2]

[1] Institute of Agricultural Economics and Rural Development GAAS, Guangzhou 510640, Guangdong, China

[2] Key Laboratory of Urban Agriculture in South China MOA, Guangzhou 510640, Guangdong, China

E-mail: wanzhong1999@163.com

Abstract: Guangdong Province is a large province of rations consumption and the abnormal fluctuations in food prices will affect its steady development. In this paper, the largest grain product in Guangdong Province - Rice is selected as the research object, and the price fluctuation characteristics is analyzed by the time series decomposition method. We try to answer the following two questions: First, what are the characteristics of price fluctuation of rice in Guangdong Province? Second, what causes the rapid rise in rice prices in the short term? The paper identifies and analyzes the fluctuation rules and characteristics of rice price in Guangdong and summarizes main influence factors for short-term drastic fluctuation in rice price by time series decomposition method based on the monthly price data in 1998-2015. The results show that the rice price in Guangdong has the following characteristics for a long time: overall rising, short-term drastic fluctuation, obvious periodic fluctuation and weakened seasonal fluctuation; random shocks drive and restrain the rice price, where external factors have serious influences on price fluctuation in some years, while the influence of random components on rice price has been obviously weakened since the second half of 2012; main factors influencing the fluctuation in rice price include market supply and demand, production cost and grain policies. In view of this, it is suggested in the paper that future price fluctuation should be predicted by utilizing the long-term trend and the periodic characteristics of the rice price, attention be paid to such factors causing abnormal fluctuation in rice price, market monitoring be strengthened and effective measures be taken to mitigate short-term drastic fluctuation in rice price.

1. Introduction

Rice, as the main food of Guangdong residents, presented a new trend with frequent price fluctuation and enlarged fluctuation range over the past more than ten years. During 2001-2015, the fixed base consumer price index (CPI) of rice in Guangdong rose from 97.9% to 207%. In 15 years, there were 7 years when the year-on-year index of rice consumer price was more than 5% and the index was even up to 22% in 2004; the price of indica rice surged to a historic high in 2011. Under the conditions of market economy, reasonable fluctuation of the grain price is an objective reflection of the relationship between supply and demand, and the rise in the grain price is the value regression of low price that maintaining for many years. However, the short-term price fluctuations in rice prices are far beyond consumer's psychological expectations, food security issues aroused wide concern in all sectors of community again. To this end, short-term fluctuations in rice prices are worth exploring in depth. There are many researches on grain price fluctuation in China, and the researches are focused on the fluctuation characteristics, influencing factors and the impacts on price and so on.

The results of the study based on the ARCH model by Luo and Liu [1] show that China's grain price fluctuation has significant ARCH effects as a whole, but the prices of indica rice, japonica and soybean did not present significant heteroscedastic effects. Miao [2] found that China's rice market price has obvious ARCH effect and its fluctuation present clustered, asymmetric and sustainable features by the study of the ARCH model. Huang's research [3], by the use of Global Trade and Energy Analysis Model (GTAP-E), showed that global energy prices, biomass liquid fuels, market speculation and national policy intervention are the main factors that caused the sharp rise and drop of the price of international agricultural products in the past three years. Zhang [4] thought that external impacts can expand the price fluctuation of primary agricultural products for 3~5 times in a short term. The grain industry chain of rice, wheat, etc. is mainly affected by the impacts of climate and natural disasters on industry production, and the degree of interpretation by such impacts for price fluctuation of both rice and wheat is about 95%. Su [5], by using GARCH model and VAR model, found that the risk of long-term fluctuation of grain price in China mainly comes from the change of domestic production cost and the change of international grain price. Wang and Li [6] considered that grain price fluctuations are both affected by the external factors like market supply and demand, production cost and internal factors like domestic policies, conduction of international market and so on. Deng [7], by using the ARCH model, found that the fluctuation of grain price in China has the ARCH effect and the "leverage effect", and the grain price fluctuation has a significant positive correlation with currency growth rate and depreciation of US dollar. Fang [8] thought that random shocks and cycle changes had a greater impact on rice price fluctuation before 2005 and the impact of random shocks on rice price fluctuation was significantly reduced after 2006. Shi and Hu [9] considered that the external random shocks had negative effects on grain prices and the effects were most serious during 1998-2003, and the grain price fluctuation presented a significant periodic characteristic and was affected by random factors. Huang [10], by using limited distributed-lag regression model, found that the rise of breakdown of food prices will be driven by price transmission of grain prices. The above studies were mainly conducted by using time series method, including ARCH model, VAR model, Beveridge-Nelson decomposition technology and H-P filter method etc. [11-15]. The current researches focus on the study of national and overall grain price fluctuation, but the fluctuation of grain prices is different in different regions and varieties. Guangdong Province is a large province of rations consumption and the abnormal fluctuations in food prices will affect its steady development. In this paper, the largest grain product in Guangdong Province - Rice is selected as the research object, and the price fluctuation characteristics is analyzed by the time series decomposition method. We try to answer the following two questions: First, what are the characteristics of price fluctuation of rice in Guangdong Province? Second, what causes the rapid rise in rice prices in the short term?

2. Fluctuation trend of rice price in Guangdong

2.1. Data sources

In this paper, rice price of Guangdong is selected as the research object. As the consumption of indica rice accounts for 70% of rice consumption in Guangdong Province, the early indica and late indica are selected as the representative study objects and the monthly price (¥/t) (the data comes from China Grain Network

Data Center)are used. The sample range is from January 1998 to December 2015 due to limited data availability.

2.2. Descriptive statistical analysis of rice price fluctuation
The price fluctuation rate is indicated by the sequential growth rate of rice prices in neighboring months, the formula is: $R_t=(P_t- P_{t-1})/P_{t-1}\times 100\%$, Pt and Pt-1 presents the rice price of month T and month T-1 respectively. It can be seen from Table 1 that the price fluctuation of early indica rice and late indica rice in Guangdong is generally higher than the national average in the sample range. Among which, the average value of the price fluctuation rate of early indica rice and late indica rice was 0.4622% and 0.5491% respectively and the standard deviation was 3.89299% and 4.37975% respectively, both were higher than the average and standard deviation of the price fluctuation rate of the national early rice and late indica rice.

Table 1. Basic statistical characteristics of price fluctuation sequence data of early indica rice and late indica rice.

Variable	Area	Min	Max	Mean	Std. Deviation	Skewness	Kurtosis
Price fluctuation rate of early indica rice	Guangdong	-10.05%	23.93%	0.4622%	3.89299%	1.513	7.661
	National	-8.69%	19.79%	0.4184%	2.74848%	2.904	17.200
Price fluctuation rate of late indica rice	Guangdong	-10.00%	24.25%	0.5491%	4.37975%	1.760	7.077
	National	-8.00%	17.44%	0.4199%	2.67219%	2.779	15.487

2.3. Fluctuation characteristics of rice price in Guangdong

2.3.1. The rate of increase of rice price is greater than the rate of decline. It can be seen from Table 1 that the maximum fluctuation rate of early indica rice and late indica rice in Guangdong is 23.93% and 24.25% respectively, and the minimum value is -10.05% and -10.00% respectively, which indicates that the increase rate of rice price is greater than the rate of decline. This is because the national food protection policy has a restrained effect on the degree of decline in rice prices to a large extent.

2.3.2. The rice price presented a trend of fluctuated rising with obvious fluctuation in some years. The price of early indica rice in Guangdong increased from 1836.82 ¥/t in 1998 to 3997.78 ¥/t in 2015, with an average annual growth rate of 4.68%, which is slightly higher than the national average rate (4.53%)of early indica rice; The price of late indica rice increased from 1971.18 ¥/t to 4964.5 ¥/t, with an average annual growth of 5.58%, which is 1% higher than that of national average price over the same period. Among the sample range, there were 27 months when the price fluctuation rate of early indica rice is more than 5% and 7 months when the price fluctuation rate is more than 10%, and the rate was 23.93% in March 2004. And there were 31months when the price fluctuation rate of late indica rice is over 5% and 8 months when the price fluctuation rate is over 10%, and the rate was 24.25% in January 2005. The price fluctuation rate of late indica is slightly higher than that of early indica rice.

2.3.3. The trend of rice prices is basically in line with that of the whole country, but the price is generally higher than the national average over the same period. It can be seen from Figure 1, price fluctuation in rice market in Guangdong presented a similar trend to that nationwide - fluctuating while rising. During 1998-2003, the prices of early indica rice and late indica rice in Guangdong were low, fluctuating around 1500 ¥/t, which was very slightly different from the national average; in 2004, the prices of early indica rice and late indica rice broke through the barrier of 2000 ¥/t, respectively being 2370.04 ¥/t and 2394.53 ¥/t, with a respective year-on-year increase of 50.01% and 39.95%; after 2004, the price of indica rice remained high, in particular, the price of late indica rice was much higher than that of early indica rice and the national average in the same period; to 2015, the prices of late indica rice and early indica rice in Guangdong were respectively 4964.5 ¥/t and 3997.78 ¥/t which were respectively 18.96% and 4.41% higher than the national average prices (4173.36 ¥/t and 3828.93 ¥/t).

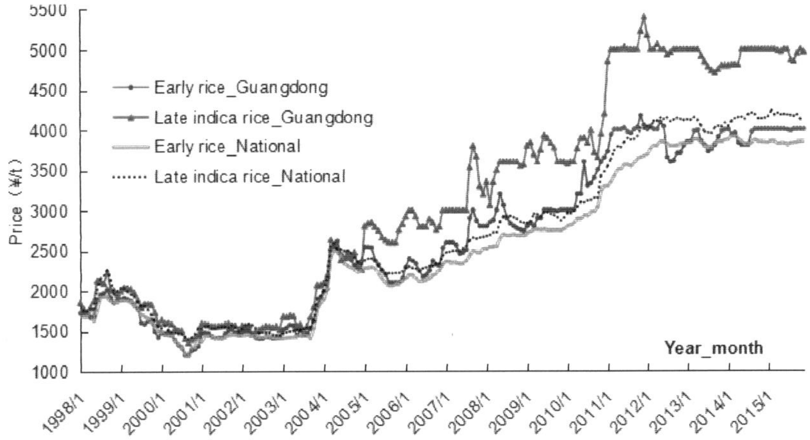

Figure 1. The Trend of indica rice price in Guangdong from January 1998 to December 2015.

3. Fluctuation decomposition of rice price in Guangdong

3.1. Time series decomposition principle

Changes of economic time series are mainly affected by four factors: trend, periodic factors (Cycle), seasonal factors (seasonal) and irregular components (irregular).Time series decomposition is designed to separate the seasons, trends, cycles and irregular components in the time series, and then analyze the statistical characteristics of the components. Seasonal fluctuations in the economic time series of monthly or quarterly that selected as time observation units show a significant trend and they tend to cover or confuse other objective changes in economic development. Therefore, when using the monthly time series to analyze the economic phenomenon, the original time series data need to be adjusted seasonally and eliminate the influence of the seasonal fluctuations (caused by the climate, social system and customs, etc.) to show the hidden trend cycle components. In this paper, the time series (Y_t) are decomposed into trend cycle components (TC), seasonal components (S) and irregular components (I) by the CensusX12 season adjustment method first. Based on this, the H-P filtering method is used to separate the trend cycle components to obtain the long-term trend factors and the short-term fluctuation cycle factors. The fluctuation regularity and characteristics of rice price are identified and analyzed. In this paper, the multiplication model is used to adjust the time series of rice price seasonally. In order to facilitate the analysis, the multiplication model is converted into the additive model, that is $\ln Y_t = \ln T_t + \ln C_t + \ln S_t + \ln I_t$.

3.2. Results analysis

Seasonal fluctuations in rice prices are weakening. From the isolated seasonal factor series (Figure 2), it can be seen that the prices of early Indica rice and late Indica rice present seasonal fluctuation characteristics, and the effect on the price of early indica rice is significantly greater than that of late indica rice. The prices of early indica rice and late indica rice in Guangdong show the characteristic of seasonal fluctuation, which has obviously greater influence on the price of early indica rice than that of late indica rice. As shown by seasonal factors, in each year, January ~ June is the high-sales season of indica rice, where the peak appears around March; July ~ December is the low-sales season of indica rice, where the prices of early indica rice and late indica rice drop to the lowest points respectively around August and October. After 2007, the effect of seasonal components on the price of Indica rice reduced significantly. Figure 3 shows the series of indica rice after the logarithm and the trend of the price sequence after the seasonal adjustment, which can be seen that seasonal factors has a little effect on the price of indica rice.

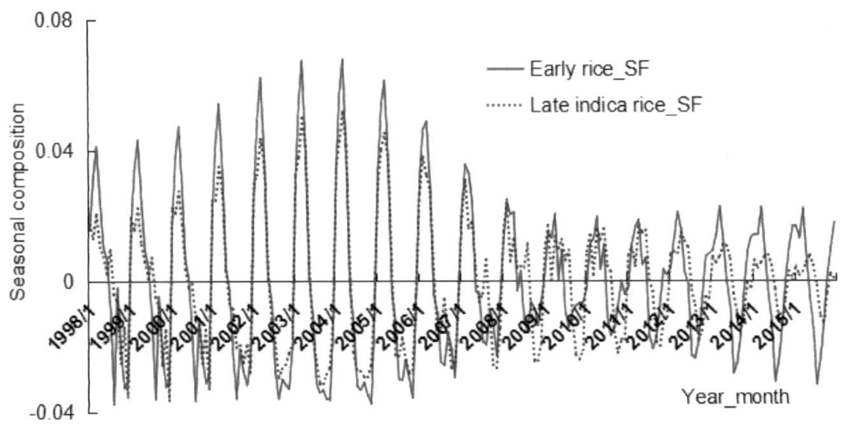

Figure 2. Seasonal factor series of early indica rice and late indica rice.

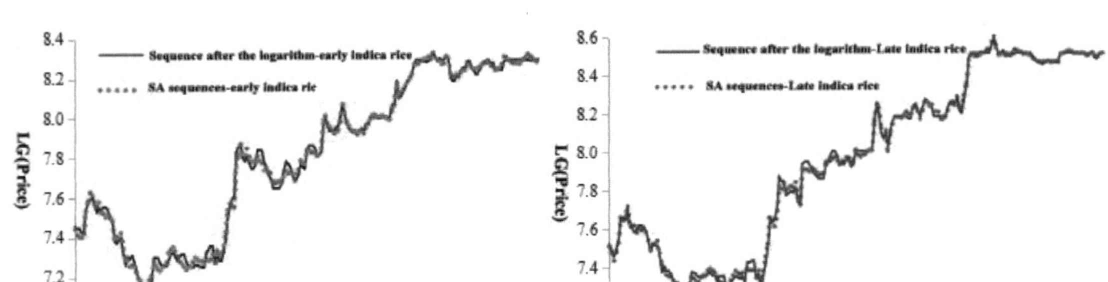

Figure 3. Price sequence trend of early indica rice and late indica rice.

3.2.1. *The long-term trend component of rice price took a dominant place.* Figure 4 shows the trend sequence (Trend) and the cycle component (Cycle) that has been decomposed by H-P filter. According to the trend sequence diagram (Trend), since 1998, the prices of early indica rice and late indica rice in Guangdong have shown the following long-term trend: the price showed a downward trend from 1998 to the first half of of 2001 and showed a long-term trend of continuous rise from the second half of 2001. This is closely related to the rigid demand of indica rice consumption of Guangdong residents and the change of population growth. In order to facilitate the analysis, this paper will convert the calculated monthly data of long-term trend factors to annual data, and calculate the ratio of long-term trend components to the price. The value of a trend component may be higher than or lower than the actual price of current period as the actual price includes the impact of random shocks and cycle component changes. When C>100%, it is indicated that random shocks and cycle components have restrained effect on price, and when C<100%, it is indicated that random shocks and cycle components have a driven effect on price. It can be found from Table 2 that the value of C was close to 100% in most years of the sample range, and the price of indica rice was mainly dominated by the trend component; The C value of the individual years such as 2000, 2003 and 2010 was obviously higher than 100%, and the random shocks and cycle components restrained the price of indica rice; The C value of 2011 was obviously lower than 100%, and random shocks and cycle components contributed to the rise of indica rice prices.

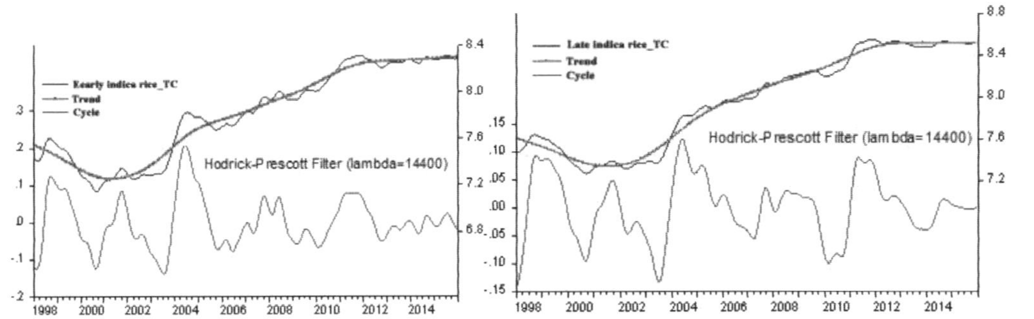

Figure 4. H-P Filter decomposition of pice trend cycle sequence of early indica rice and late indica rice.

Table 2. Long-term trend composition and proportion of early indica rice and late indica rice.

Year	Price of early indica rice (y_{1t})			Price of late indica rice (y_{2t})		
	Lg (y_{1t})	T_{1t}	$C_{1t} = T_{1t}$/ Lg (y_{1t})	Lg (y_{2t})	T_{2t}	$C_{2t} = T_{2t}$/ Lg (y_{2t})
1998	7.514	7.500	99.82%	7.583	7.570	99.82%
1999	7.442	7.390	99.30%	7.540	7.477	99.16%
2000	7.205	7.288	101.15%	7.315	7.381	100.89%
2001	7.289	7.257	99.57%	7.360	7.340	99.72%
2002	7.271	7.311	100.54%	7.338	7.381	100.58%
2003	7.362	7.447	101.16%	7.440	7.515	101.01%

2004	7.767	7.616	98.05%	7.779	7.706	99.06%
2005	7.728	7.726	99.98%	7.911	7.868	99.45%
2006	7.747	7.795	100.62%	7.965	7.978	100.17%
2007	7.892	7.868	99.70%	8.070	8.068	99.98%
2008	7.965	7.943	99.72%	8.159	8.153	99.92%
2009	7.984	8.024	100.50%	8.227	8.234	100.08%
2010	8.104	8.125	100.27%	8.244	8.329	101.04%
2011	8.290	8.217	99.12%	8.525	8.434	98.93%
2012	8.255	8.260	100.06%	8.519	8.496	99.73%
2013	8.261	8.272	100.13%	8.488	8.511	100.27%
2014	8.275	8.280	100.07%	8.503	8.511	100.09%
2015	8.293	8.291	99.97%	8.510	8.510	100.00%

3.2.2. The rice price presents a trend of obvious periodic fluctuation. During the 18 years since 1998, the prices of early indica rice and late indica rice in Guangdong have experienced two most obvious price increases. The first time: From July 2003 to May 2004, Guangdong early indica rice prices rose from 1456.67¥/t to 2620¥/t with an increase of 79. 86%; From July 2003 to March 2004, Guangdong late indica rice prices rose from 1460¥/t to 2635¥/t, or with an increase of 80.48%. The second time: From March 2010 to November 2011, early indica rice prices rose from 3,000¥/t to 4175¥/t with an increase of 39.17%; From March 2010 to December 2011, late indica rice prices rose from 3600¥/t to 5400¥/t, with an increase of 50%. According to the periodic determination of the principle of "trough-trough", the price fluctuation of early indica rice and late indica rice in Guangdong during the sample range can be divided into five complete cycles (Table 3 and Table 4). The average cycle length is 35 months and 38 months.

Table 3. The price fluctuation period of indica rice from January 1998 to December 2015.

Cycle	Time Span	Duration (month)	Rise (month)	Peak	Decrease (month)	Trough
First	Feb.1998- Aug.2000	31	8	Sept. 1998	23	Aug. 2000
Second	Sept. 2000- Jul.2003	35	14	Oct. 2001	21	Jul. 2003
Third	Aug.2003- Jun.2006	35	11	Jun. 2004	24	Jun. 2006
Fourth	Jul. 2006- Feb.2010	44	16	Oct. 2007	28	Feb. 2010
Fifth	Mar.2010-Sept.2012	31	19	Sept. 2011	12	Sept. 2012

Table 4. Price fluctuation cycle of late indica rice from January 1998 to December 2015.

Cycle	Time Span	Duration (month)	Rise (month)	Peak	Decrease (month)	Trough
First	Jan. 1998- Sept. 2000	33	10	Oct. 1998	23	Sept. 2000
Second	Oct. 2000- Jul. 2003	34	12	Sept. 2001	22	Jul. 2003
Third	Aug. 2003- Mar. 2007	44	10	May 2004	34	Mar. 2007
Fourth	Apr. 2007- Feb. 2010	35	6	Sept. 2007	29	Feb. 2010
Fifth	Mar. 2010- Jan. 2014	47	14	Apr. 2011	33	Jan. 2014

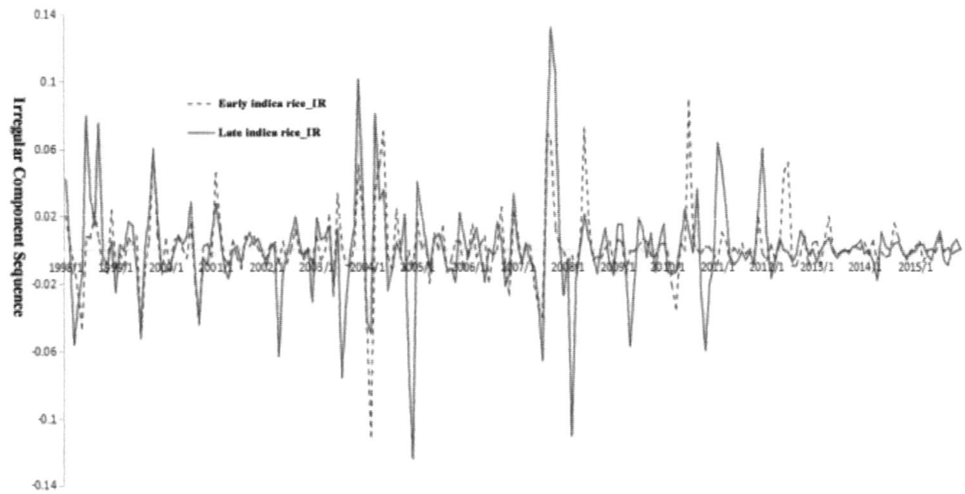

Figure 5. Irregular component sequence of the price of early indica rice and late indica rice.

3.2.3. Random factors have serious impacts on the price of indica rice in some years. From the irregular component sequence of the price of early Indica rice and late Indica rice (Figure 5), it can be seen that the value of irregular component is around 0 with frequent random fluctuations. And the fluctuation degree was much higher in some years. This shows that the random shocks have both positive (driven) and negative (restrained) effects on the price of Indica rice, and random factors have a severe impact on the price of indica rice. From different time periods, random factors have obviously driven effects on the price of late indica rice in June 1998, November 2003, March 2004, September~October 2007, and the proportion of random components accounted for 1.04%, 1.33%, 1.03%, 1.6% and 1.29% respectively. Random factors have obviously restrained effects on the price of late Indica rice in July 2003, December 2004 and February 2008, and the proportion of random components accounted for -1.04%, -1.61% and -1.38%. As for the price of early indica rice, there were fewer months when the irregular components presented obvious fluctuation with slight fluctuation degree. There were two years when the the proportion of random components accounting for more than 1% of the original sequence value and the proportion was -1.46% in February 2004 and 1.08% in June 2010. Since the second half of 2012, with the country's emphasis on food security issues, the impact of random components on the price of indica rice has been significantly weakened, with a fluctuating value of around 0.

4. Factors analysis on abnormal fluctuations of rice price

4.1. Supply shock

Since October 2003, Guangdong rice prices rose rapidly on the basis of low prices over a long period of time and the price of early indica and late indica rose to 1815 ¥/t and 2070 ¥/t respectively in November 2003, with a growth rate of 11.61% and 15% respectively compared to that of October in the same year. Generally, successive years of rice production reduction and tightening relationship of supply and demand contributed to the rise of rice price .Under the influence of a variety of factors like the decline of planting area and drought etc., the rice production in China has been reduced for 6 consecutive years. The production reduced from 200.735 million tons (peak) in 1997 to 160.656 million tons in 2003, with a reduction of 24.95%. The inventory fell to the lowest level since the reform and opening up. The rice production in Guangdong decreased by 25.95% from 16.8853 million tons in 1998 to 12.5038 million tons in 2003. The rice price in Guangdong remained high in 2007, with a slight rise, which was adjusted in August ~ September. The main reason was: early indica rice in main producing provinces of indica rice, namely Jiangxi, Hunan and Sichuan, suffered from drought and production reduction, which resulted in relatively tight supply of indica rice, and the domestic rice market continuously presented the characteristic of "indica rice market being more robust than japonica rice market". In 2011, a severe drought broke out in the middle and lower reaches of Yangtze River, followed by flood disaster, which caused much more concern about production reduction of summer grain crops in the market and increased the expectation for grain price rise. During November ~ December that year, the prices of early indica rice and late indica rice in Guangdong rose to the historical high points. Therefore, the price fluctuation of rice in Guangdong is affected by rice supply shocks in the recent decade.

4.2. Cost driven

In 2004, Guangdong rice prices rose first and then fall and the price of early indica rice and late indica rice in March reached the highest level in the whole year, with an increase of 23.93% and 19.77% respectively compared with that in February. Generally, the rise in rice prices is still recovering. The root causes of the rise of rice prices in the 2004 were cost pull, rice production and the increasing contradiction of the supply structure among regions and varieties. From the microscopic supply factors, the rising production costs of seed, pesticide, fertilizer, machinery operation and labor have been supporting the rice price since 2004. In 2004, the production cost of early indica rice per mu in Guangdong and the country was 376.64 yuan and 383.82 yuan respectively, which was 26.87% and 32.18% higher than that in 2003. The production cost of late indica rice in Guangdong and the country per mu was 377.91 yuan and 382.13 yuan respectively, with a year-on-year rise of 25.15% and 31.39% (Table 5).

Table 5. Changes of production costs of indica rice in Guangdong and nationwide in 2004.

Year	Early indica rice		Late indica rise	
	Guangdong	Nationwide	Guangdong	Nationwide
2003	296.88	290.37	301.96	290.84
2004	376.64	383.82	377.91	382.13
Rise	26.87%	32.18%	25.15%	31.39%

Note: Data come from *Collection of National Agricultural product cost benefit of Y2004-2005*

4.3. Policy intervention

In April 1998, our country launched the "three policies" of grain reform(that is, purchase surplus grain from farmers without limit at protective prices, sales at profitable price and closed operation of purchasing funds), The prices of Guangdong early indica rice and late indica rice rose to 1884 ¥/t and 2133 ¥/t in June of the same year, with an rise of 11.6% and 13.23% compared with that in May In 2004, after the liberalization of the grain purchase market and the purchase price in China, the minimum purchase price policy for paddy rice was put forward from 2004 to protect farmers' interests and grain cultivation enthusiasm. The minimum purchase price has been raised for 6 consecutive years since 2006. The purchase price increased to 1.39 yuan per catty with an increase of 92%. The policy on the market driven and support efforts continue to increase.

5. Conclusion and policy implications

Based on the empirical analysis on the price fluctuation of early indica rice and late indica rice in Guangdong Province, it can be concluded that: (1) The price of rice in Guangdong presents a trend of overall fluctuating rise and the growth rate is generally higher than the national average over the same period, which has a effect on the stability of rice market. (2) Seasonal factors have limited impact on the rice prices and seasonal fluctuations have been significantly weakened in recent years. This shows that the consumption of rice rations of Guangdong residents tends to be stable and the "effect of slack-peak seasons" for consumption is gradually reduced while the strong support of the lowest purchase price policy offsets the impacts of seasonal supply on low rice prices. (3) It can be indicated that the fluctuation of rice price is predictable to some extent from the cyclical fluctuation characteristics and the long-term rising trend of rice price. (4) Random shocks have both positive (driven) and negative (restrained) effects on the price of indica rice and random factors have a severe impact on the price of indica rice in some years. (5) The rice prices are mainly affected by market supply and demand, production costs and food policy intervention (The main factors affecting the fluctuation of rice price include market supply and demand, production costs and food policy intervention). The sharp abnormal fluctuation of rice prices will affect the development of rice industry, residents' life and national economy. Therefore, the relevant departments should attach great importance to the factors that cause abnormal fluctuations in rice prices and take effective policy measures in practice to achieve long-term stability of the supply of grain market and reasonable fluctuations in grain prices in Guangdong through protecting effective land area, improving grain yield, constructing grain source base, building grain supply system and strengthening market monitoring and early warning etc.

Acknowledgments

We thank the supported by grants from the Guangdong Science and Technology Program: Reform and innovation project of Guangdong provincial scientific research institutions-Monitoring & early warning platform for modern agriculture industry economy and circulation (2014B070706015).

References

[1] Luo W C and Liu R, 2010, Analysis on grain price fluctuation in China: based on ARCH Model. *Chin. Rural Econ.*, **4**: 30-7.
[2] Miao S S, 2014, Price fluctuation of rice market and periodic characteristics. *J. Huazhong. Agri. Uni. (S. S. E.)*, **2**:68-73.
[3] Huang J K,Yang J and Qiu H G, 2009, The fluctuation of grain price in this round: main reasons and future trend. *Management World*, **1**: 72-8.
[4] Zhang L X and Zhang X C, 2011, Impact of external shocks on price fluctuation of agricultural products in China--based on the view of agricultural industry chain. *Management World*, **1**: 71-81.
[5] Su Z F, Wang X and Chen C N, 2012, Study on the influence factors of low-frequency fluctuation of grain price in China: based on panel VAR model. *J. Agrotech. Econ.*, **10** :22-30.
[6] Wang S J and Li S Y, 2013, Causes of grain price fluctuation and its control countermeasures.

Econo. Review, **2**: 60-4.
[7] Deng H L and Huang T Y, 2013, An empirical analysis of grain price fluctuation in China. *Statist. Decis.*,**24**: 91-5.
[8] Fang W, Lin W J and Wan Z, 2013, Economic evaluation on whether the fluctuation of rice market price is abnormal--adjustment and calculation based on X-12 and P method. *J. China Agri. Uni.*, **18**: 211-9.
[9] Shi Z Z, 2015, Impact of random shocks on grain price fluctuation in China. *J. Huazhong Agri. Uni. (S. S. E.),* **2**:33-40.
[10] Huang C Y, 2015, The influence of grain price on urban residents' life: function mechanism and empirical test. *J. Agrotech. Econ.*, **9**: 77-84.
[11] Liu D B and Zhou H H, 2017, Research on rice price fluctuation based on multiple linear regression and VAR model. *Price Monthly*, **9**: 1-6.
[12] Yang W J and Liu Q, 2017, Study on the fluctuation characteristics of international grain price and its random impact effect: taking rice price as an example. A*gri. Modern. Res.*, **38**: 589-96.
[13] Li X G, 2017, Analysis of the characteristics of international grain price fluctuation based on ARCH family model.*Chin. Forei. Entrepren.*,**16**: 1-3.
[14] Wu Z M, 2017, An empirical study on the transmission effect of grain price under the "market support policy" . *Nanjing Uni. Finances Econ.*
[15] Zhou Z and Li G S, 2016, Empirical analysis of the fluctuation characteristics and influencing factors of rice prices in China. *Theo. Pract. Price*, **3**: 93-6.

Decomposition and measurement for conduction and interaction of pig price fluctuation in Guangdong

W Fang[1,2,3], J F Liang[1,2], L Kang[1,2] and M T Lin[1,2]

[1] Institute of Agricultural Economics and Rural Development GAAS, Guangzhou 510640, Guangdong Province, China

[2] Key Laboratory of Urban Agriculture in South China MOA, Guangzhou 510640, Guangdong Province, China

E-mail: fangwei9103@163.com

Abstract: The price conduction and interaction in the circulation of agricultural products affect the development of the whole swine industry. In recent years, the imbalance between the conduction and interaction of the price of pork has caused the contradiction between residents' consumption and pig breeding enterprises. In this paper, Granger causality test, pulse function test and variance decomposition test were conducted on price fluctuations in the processes of production, wholesale and retail based on VAR model, it was found that the price fluctuation in the production, wholesale and retail of pig market in Guangdong obviously presents the characteristic of "demand driving and the tendency of "cost push" is not statistically significant.

1. Introduction

At present, pig breading in Guangdong has a large scale and a high degree of intensiveness, with breeding by large enterprises gradually becoming the mainstream in the market. The proportion of large-scale breeding of pigs is up to 86%, being 20% higher than the national average level, and ranking among the top in the country [1]. According to the market structure theory, as pig breeding in Guangdong has presented the large-scale intensive development trend, and large breeding enterprises have become the main market players [2], the cost-driven mechanism of "production-circulation (wholesale)-retail" will inevitably formed. In other words, the rising of breeding cost will inevitably result in the rising of price during circulation and retailing [3]. In the respect of pork circulation, the proportion of pork circulating from the wholesale market to the retail market accounts for over 90%, while that of the pork directly sold from the pig breeding enterprises to the pork retail market accounts

IOP Conf. Series: Earth and Environmental Science **185** (2018) 012038 doi:10.1088/1755-1315/185/1/012038

for less than 10%. In addition, only a small amount of pork from Guangdong Yihao Pig Co., Ltd. and WENS Group, etc. is circulating through such direct selling modes as independent production and marketing, and farm-supermarket cooperation; moreover, the degree of self-sufficiency of pork in Guangdong is only 60%, with the remaining part being allocated and transported from other places. [4] The above two factors directly make pig breeding enterprises in Guangdong be confronted with a highly competitive external market. Based on the market structure theory, the highly competitive external market will greatly weaken the breeding enterprises' control ability on the retail market price, thus forming the demand-driven mechanism of "retail-circulation (wholesale)-production". That is, as the pork price fluctuates ((drops in particular), enterprises can only accept the price transferred to the production link due to their weak ability of controlling the market price. Moreover, pig breeding enterprises have strong monopoly power on the breeding market, which will inevitably lead to depression of price of breeding materials by breeding enterprises and finally subject the price of breeding materials to a downward trend [5-12].

2. Framework for analysis of conduction and interaction of pig price fluctuation in Guangdong

The large-scale production and large market of pigs in Guangdong directly result in two completely opposite mechanisms of action: "production-circulation (wholesale)-retail" cost-driven mechanism and "retail-circulation (wholesale)-production" demand-driven mechanism. The superiority between the two will directly determine the price conduction and interaction of "production-wholesale-circulation". To study this problem in depth, VAR model and its related theories are used in this paper to conduct empirical analysis on price conduction and interaction among the production, wholesale and retail links of the pig market in Guangdong, which is based on the monthly evidence of pork in Guangdong from 2012 to 2015, trying to reveal the economic mechanism of price conduction and interaction among the production, wholesale and retail links. The VAR model is composed of benefit change (cost) during production, benefit change during wholesale (wp) and benefit change during retail (rp), with the specific form as below:

$$Y_t=A_0+A_1Y_{t-1}+A_2Y_{t-2}+\ldots+A_kY_{t-k}+u_t$$

Among them, $Y_t=(cost_t, wp_t, rp_t)^T$, A_0 is a constant item, K is the delay order number, the specific data need to be determined according to the Lag Order selection criteria table. The test methods for determining the delay order number are: Likelihood ratio test, AIC information criterion and SC information criterion.

3. Variable selection and data analysis

As the feed cost accounts for over 70% of the pig breeding cost, and corn cost accounts for over 60%-70% of the feed cost [5], in light of data availability and representativeness, change in the transaction price of corn in Guangdong ports is considered as the benefit change during production; change in the marketing price of live pigs is considered as the benefit change during wholesale; change in the price of the skinless bone-in pork is considered as the benefit change during retail. The time span for the data is from November 2012 to June 2015. Unit root test is required in this paper in order to test the stability of each variable data. The most common ADF test is used herein. Differential treatment will be made if this variable data is not stable, then unit root test will be performed until it becomes stable (see Table 1), and then descriptive statistical analysis will be performed on the stable data obtained (see Table 2). Finally, quantitative analysis will be made.

Table 1. ADF unit root test for price change data during production, wholesale and retail.

Variable	COST (Benefit change during production)	RP (Benefit change during retail)	WP (Benefit change during wholesale)
ADF test t value (concomitant probability p)	-7.2866 (0.0000)***	-3.9928 (0.0045)***	-3.5830 (0.0123)***

Note: *** indicates 1% significance level; ** indicates 5% significance level; * indicates 10% significance level.

Table 2. Statistic analysis on price change during production, wholesale and retail.

	COST (Benefit change during production)	RP (Benefit change during retail)	WP (Benefit change during wholesale)
Mean	0.0094	0.0109	0.0079
Median	0.0031	0.0067	0.0048
Maximum	0.0871	0.1055	0.1509
Minimum	-0.0546	-0.0721	-0.1380
Std.Dev.	0.0278	0.0410	0.0584
Skewness	0.2403	0.5211	0.2800
Kurtosis	3.8514	3.1826	3.8901
Observations	31	31	31

4. Granger causality test based on VAR model

Results of the Granger causality test show that, under the 1% significance level, price change during retail process is the change in the production cost; under the 10% significance level, price change during retail is the Granger causality of the price change during wholesale; under the 1% significance level, price change during wholesale is the price change during production. This indicates that the interaction strength of the demand-driven mechanism is greater than that of the cost-driven mechanism, in other words, price fluctuation during retail will affect the price during wholesale, and price fluctuation during wholesale will further affect pricing during production.

Table 3. Results of granger causality test based on VAR model.

Parameter	Cost	WP	RP
Cost	—	2.9506 (0.0625)*	22.5944 (0.0000)***

WP	0.6675 (0.4139)	—	2.7192 (0.0898)*
RP	1.4872 (0.2226)	0.8665 (0.3519)	—

Note: *** indicates 1% significance level; ** indicates 5% significance level; * indicates 10% significance level.

5. Test of impulse response function based on VAR model

Under the impulse of one standard deviation in the price fluctuation during retail, a positive response of 0.031 is generated on itself within the first month, and the accumulative response amplitude reaches 0.048 within 6 months; the accumulative response to the price fluctuation during wholesale is a positive response of 0.035 in the first month, and the accumulative response is a positive response of 0.053 within 6 months; the accumulative response to the price fluctuation during production is a positive response of 0.0019 in the first month, but it turns to be a negative response of 0.0017 in the second month, and the accumulative response is a negative response of 0.0021 within 6 months. From the "demand driving" perspective, the rise of retail price leads to significant rise in the wholesale price and thus raises the breeding cost obviously at the initial stage. This fully reflects the very obvious "demand-driven" mechanism in the pig market in Guangdong, which is completely consistent with results of the Granger causality test.

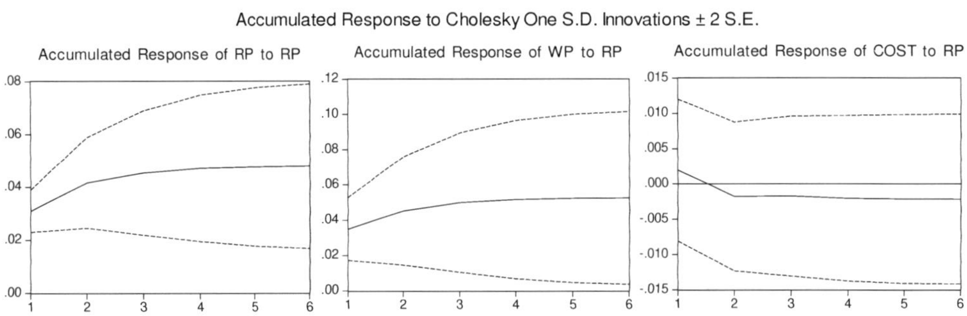

Figure 1. Impact of price change during retail on itself and on price change during production and wholesale.

Under the impulse of one standard deviation in the price fluctuation during wholesale, a positive response of 0.042 is generated on itself within the first month, and the accumulative response amplitude reaches 0.071 within 6 months; the accumulative response to the price fluctuation during retail is a positive response of 0.026 in the second month, and the accumulative response is a positive response of 0.033 within 6 months; the accumulative response to the price fluctuation during production is a positive response of 0.0078 in the first month, its intensity drops gradually in the second month and it turns to a positive response of 0.0054 in the third month, and the accumulative response is a positive response of 0.0061 within 6 months. From the "intermediate diffusion" perspective, on the one hand, rise in the wholesale price will cause the rise in the retail price, which tends to be stable very rapidly; on the other hand, the breeding cost will also take on a trend of significant rise.

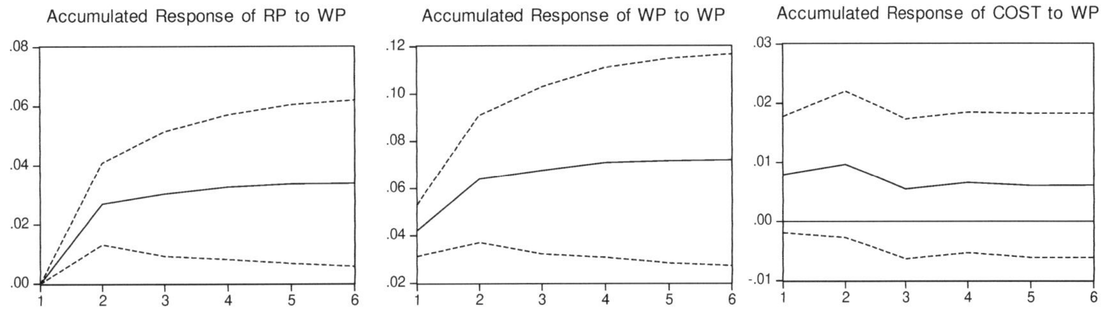

Figure 2. Impact of price change during wholesale on itself and on change during retail and production.

Under the impulse of one standard deviation in the price fluctuation during production, a positive response of 0.026 is generated on itself within the first month; the accumulative response amplitude drops significantly in the second month, being 0.017; the accumulative response amplitude within 6 months reaches 0.020. The accumulative response to the price fluctuation during wholesale is a negative response of 0.014 in the second month, and the accumulative response is a negative response of 0.017 within 6 months. The accumulative response to the benefit fluctuation during retail is a negative response of 0.0074 within the second month, and the accumulative response is a positive response of 0.012 within 6 months. From the "cost driving" perspective, with the rise of breeding cost, there is no positive growth in both the prices during wholesale and retail; instead, there are negative changes. All of these indicate that the "cost-driven" mechanism in the pig market in Guangdong is not so obvious.

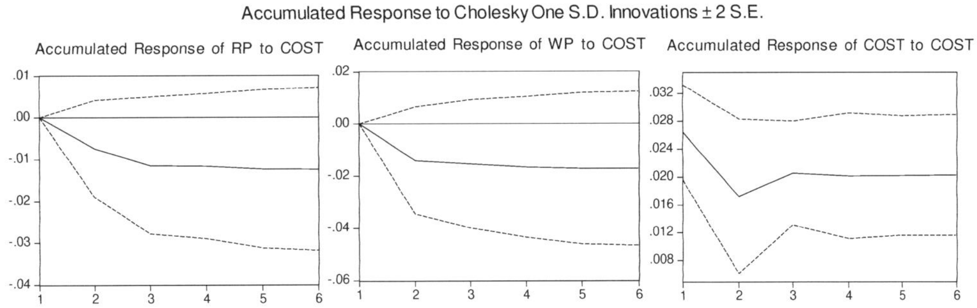

Figure 3. Impact of price change during production on itself and on benefit change during wholesale and retail.

6. Conclusions and Suggestions

Based on the above three analyses, it can be concluded that price change fluctuation during retail is relatively independent, subject to certain (but not obvious) influence from the wholesale and almost no influence from the production; price change during wholesale is greatly affected by the retail (approximately 40%), subject to almost no influence from the production; price change during production is significantly affected by the wholesale (approximately 40%), subject to little influence from the retail. This fully reflects that there is very obvious "demand-driven" mechanism, rather than "cost-driven" mechanism, in the pig market in Guangdong, indicating that price fluctuation during production, wholesale and retail of the pig market in Guangdong is remarkably characterized by being "demand-driven" instead of "cost-driven".

Acknowledgments
We thank the supported by grants from the Guangdong Science and Technology Program: Reform and innovation project of Guangdong provincial scientific research institutions—Monitoring & early warning platform for modern agriculture industry economy and circulation (2014B070706015).

References

[1] Xia S and Zhang C. 2008 The cost composition and benefit distribution of agricultural products circulation in China--a case study on vegetable circulation in Dalian. *Issues Agri. Econ.*, **2**:39-48.
[2] Li S J. 2010 An empirical analysis on benefit distribution mechanism of agricultural products during circulation. *J. Agrc. Tech*, **11**:108-14.
[3] Liang J F and Ma W 2011 Development situation and suggestion for Guangdong pig industry in the first half of . *Guangdong Agric. Sci*, **19:**21-24.
[4] Zhang M and Xie J Z. 2012. Cost-driven, external input and the price rise of Chinese agricultural products. *J. Agrc. Tech*, **5:** 13-9.
[5] Ji C J, Hu Y J and Tang B J. 2018..Research on carbon market price mechanism and influencing factors: a literature review. *Nat. Haz*, **92 (2)**:761-782.
[6] Wang G Y, Wang X H and Li C X. 2018. Study on the stabilizing effect of capitalization on the price fluctuation of pig - based on the empirical analysis of Chinese panel data. *Chin. rural Econ*, **06**: 55-66.
[7] Wang Y, Zhang Y and Ai Y H. 2018. Study on the influencing factors of hog price fluctuation in China based on equilibrium price theory . *Pri. Mon*, **04**: 62-65.
[8] Fu L L, Yang N and Zhang Z W. 2018. Asymmetric analysis of Hog Price Fluctuation: Based on Markov zone transfer VAR model . *Chin. J. Anim. Hus*, **54 (04)**: 121-126.
[9] Yang H C, Shen C, Wu J Z and Kong F T.2018. Analysis and Prospect of price transmission mechanism of pork market in Beijing City. *Agric Outlook*, **14 (03)**: 22-29.
[10] Huang B. 2018. Based on the monthly ratio of annual average moving average pig to grain ratio combination index to determine the turning point of pig cycle . *Fina. Theo. Pra*, **03**: 89-93.
[11] Xue Y H, Li Z and Pan F H. 2018. Study on the periodic fluctuation of live pig prices in Heilongjiang Province Based on MS-AR model. *Heilongjiang Anim. Hus. Vet. Med.* **02**: 1-4.
[12] Wang G Y, Wang X H and Li H S. 2018. Study on the spatial spillover effect of Chinese pig price -- spatial econometric analysis based on synchronization coefficient matrix. *Agric. Mod*, **39 (01)**: 105-112.

Study on the ecological construction of livestock husbandry at county level

S H Chai[1]

Longxi Livestock Husbandry and Veterinary Technical Service Center, Gansu Province, 748100, China

E-mail: 94792702@qq.com

Abstract: In order to find a balance between livestock husbandry and environmental protection, a feasible, mutually beneficial, inclusive and win-win development system should be established. Taking Longxi County in Gansu Province for instance, the paper demonstrates how to establish livestock husbandry ecosystem in all directions by analyzing and studying the current situation of livestock husbandry development, environmental pollution and the relationship between them. The results show as follows: 1) The dialectical relationship between the total output of pollutants produced by livestock and the carrying capacity of the environment is the basis for formulating relevant macroeconomic policies in county administrative regions. The balance and unification of "water-land-grass-livestock" is the prerequisite for sustainable development of livestock husbandry; 2) Livestock husbandry should adjust the development direction within the safety range of critical points of environmental protection; 3) Livestock husbandry should adjust livestock species, breeding quantity, supporting anti-pollution hardware facilities according to local geographical environment and external carrying capacity, and increase input intensity in soft environment, so as to achieve the aim of reducing emissions. The results of this study can provide a powerful theoretical reference and administrative basis for the coordination of livestock husbandry and environmental protection in county administrative regions.

1. Introduction

With the rapid economic development, China is facing two major pressures - population growth and environmental damage. Therefore, the concept of sustainable development has drawn more and more attention of the whole society [1-2]. As the direction of future agricultural development, ecological agriculture has become an inevitable choice for the sustainable development of agriculture in China. Recently, Livestock husbandry has been developing rapidly, which promotes rural area economic development and increases farmers' income. However, large amount of livestock and poultry manure, which cannot be applied to farmland in time, accompanied by the poor operation of the facilities for the disposal and utilization of manure, has become an important source of non-point source pollution [3]. In addition, livestock husbandry has a tremendous negative impact on grassland vegetation, disease transmission, greenhouse gas emissions, land occupation and water resources. At present, a lot of studies have been done in this field in China and the management of scope of sub-item index is strictly demonstrated from different angles of grassland carrying, fecal disposal, gas disposal, gas emissions and so on. This article analyzes and evaluates how the industrial layout and mode of production of livestock husbandry conform to the environmental protection work based on macroeconomic regulation and control. Then establishes the concept of "three determinations" (means water, land and grass determines the development of livestock husbandry), which provides a theoretical reference to policy making and livestock husbandry development.

2. Longxi livestock husbandry development status and sewage discharge

2.1. Longxi livestock husbandry development status

In 2017, the total number of livestock and poultry raised in the county was 4,629,100. Among them, cattle, sheep, pigs and poultry were 71,500; 989,100; 444,100 and 3,124,400 respectively; the total number of livestock and poultry for sale was 2,241,600. Among them, cattle, sheep, pigs and poultry were 21,600; 385,900; 235,700 and 1,598,400 respectively; the total production of meat, eggs and milk was 55,500 tons. Among them, meat, eggs and milk were 39200tons, 10200 tons and 6100 tons respectively; the gross output value of grass and livestock husbandry reached 2.507 billion yuan, and the per capita net income was 2048 yuan [4]. From the above statistics got at Longxi County level, we know that livestock husbandry is ingrained in the county. The county is known as "a hometown of Chinese bacon", where the processing of livestock products has a long history. There are 7 meat processing enterprises with over 1000 tons of meat products, and 18000 tons of meat products are processed annually. There are 209 livestock husbandry professional communities and 16631 large scale farmers in the county, of which 160 large sheep farmers with more than 100 sheep, 23 large cattle farmers with more than 50 cattle and 36 large chicken farmers with more than 1000 chickens. There are 51 leading companies including Gansu Longyuan Zhongtian Sheep Industry Co., Ltd. etc [5]. In the past two years, the county has vigorously implemented "5532" and "5522" developing models (that is to say, farmers loan 50,000 yuan, plant corn or grass of 5 acres, then raise 3 cattle, they will have annual income of 20,000 yuan or more; or farmers loan 50,000 yuan, plant corn or grass of 5 acres, then raise 20 sheep, they will have annual income of 20,000 yuan or more) among farmers who are identified as precision poverty alleviation filling card households (the local policy aims to alleviate the poverty). It is a good way to increase farmers' income and becomes indispensable for local economic development. However, it will inevitably become a major target for the eradication of pollution.

2.2. Livestock husbandry pollution investigation

2.2.1. Livestock waste discharge pollution. The rapid development of aquaculture plays an important role in the economic development of rural areas and the increase of farmers' income. However, with the intensification of livestock and poultry farming, the shift of agricultural labor force and the application of fertilizers which has changed from organic fertilizer to chemical fertilizer, livestock and poultry feces changed from treasure to waste, which brought about the problem of environmental pollution [6]. Pigs, cattle, sheep and chickens are the four big livestock and poultry breeding according to the data of the county by the end of 2017. According to the yield and discharge coefficient of individual livestock and poultry, the amount of excrement, chemical oxygen demand(COD) production and total nitrogen are showed as follows[7] (see Table 1, because of the geographical relationship of the object of the argument, this paper only extracts the data of "Northwest region").

Table 1. Pollution production coefficient of livestock and poultry.

	Amount of excrement/$Kg{\cdot}d^{-1}$	COD production/$g{\cdot}d^{-1}$	Total nitrogen production (TN)/$g{\cdot}d^{-1}$
Pig	3.54	334.55	31.73
Cattle	20.42	2235.21	104.10
Sheep	0.87	0.46	2.15
Chicken	0.18	0.18	1.85

In Table 1, the livestock and poultry breeding period is calculated according to the data published by the State Environmental Protection Administration when necessary [8]. That is to say, the breeding period of the pig is 199 days, the cow is 365 days, the sheep is 365 days, and the chicken is 210 days. The formula for calculating the total fecal amount, COD, total nitrogen (phosphorus) is as follows [7].

$$Q = \sum_{i=1}^{4} N_i \times T_i \times P_i \tag{1}$$

In the formula, Q is feces, COD, total nitrogen (phosphorus) production components (ten thousand tons), N_i is the breeding quantity of i species of livestock and poultry. T_i is the feeding period (d); P_i is the coefficient of pollutant production (kg/d or g/d), and i is the i-th kind of livestock. Taking pigs as an example, the annual total amount of sewage discharged from this type of livestock in this county is (due to the almost uninterrupted livestock and poultry farming in large-scale farms, the calculation result is converted into the annual output of sewage):

$$Q{=}44.41\times199\times(3.54+334.55\times1000^{-1}+31.73\times1000^{-1})\times365/199/1000 = 633200\ \text{t} \tag{2}$$

In the same way, the total amount of sewage discharged from cattle, sheep and chickens was 553,900 tons, 315,000 tons and 205,700 tons respectively. The annual discharge of livestock in the county was 1,707,800 tons, of which the total nitrogen and phosphorus emissions were 63,800 tons, the excrement and urine is about 1,644,000 tons, the latter accounting for 96.26% of the total amount of discharged pollutants, which is the largest source of pollution.

2.2.2. Evaluation of the external environment capacity. The external environment includes land, forage grass and water resources. As a consumptive resource, they are the basis for the development and growth of livestock husbandry. From the perspective of environmental protection, the carrying capacity of various items must be kept within the bearing range.

i) Land carrying capacity

In accordance with the "Survey on Pollution and Countermeasures of Large Scale Livestock Farming in China", it is suggested that the livestock manure per hectare of cultivated land can carry about 30 tons [9]. The county has an area of 134,040 mu (=0.0667 hectares) of cultivated land [10], and it can be seen that the peak value of the livestock manure in the county is about 4,021,200 tons, and the total amount of the actual discharge pollution is about 40.88 % of the bearing capacity. In the case of this data, it seems to be within a reasonable range. However, according to the latest study "Technical Guide for Estimating Carrying Capacity of Livestock Contaminated Land" [11] released by the Office of Ministry of Agriculture, the total nitrogen and phosphorus nutrient demand of all kinds of plants in this county is much smaller than the total amount of nitrogen and phosphorus emission which exceeds the carrying capacity by 52.63%. In Longxi County, there are 200,0000 mu (=0.0667 hectares) of cultivated land. Among them, about 150,000 mu is planted with wheat, 650,000 mu of corn, 50,000 mu of alfalfa, 450,000 mu of potatoes and 100,000 mu of vegetables, 350,000 mu of all kinds of Chinese herbal medicines and 250,000 mu of fruit trees. The yield per mu of above crops are wheat 400Kg/ mu, corn 750Kg/ mu, alfalfa 1500Kg/ mu, potato 3200Kg/ mu, vegetable 1500Kg/ mu, Chinese herbal medicine 800Kg/ mu, forest and fruit 2500Kg/ mu. According to the above index, 100Kg yields in different plants need recommended value amount of nitrogen and phosphorus as shown in Table 2.

Table 2. 100kg of yield formation in different plants need recommended value amount of nitrogen and phosphorus.

	Nitrogen N/Kg	Phosphorus P/Kg
Wheat	3.0	1.0
Corn	2.3	0.3
Alfalfa	0.2	0.2
Potato	0.5	0.088
Vegetables	0.36	0.088
Chinese herbal medicine	2.32	0.323
Fruit trees	0.3	0.08

Note: Because the vegetables are widely planted, they are calculated by the average value of 8 kinds of vegetables, the Chinese herbal medicine is calculated by astragalus, and the fruit tree is calculated by the apple trees which are widely planted.

$$Q = \sum_{i=1}^{7} P_i \div 100 \times A_i \times a_i \tag{3}$$

In the formula, Q is the demand of regional plant nitrogen and phosphorus (10,000 Kg), P_i is the yield per mu of i crop (Kg/mu), A_i is the total area (mu) of planting i crop, a_i is the recommended value of nitrogen and phosphorus for i crop to form 100Kg yield (kg). It can be obtained that the total nitrogen and phosphorus nutrient demand of all kinds of plants in Longxi County is about 41,849.6 tons, of which the total nitrogen nutrient demand is 36,833.5 tons, and the total phosphorus nutrient demand is 5,016.1 tons. It can be seen that the total nitrogen and phosphorus nutrient demand of all kinds of plants in this county is 41,800 tons, which is much smaller than the total amount of 63,800 tons of nitrogen and phosphorus emission, and the nitrogen and phosphorus emission exceeds the carrying capacity by 52.63%.

According to the results of (1) and (2), combined with the concept of pig equivalent, 100 pigs are equivalent to 15 cows, 30 beef cattle, 250 sheep and 2500 poultry [11]. The number of livestock raised in the county is as follows: the pig equivalent is 444,100 pigs equivalent, cattle 238,300 pig equivalent, sheep 395,600 pig equivalent, poultry 125,000 pig equivalent, and the total discharge of pig, cattle, sheep and poultry is 633,200 tons, 553,900 tons, 315,000 tons, and 205,700 tons respectively. The gross pollution and pig equivalent table is as follows (Table 3).

Table 3. Equivalent and produce pollution gross livestock and poultry pig Longxi County.

	Pig/ten thousand	cattle/ten thousand	sheep/ten thousand	chickens/ten thousand
Total output of Sewage/ million t	63.32	55.39	31.50	20.57
Equivalent to pig equivalents/ million	44.41	3.83	39.56	12.5
Sewage yield rate of per pig equivalents	1.42	2.32	0.79	1.64

As can be seen from the above table , the discharge capacity of pigs is the largest , accounting for 37.08 % of the total discharge amount ;followed by cattle and sheep, chickens with the least amount of 205,700 tons, accounting for 12.04% of the total. It can be seen that livestock and poultry farming in the county is mainly concentrated on pig farming and cattle breeding.

ii) Grassland carrying capacity

As a result of global warming and the influence of natural and human factors such as grassland reclamation, overgrazing, and increasing range of human activities, grassland vegetation has been seriously destroyed, resulting in grassland degradation, desertification, salinization, etc. The area of rocky desertification continues to expand and the carrying capacity of grassland decreases seriously. At present, 90% of the available natural grassland in the country has been degraded to some different degrees. According to statistics, the livestock overloading rate of grassland in deserted areas is from 50% to 120%, and in some areas it is as high as 300%. [12]. Longxi County is located in southeastern Gansu Province and in the central part of Dingxi City, with a total area of 2,408 square kilometers and a total population of 519,200. Among them, the agricultural population is 433,000. The county is from 1612 meters to 2762 meters above sea level, the average annual temperature is 8.1 °C, the sunshine hours are 2,210 hours, and the rainfall is 415mm.The frost-free period is 160 days, which belongs to the Loess Plateau area [13]. The county has natural grassland of 630,000 mu, among which 578,100 mu can be used as grassland. It is the largest terrestrial ecosystem and also a water conservation area [14].

Due to the local corn silage, the balance of grass and livestock is used to characterize the grassland carrying capacity [15].

$$Fr=Fp \quad (4)$$

$$Fr=Ns\times CSN\times T\times a \quad (5)$$

$$Fp=Fng+Fs+Fig \quad (6)$$

F_r is the forage requirement , Fp is the total forage yield , CSN is quantitative for livestock and poultry daily needs, Ns is the number of raised sheep , T is the number of breeding days , Fng is the grassland grass yield , Fs is the straw available amount, Fig is yield grass for irrigated artificial grass .

Combined with the local conditions in Longxi, the sum of F_{ng} is 48,000 tons, F_{ig} and F_s is 1,426,000 tons, and F_p is 1,474,000 tons.

According to the conversion relationship of "1 cow = 5 sheep units, 1 pig = 1.5 sheep units and 20 chickens = 1 sheep unit", the total number of livestock breeding in this county is 2,169,000 sheep units. Due to the county feeding level, the value of CSN is 2.5Kg, and the coefficient of livestock age is 0.8, so the formula (2) is calculated:

$$F_r = 2.5\text{Kg}\times 365\text{d}\times 2{,}169{,}000\text{sheep units}\times 0.8=1{,}583{,}400 \text{ t} \quad (7)$$

$$F_r - F_p = 109{,}400 \text{ t} \quad (8)$$

It can be seen from the above that the carrying capacity of artificial and natural grasslands in this county is seriously “overloaded”, exceeding 5% of the basic group. Reasons for this are as follows: 1) In recent years, the rapid development of livestock husbandry has led to the over-expansion of basic mother livestock without policy guidance; 2) Less annual precipitation (see Table 4 [16]), is restricted for forage growth. In addition, the "warm winter" climate reduces the snow cover in winter and prevents the grass from turning green and growing fast in spring. 3). The appearance of idle farmland in rural areas led to the further expansion of straw crops. At the same time, because livestock demand is greater than supply in the market, many farmers sell off the livestock when they don’t grow the best condition. What’s more, the benefit of livestock farmers has not been maximized simply because of the shortage of forage.

Table 4. Precipitation of Longxi County in 2017.

Month	1	2	3	4	5	6	7	8	9	10	11	12
Precipitation/ mm	4	5	11	23	40	48	67	58	38	26	5	2

iii) Water resources carrying capacity

Water resources are basic and strategic resources and one of the basic conditions for maintaining the steady development of the national economy and the sustainable development of cities [17]. The total amount of water resources in Longxi County is 1,707,620,000 cubic meters, of which available water resource is about 71,800,000 cubic meters. The annual water consumption of industry, agriculture and residents is about 61,100,000 cubic meters. Meantime, the branch canal and supporting projects of the first phase of Taohe River have water supply and irrigation projects in the county. It is planned that by 2020, 66,400,000 cubic meters of water will be transferred [18]. According to the fact that "the sheep need 2 to 3 kilograms of drinking water for per kilogram of forage in an environment of minus 17 degrees Celsius to 27 degrees Celsius" [19], when the peak value of 3Kg for per kilogram of forage water is taken, combined with the formula CSN=2.5Kg, the daily water requirement of a sheep is 7.5 kg. Then total Water requirement of livestock Q is as follows:

$$Q = a \times b \times 7.5 \times 356 = 0.8 \times 2{,}169{,}000 \times 7.5K \times 365d = 4{,}750{,}000\ t \quad (9)$$

In the formula: the coefficient of adult livestock a is 0.8, the sheep units b is 2,169,000.

It can be seen that the livestock water demand in Longxi accounts for 6.62% of the available water resources(this result does not include the wasteful water resources, the increase of drinking water due to seasonal factors and the water for forage irrigation). Therefore, we can make a preliminary conclusion that the water resources in this county can carry the development of livestock husbandry.

3. Livestock husbandry and the relationship between the environment

3.1. Effects of livestock husbandry on the environment

"Long Shadow Environment Problems and Solutions of Livestock Husbandry" [20] issued by the UN Food and Agriculture Organization of the United Nations in 2006 pointed out that if the changes in land use caused by land occupation in livestock husbandry and livestock farms are taken into account, the global livestock industry accounts for 9%, 65%, 37% and 64% of the total emissions of CO_2, N_2O, CH_4 and NH_3 from human activities respectively. Greenhouse gas emissions of livestock husbandry account for 18% of the total emissions from human activities, thus livestock husbandry has become a great threat to climate change. According to the data released by China's initial national communications on climate change, in 2004, the CH_4 emissions from livestock enteric fermentation and livestock manure management system accounted for 59.21% and 5.04% respectively of the agricultural emissions, and the two accounted for 29.70% and 2.53% respectively of nationwide emissions. Livestock husbandry has become the source of the largest CH_4 emissions in the field of agriculture in China [21]. At the same time, most livestock farms are built in rural areas because of land prices, grass sources and other reasons. The excrement discharge from livestock husbandry has not had a small impact on the surrounding residents, which is contrary to the construction of a new socialist beautiful countryside. In addition, the harmful gas in the enclosure can seriously harm the livestock.

3.2. The environmental impact on livestock husbandry

In recent years, environmental protection, guided by sustainable development strategies and new development concepts, has become a vane for the development of various industries .Only on the premise of not violating the relevant laws and regulations of environmental protection and only by opening up and innovating boldly, can we walk out of an industrial road that conforms to the county's conditions. As far as environmental protection is concerned, the promulgation of various policies has forced the livestock husbandry to reshuffle and prepare for new start. The major impacts on livestock husbandry include: Firstly, breaking the existing pattern of breeding, it is mainly manifested in the delineation of prohibited areas and restricted areas. According to "Longxi County Livestock and Poultry Farming Area Division Program" [22], 18 farms in this county are in the prohibited area and have to be converted or relocated, and the prospects of the farms in the restricted area are also not optimistic. The restriction of the amount of breeding has undoubtedly blocked the way to expand the reproduction of livestock industry. Secondly, it is urgent to replan the layout of the industry. The county administer 17 townships, each of which has different competitive power of livestock husbandry, and the corresponding pollution degree is different, which weakens the government's macro-control to a certain extent. Lastly, the new policy of environmental protection puts forward new requirements for industry development. As far as the livestock is concerned, it is necessary to change from pursuing the greatest economic benefits in the past to set up the green production system under the new normal condition by supporting the related environmental protection facilities.

4. Conception of unified and coordinated development

At present, agricultural environment in China is under the double pressure of exogenous and endogenous pollution, which has increasingly become the bottleneck of agricultural sustainable development. What's more, agricultural non-point source pollution is becoming more and more serious due to over-exploitation of agricultural resources [23-24]. As an important part of agriculture, livestock husbandry is also facing innovation. In the paper the author makes a tentative plan on how to coordinate the development of livestock husbandry and environmental protection based on Longxi County, Gansu Province.

4.1. Adjust the layout for environmental protection management

In order to adjust the layout of livestock husbandry industry, the following facts should be followed:

First, consider the bearing capacity carefully. Take Longxi County for instance, Wenfeng Town has a total area of 257.6 square kilometers and an area of 110,400 mu of cultivated land, while Shuangquan Town has a total area of 75 square kilometers and the area of 32,300 mu of cultivated land. On the basis of actual situation ascertained in March 2017, there are 1,379 farmers in Wenfeng Town and 1,318 households in Shuangquan Town. As far as the carrying capacity of cultivated land is concerned, these two towns are not equal. Therefore, the indicator is an important reference for the overall layout of the industry.

Second, consider the altitude and monsoon factors. Viewed the map of Longxi County, Mahe Town has an average elevation of 1,800m in the north, Wenfeng Town 1,682m at the southernmost end, Shuangquan Town 2,116m at the westernmost end, Weiyang Town 1,850m at the easternmost end, and Yuntian Town 1,976m in the central part. As we all know, the molecular weight of air is about 29, and the proportion of harmful gases in livestock is relatively large. In addition, the humidity of livestock enclosure is high, and there are more dust and microorganisms. Therefore, it is generally considered that the proportion of air in the enclosure is larger than that of air, and the mobility is relatively slow. Because Longxi is a temperate continental monsoon climate, the breeding population should be in the lower tuyere at low altitude, and the air pollution treatment of the livestock enclosure is mainly "upper ventilation, lower adsorption", which is also an important basis for adjusting the layout.

Third, adjust the composition of livestock species; the main way is to advocate the raising of low sewage livestock. According to the Table 3 in this paper, "pollution rate for per unit pig", the amount of sewage discharged by one head of four major livestock breeds is from large to small, which is cattle > pig > chicken > sheep, and the livestock species are also affected by market price, dietary habits, breeding habits and so on. But as far as county environmental protection is concerned, it is also an important reference for macro-control.

4.2. Reduce the environmental load balance capacity

As far as Longxi County is concerned, livestock husbandry has a large amount of fecal waste and excessive carrying capacity of cultivated land, which is due to the fact that the quantity of aquaculture is too large and leads to the status quo of environmental overloading, so it is imperative to reduce emissions by reducing the amount of manure.

First, the elimination of lower feed utilization rate and the promotion of livestock and poultry with high rates of return have always been the source power for the sustainable development of livestock husbandry, which is particularly important today when the environmental protection problem is increasingly prominent. It is also a necessary means of reducing basic groups.

Second, popularize high quality forage planting. The productivity of grassland per hectare in China is only 7 livestock production units, and the meat yield per hectare of grassland is only 30% of the world average [25]. This method is a necessary way to increase the total amount of forage grass. Increasing yield within a fixed unit area can not only guarantee the basic needs of livestock husbandry development, but also an important means to increase farmers' income.

Third, upgrade downstream processing production capacity. This mainly refers to the meat processing industry. As the "intermediate" in the docking market, the greater the production capacity of this link, the shorter the "stagnation" time of adult livestock and poultry in the enclosure, one is to reduce the raising cost of breeding owners, the other is to reduce the discharge of livestock and poultry manure, and the third is to effectively revitalize the entire industry chain of livestock husbandry, which can be described as "kill three birds with one stone".

4.3. Strengthen configuration and Advocate green production

At the county level, it is necessary to establish livestock carcasses innocuous treatment sites, organic fertilizer production line, and so on. At the same time, livestock farmers, as the most advanced individuals, should take their own improvement as their own responsibility. Livestock husbandry authorities at the county level should inculcate breeding owners through on-the-spot teaching, observation, etc. Let the farmers truly realize that ecosystem farming technique is a great event that benefits other people as well as themselves and it is the critical point of improving quality and increasing efficiency [26]. At the same time, the finance branch at the county-level should actively take care of the project and finance, starting with leading enterprises and setting up ecological circulation operation system which take "self-discharge and self-elimination" as the appeal to reduce the amount of external discharge in the region. What's more, farming and livestock husbandry can cooperate with each other through "forest culture" models, and then bring the farmers around to fully participate.

5. Conclusion

The livestock husbandry in China is at the crossroads and is changing from the traditional livestock husbandry to the modern livestock husbandry. With regard to the general direction of the transformation of Chinese livestock husbandry, the CPC Central Committee has pointed out that it is to build a resource-saving, environment-friendly livestock husbandry, to build harmony between man and nature and human-oriented healthy livestock husbandry and to build circular and sustainable development livestock husbandry in accordance with the scientific development concept and the requirements of building ecological civilization [27]. Based on the analysis of the development of livestock husbandry in Longxi County, Gansu Province, the author draws the following conclusions on how to establish livestock husbandry ecosystem in an all-round way:

First, the total amount of sewage and environmental carrying capacity of livestock are prerequisite for the harmonious development of livestock husbandry and environmental protection and it is also the basis for the formulation of relevant macro policies in county administrative regions. "Water-land-grass-livestock" is the main line of balance and the basis for coordinated development. The cultivated land and grass resources in this county meet the carrying capacity requirements of the existing livestock and poultry quantity.

Second, it is necessary to grasp the relationship between livestock husbandry and environmental protection, and adjust the development direction within the safety range of the critical point.

Third, the development of livestock husbandry should be based on the local geographical environment, external bearing capacity, timely adjustment of livestock species, feeding quantity, and supporting anti-pollution hardware facilities. It is also indispensable to increase input in soft environment to achieve the purpose of emission reduction.

References

[1] National Ecological Agriculture Construction Leading Group Office. 1996 China ecological agriculture. Beijing, China, China Agricultural Science and Technology Press.

[2] Rahman G and Sun Z J. 2004 Sino-German research collaboration sustainable biomass energy farming. Beijing, China, China Agricultural University.

[3] Dong H M, Zhu Z P and Huang H K. 2011 Pollutant generation coefficient and discharge coefficient in livestock production . *J. Agri. Engin.*, **27**(1): 303-308.

[4] Report on the work summary of the 2017 livestock industry and the plan of work in 2018. Longxi County, Gansu Province, China, Longxi County Livestock Husbandry and Veterinary Technology Service Center.

[5] People's Government of Longxi County. 2016 Longxi County Situation in 2016. Longxi County, Gansu Province, China, People's Government of Longxi County.

[6] Zhao L X. 2009 Q&A for rural biogas engineering. Beijing, China, Chemical Industry Press.

[7] Geng W, Hu L and Cui J Y. 2013 Biogas energy potential for livestock manure and gross control of livestock feeding in region level of China. *J. Agri. Engin.* , **29**(1): 171-9.

[8] The State Environmental Protection Administration. Notice on the reduction of sewage charges and other issues. http://www.zhb.gov.cn/epi-sepa/zcfg/w3/fa2004-43.htm2004/3/15/2012/1219

[9] Department of Natural Ecological Protection of Environmental Protection Bureau, 2002 Investigation on the pollution of large-scale livestock and poultry breeding and its countermeasures in China. Beijing, China, China Environmental Science Press: 77-78.

[10] People's Government of Longxi. 2010 General plan of land use in Longxi County (2010-2020). Gansu Longxi County: People's Government of Longxi.

[11] The General Office of the Ministry of Agriculture. 2018 Guide to the measurement of the bearing capacity of livestock and ooultry manure (Agricultural Office, No. 1). Beijing: General Office of the Ministry of Agriculture.

[12] Li S. 2006 The decrease of the capacity of grassland in China. China Meteorological Report.

[13] Longxi County People's Government. County Profile\http://www.longxi.gansu.gov.cn/showxw.asp?id=5561

[14] Longxi County Livestock Husbandry and Veterinary Technical Service Center. 2016 Longxi County construction western grass development planning (2016-2020). Gansu Longxi: Longxi County Livestock Husbandry and Veterinary Technical Service Center.

[15] Lu H Y, Li H P and Gao Z Y. 2016 Water and land resources allocation model of pastoral area based on grassland conservation. *J. Agri. Engin.*, **32**(23): 123-130.

[16] Weather Network. http://www.tianqi.com/qiwen/city-longxi/

[17] Shao T. 2008 Research on water environment metabolism in eastern China. Shanghai: Fudan University.

[18] People's Government of Longxi. 2016 Longxi County Industrial Poverty Alleviation and Development Plan 2016-2020. Gansu Longxi: People's Government of Longxi.

[19] China Agricultural Information Network. [EB / OL]. Http://www.agri.cn/kj/syjs/201512/t20151202_4921745.htm

[20] Steinfeld H, Gerber P and Wassenaar T. 2006 Livestock's Long Shadow: environmental issues and options. Rome: FAO.

[21] National Development and Reform Commission. Initial national communications of the People's Republic of China on climate change [EB / OL]. Http://www.ccchina.gov.cn/file/source/aa/ga2004110901.pdf [20050110].

[22] People's Government of Longxi County. 2017 Report on the division of livestock and poultry breeding in Longxi County 2017. No.132. Gansu Longxi: People's Government of Longxi County.

[23] Guoxin Office Held a Press Conference on the Prevention and Controlof agricultural non-point source pollution (2015-04-14) http://www.scio.gov.cn/xwfbh/xwbfbh/wqfbh/2015/ 20150414 / index.htm.

[24] Liu Q P, Lin Z S and Zhou L. 2015 Spatial-temporal differentiation and environmental risk assessment of fertilization in Shandong Province, China. *J. Agri. Engin.* **31**(7): 208-214.

[25] Yan H P. 2013 Standardized management guide for livestock husbandry and veterinary medicine in the new era. Shanxi: Shanxi Publishing Media Group.
[26] Chai S H. 2017 The development prospect of ecological aquaculture in Longxi County. Livestock Husbandry and Veterinary Medicine of Gansu Province. **46**(9).7-10.
[27] Livestock industry development monitoring analysis and development trend forecast report (2018-2025). China Industry Research Network. 2018.

IOP Conf. Series: Earth and Environmental Science **185** (2018) 012040 doi:10.1088/1755-1315/185/1/012040

Accuracy evaluation of the CCI remote sensing soil moisture for revealing drought in Northeast China

X L Yao[1,2], S Y Sun[2], X J Li[1,3] and R Liu[2]

[1] College of Resource Environment and Tourism, Capital Normal University, No. 105 XiSanhuan North Road, Beijing 100048, China

[2] China Science Map-universe Technology Co., Ltd, No., Jia 11 Anxiangbeili, Beijing 100875, China

Email: lixiaojuan@cnu.edu.cn

Abstract. The validation research is the basis of application of remote sensing data. This study using soil surface measured data for 10 cm in Northeast China in 1992-2013 to verify the European space agency soil moisture data set (ESA CCI SM). In this study, two methods evaluating CCI data for revealing drought was used: common Pearson coefficient method and drought judgement method for coincidence degree of drought. This verification study is located in the main grain producing areas of Northeast China, which is dominated by rain-fed agriculture. After the result analysis, CCI remote sensing soil moisture is feasible for assessment research on agricultural drought. Through the evaluation of this study, this remote sensing soil water can be used as an effective indicator of agricultural drought in the main grain-producing region of Northeast China.

1. Introduction

The climate drought has a significant impact on food production. Drought affects crop growth to a certain extent, especially in large areas of rain-fed agriculture. Therefore, drought monitoring on large scale and timely manual intervention should be carried out during the critical period of crop growth in the food production base to ensure food production and food security. Because of the complexity of the drought, the traditional monitoring method is difficult to fully grasp the drought information completely. Satellite remote sensing technology overcame the scarcity of traditional monitoring sites, low level of automation and data processing technology, etc. This technology can get continuous surface information quickly and efficiently, and it has obvious advantage in space representation and sampling period. Remote sensing technology also can obtain the information of surface soil, vegetation, surface temperature and meteorological aspects, and monitor drought from different angles. Soil moisture, as a direct and timely indicator of drought monitoring, is the essence of agricultural drought. With the development of microwave remote sensing technology, the accuracy of remote sensing soil moisture has improved further. A variety of global remote sensing soil moisture data have been produced, and a series of study about remote sensing data accuracy have been carried out [1-2]. Among them, the research on the relationship between soil surface and deep water content is the theoretical basis for remote sensing of soil moisture to characterize the soil moisture change in deep roots and to carry out practical application.

According to the root depth study of various crops, 0~50cm is generally regarded as the depth of the root layer of crops [3]. Spring corn is main rain-fed crop in northeast China, followed by soybean. The root of spring corn distributes in less than 40 cm soil layer at seedling stage, will grow to 160 cm at flowering period, and right along grow to 180 cm at ripening period. However, the main root system that absorbed water in the whole reproductive stage always distribute in the soil layer of 0~40 cm deep [4]. The main root system of soybean is within 0~20cm underground depth [5]. Within a certain depth of soil, soil water content has good continuity and removable property, and is a typical regional variable. When there is little water content in the shallow layer, the deep soil water will supply shallow layer water content, so there is a certain correlation between the soil moisture content in the adjacent areas. Biswas and Dasgupta [6] firstly proposed a linear expression for estimating deep water based on surface soil moisture, which was widely used as an empirical formula for the relationship between soil moisture in different layers. Combined different land use and hydrological and meteorological conditions, Mahmood and Hubbard [7] revealed a certain correlation between surface and deep soil water in Nebraska area of the United States. Ragab [8] estimated deep soil moisture using remote sensing surface soil water based on the relationship between the soil water of surface layer and each layer within 50cm. In the study about the relationship between different depth data of 31 soil moisture observation sites of different land use types in China, results show that the surface soil moisture and soil moisture in 0~100 cm soil deep has good correlation, and this correlation has a certain reduce with the increase of depth [3]. Estimating deep soil water content by surface remote sensing of soil water directly may lead to error. For remote sensing data analysis, the analysis of relative dynamic change is more important than the absolute value change [1]. The long time series of remote sensing of soil water can reflect the change of soil water in the deep roots to some extent. With the studies on relationship between soil moisture at different soil depth, remote sensing soil water has gradually become an effective means for large-scale agricultural drought monitoring research.

The key to the application of remote sensing soil moisture in agricultural drought assessment is the correlation between remote sensing soil water and measured soil moisture during the process of soil water reduction, that is, the accuracy of revealing drought. Because of the inconsistency between measuring methods and measuring depth, it is a great challenge to evaluate the remote sensing data by using the field measured data. The scarcity of measuring site in China adds the difficulty in evaluation.

First of all, the measured soil water is expressed by field capacity in China which is different from the international remote sensing data. The conversion between these two kinds expression is related to regional soil texture (density, porosity, etc.) information. Secondly, remote sensing data is the average moisture in a certain thickness of soil surface, but the measured data by using sensor probe in China represent the moisture of current certain depth soil. Thirdly, remote sensing data is the instantaneous data at a certain time, which is not consistent with the continuous monitoring data of the actual measurement. Fourthly, remote sensing data in each grid reflects different meteorological conditions, land cover, soil types, and the soil moisture conditions, terrain conditions. The sparse monitoring site data cannot represent a grid within the scope of the overall situation [9]. In order to assess the expressing ability of CCI, this study introduced four quantitative evaluation indexes to reflect the judgment accuracy of the drought. This method can directly analyse the times of drought events by CCI and measured data instead of the absolute value comparison. This research has important theoretical and application value on remote sensing evaluation of drought.

As one of the three black earth regions in the world, Northeast China is an important commodity grain production base and a large granary for ensuring national food security. In the context of global climate change, Northeast China, as one of the most sensitive regions to global warming, will continue to show a trend of warming and drying, especially in summer. The rain and heat resource allocation in Northeast China is accordance with crop growth cycle. Therefore, rain-fed agriculture is the main planting pattern in there. In Northeast China, with the backwardness of agricultural infrastructure, crops growth is dependent on natural precipitation. In recent years, agricultural drought occurred frequently, which have greatly affected the agricultural production. Due to wide distribution of farmland, flat terrain, and small differences between soil types, the underlying surface condition is relatively simple. Northeast China has favorable conditions to carry out agricultural drought remote sensing evaluation. In this study, accuracy evaluation of CCI remote sensing soil moisture for revealing agricultural drought was carried out in Northeast China.

2. Data and pre-treatment

In this study, the accuracy of CCI remote sensing soil moisture revealing drought was evaluated in 39 counties and 6 municipal districts in the west of Northeast China where is urgently needed for agricultural drought monitoring for frequent drought disasters in recent years. In there, spring corn is the main crop, and its planting proportion increase continuously. There are 8 monitoring sites called Keshan, Fuyu, Helen, Tailai, Anda, Qianguo, Harbin and Changling. The distribution of the study area and monitoring site is shown in Figure 1.

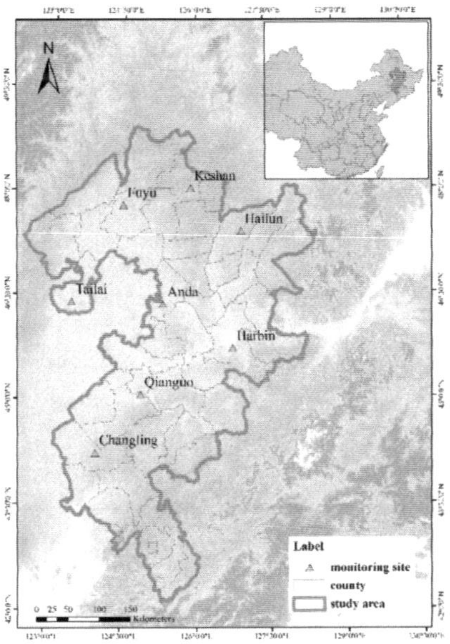

Figure 1. Study area and monitoring site distribution map.

2.1. *CCI remote sensing data*

CCI remote sensing soil moisture data is a set of global soil moisture data products based on active and passive microwave sensors including SMMR, SSM/I, TMI, AMSR-E, Windsat, AMSR2, ERS-1/2 etc.. This combination of C-band scatterometer and multi-frequency radiometer data makes the product highly suitable for soil moisture inversion and long-term technical inheritance. As a set of soil moisture data products with a long time series, CCI data has been widely concerned since its release and has been verified and applied in many regions of the world [10-12].These studies indicated that CCI data are in good agreement with measured data from sites in many regions of the world, and can accurately represent soil moisture and its spatial and temporal changes. In this study, daily CCI data covers 1979-2013 in the latest version (v02.2) was used, and its spatial resolution is 0.25.

The original CCI data is in NetCDF format. After downloading, the data is stored in ASCII format with one scene per day. Due to the scarce site in study area and the offset of remote sensing data itself, the average of several remote sensing grid data nearby the site was be compared with the site monitoring data in this study. Considering the cumulative effect of soil water content, CCI soil moisture is treated to the average for 10 days.

2.2. *Measured data*

This study chooses relative soil moisture every ten days in 10 centimeters (most shallow) issued by the China meteorological data network, to evaluate soil moisture data of CCI. It is from 1992 2013 with 8 stations of measurement. The measured data were converted into the same volume water content as the remote sensing soil water firstly.

3. Evaluation methods

3.1. Pearson coefficient method

Pearson correlation coefficient is a common correlation analysis method, which can be used to describe linear correlation between two different random variables. In the correlation analysis, the correlation between the two variables can be intuitively shown by coefficient value and scatter diagram.

Pearson correlation coefficient analyses the correlation between the two, and the formula is as following:

$$\rho_{XY} = \frac{Cov\left(X, Y\right)}{\sqrt{Var(X)Var(Y)}} \tag{1}$$

where, Var represents the variance of variables, and Cov represents the covariance of variables. The domain value of the correlation coefficient is [-1, 1], where 1 indicates the complete positive correlation between X and Y, -1 indicates the complete negative correlation, 0 indicates the independence, and the greater the absolute value indicates the stronger the correlation. The correlation coefficient is also usually represented by the symbol R [13].

3.2. Drought judgement method

The purpose of this research is to assess the accuracy of CCI remote sensing data in revealing agricultural drought. This study introduced four quantitative evaluation indexes: Hit ratio, False alarm rate, Equitable Threat Score and deviation coefficient Bias to reflect the judgment accuracy of the drought. Among them, Equitable Threat Score (ETS) is used to test the coincidence degree between remote sensing soil moisture data and measured data [14].The higher the ETS value, the higher the Hit rate, the lower the False alarm rate, indicating the closer between the remote sensing data and the truth. Deviation coefficient Bias, namely, the ratio of number of drought judged by remote sensing soil moisture and by the measured soil moisture data. When the Bias value is greater than 1, this value represents the drought event number remote sensing overestimated. When the Bias value is less than 1, this value represent the drought event number remote sensing underestimated. The comprehensive result of ETS and Bias can evaluate the accuracy of remote sensing soil moisture revealing drought.

These four indexes are calculated by a, b, c and d. The drought event number detected both by the remote sensing data and measured data is a. The drought event number detected only by the remote sensing data is b. The drought event number detected only by the measured data is c. The drought event number neither detected by the remote sensing data nor detected by the measured data is d.

The calculation formula of hit ratio H is:

$$H = \frac{a}{a + b} \tag{2}$$

where, the value range of H is 0~ 1, 1 is the best.

The calculation formula of false alarm rate F is:

$$F = \frac{b}{b + d} \tag{3}$$

where, F value range 0~ 1, 0 is the best.

The calculation formula of Equitable Threat Score is:

$$ETS=\frac{a-a_{ref}}{a-a_{ref}+b+c} \tag{4}$$

where, ETS value range -1/3~1, 1 is the best. The a_{ref} is calculated as:

$$a_{ref}=\frac{(a+b)(a+c)}{(a+b+c+d)} \tag{5}$$

The calculation formula of deviation coefficient Bias is:

$$Bias=\frac{a+b}{a+c} \tag{6}$$

where, the value range of Bias is 0~ infinity, and 1 is the best [15].

4. Results analysis

For correlation analysis, both remote sensing and measured data were selected from May to September covering 1992 to 2013.

4.1. Pearson correlation analysis of CCI and measured data

The following Table 1 shows the Pearson correlation of 8 monitoring sites and remote sensing data. Pearson correlation coefficients in Keshan, Tailai, Harbin and Changling exceeded 0.5. Pearson correlation coefficients in Fuyu exceeded 0.4. The Pearson correlation between remote sensing and measured data of Hailun, Anda and Qianguo site was poor. Considering the remote sensing feature and the different mechanism, Pearson correlation of absolute value comparison cannot fully express the application performance of remote sensing data in agricultural drought. Therefore, this study proposed another new assess method to evaluate the accuracy of the CCI remote sensing soil moisture for revealing drought.

Table 1. Pearson correlation coefficient between monitoring site data and remote sensing data.

County	ρ_{xy}
Fuyu	0.428
Keshan	0.517
Hailun	0.268
Tailai	0.581
Anda	0.379
Harbin	0.531
Changling	0.476
Qianguo	0.364

4.2. Analysis of drought coincidence degree between CCI and measured data

Hit ratio H, false alarm rate F, Equitable Threat Score (ETS) and deviation coefficient Bias may express drought coincidence degree between CCI and measured data quantitatively. The comparison of 8 sites in the study area was calculated as shown in the Table 2. The H value of Keshan, Fuyu, Tailai, Anda, Qianguo, Harbin and Changling are all more than 0.4, which indicates that the CCI has a high hit rate in judging drought events. Among of these, H of Keshan is more than 0.6. The false alarm rate of all 8 sites was lower than 0.18. It indicates that the false alarm rate of CCI on agricultural drought events is low. The deviation coefficient Bias of all 8 stations remained between 1 and 1.16, indicating that remote sensing data had few overestimations of drought events. The comprehensive ETS value of seven sites is more than 0.1.Based on literature summary and analysis [14], these seven

Equitable Threat Scores indicate that CCI data in most areas can revealing most drought events. The drought expressing ability of the CCI was poor in Hailun city, which was consistent with Pearson result. The reason why the CCI data in Hailun city has more error in revealing drought remains to be further analyzed. Generally, it is concluded that CCI remote sensing soil moisture can revealing drought event in this study area. This method remedied the defect of Pearson absolute value comparison in terms of performance expression.

Table 2. The drought coincidence degree between CCI remote sensing data and measured data.

County	H	F	ETS	BIAS
Fuyu	0.51	0.22	0.17	1.1
Keshan	0.62	0.19	0.26	1.16
Hailun	0.36	0.28	0.04	1.01
Tailai	0.58	0.18	0.25	1
Anda	0.54	0.2	0.2	1.05
Harbin	0.47	0.23	0.13	1.07
Changling	0.43	0.24	0.1	1
Qianguo	0.46	0.23	0.13	1.05

5. Conclusion and discussion

Assessment of remote sensing data based on site measured data is challenging. The scarce monitoring sites in Northeast China are one reason of poor management for agricultural drought. Considering the characteristics of large area, simple land use, flat and open topography, the Northeast China is a favorable area to apply remote sensing data. Based on this, absolute and relative method is proposed in this study to evaluate the accuracy of CCI remote sensing data in revealing drought. Pearson correlation coefficient assessed the accuracy of CCI in absolute value. Four quantitative indexes evaluate the remote sensing data by drought events in relative mode. In Pearson absolute comparison, the correlation coefficient of CCI and monitoring data exceed 0.5 only in three sites. However, the accuracy of CCI reveals drought event in seven sites. The drought events judgment by CCI data only has a small of overestimate. Drought judgment method is more suitable for the evaluation of remote sensing data applying in agricultural drought. Based on the above analysis, CCI data is feasible for research on agricultural drought assessment. The method put forward in this study has important value in remote sensing application assessment.

To evaluate the remote sensing data by site monitoring data, two points should be considered. Firstly, the correlation of soil water between surface and deep soil; Secondly, the representativeness of the measured site. For agricultural drought, its essence is the lack of soil moisture in root region. When the relationship between surface and deep soil moisture has obvious positive correlation, the surface water reduce can indirectly reflect the deep water situation and agricultural drought. The objective of this study is to evaluate the accuracy of remote sensing soil moisture in revealing agricultural drought. Drought judgment method proposed in this study is more suitable remote sensing application research. And this method may obtain better evaluation results when applied to the dense area of the site. CCI remote sensing soil moisture data has long time series, which has advantage in frequency analysis. Due to the slow update frequency of data, it is difficult to carry out real-time drought assessment analysis, which is an obstacle in CCI data application. The fusion study of different remote sensing data with different time series characteristics is a hot topic in the application research of remote sensing data in the future.

Acknowledgments

This research was financial supported by Beijing Postdoctoral Research Foundation, and Project funded by China Postdoctoral Science Foundation.

References

[1] Liu Y Y, Parinussa R M, Dorigo W A, De Jeu R A M, Wagner W, Van Dijk A I J M, McCabe M F and Evans J P 2010 Developing an improved soil moisture dataset by blending passive and active microwave satellite-based retrievals *Hydrol. Earth Syst. Sci. Discuss.* **15** 425-436
[2] Hogg E H, Barr A G and Black T A 2013 A simple soil moisture index for representing multi-year drought impacts on aspen productivity in the western Canadian interior *Agr. Forest Meteorol.* **178-179** 173-182
[3] Liu S X, Xing B, Yuan G F, Mo X G and Lin Z H 2013 Relationship analysis between soil moisture in root zone and top-most layer in China *Chin. J. Plant Ecol.* **37** 1-17 (In Chinese)
[4] Guan J H, Liu K L and Guo X Y 2006 Advances of research on maize root system architecture *J. Maize Sci.* **14** 162-166 (In Chinese)
[5] Fu J M and Dong Z 1987 Study on root growth and relationships between root and yield in soybean *Soybean Sci.* **6** 261-271 (In Chinese)
[6] Biswas B C and Dasgupta S K 1979 Estimate of soil moisture at deeper depth from surface layer data *Mausam* **30** 40-45
[7] Mahmood R and Hubbard K G 2007 Relationship between soil moisture of near surface and multiple depths of the root zone under heterogeneous land uses and varying hydroclimatic conditions *Hydrol. Process.* **21** 3449-3462
[8] Ragab R 1995 Towards a continuous operational system to estimate the root-zone soil moisture from intermittent remotely sensed surface moisture *J. Hydrol.* **173** 1-25
[9] Jackson T J, Bindlish R, Cosh M H, Zhao T, Starks P J, Bosch D D, Seyfried M, Moran M S, Goodrich D C, Kerr Y H and Leroux D 2012 Validation of soil moisture and ocean salinity (smos) soil moisture over watershed networks in the U.S IEEE Transact. *Geosci. Remote Sensing* **50** 1530-1543
[10] Dorigo W A, Gruber A, De Jeu R A M, Wagner W, Stacke T, Loew A, Albergel C, Brocca L, Chung D, Parinussa R M and Kidd R 2015 Evaluation of the ESA CCI soil moisture product using ground-based observations *Remote Sens. Environ.* **162** 380-395
[11] Albergel C, Dorigo W, Balsamo G, Munoz-Sabater J, De Rosnay P, Isaksen L, Brocca L, De Jeu R and Wagner W 2013 Monitoring multi-decadal satellite earth observation of soil moisture products through land surface reanalysis *Remote Sens. Environ.* **138** 77-89
[12] Zhang A and Jia G 2013 Monitoring meteorological drought in semiarid regions using multi-sensor microwave remote sensing data *Remote Sens. Environ.* **134** 12-23
[13] Nicewander W A 1988 Thirteen ways to look at the correlation coefficient *Am. Stat.* **42** 59-66
[14] Yuan X, Ma Z, Pan M and Shi C 2015 Microwave remote sensing of short-term droughts during crop growing seasons *Geophys. Res. Lett.* **42** 4394-4401
[15] Wilks D S 2011 Statistical Methods in the Atmospheric Sciences, International geophysics series ed San Diego (Calif.: Academic Press) p 676

Institute of Physics
Dirac House, Temple Back
Bristol BS1 6BE UK

ISSN: 1755-1307
ISBN 978-1-5108-7024-6

CURRAN ASSOCIATES INC.
proceedings.com